The Nature of the Lunar Surface

13.50

THE NATURE
OF THE LUNAR SURFACE:

PROCEEDINGS OF THE 1965 IAU-NASA SYMPOSIUM

EDITED BY WILMOT N. HESS
DONALD H. MENZEL
JOHN A. O'KEEFE

Sponsored by IAU Commission 17 (the Moon) and NASA,
Held at Goddard Space Flight Center, April 15–16, 1965

THE JOHNS HOPKINS PRESS, BALTIMORE

Contents

PART III. Physics and Chemistry of the Lunar Surface

PART IV. Conclusions

Introduction

Under the joint auspices of Commission 17 (The Moon) of the International Astronomical Union and the National Aeronautics and Space Administration, the Conference on the Nature of the Surface of the Moon was held at Goddard Space Flight Center, Greenbelt, Maryland, on April 15 and 16, 1965. The objective of the conference was to bring about a confrontation between various theories of the lunar surface and to promote mutual criticism so as to bring about progress toward a solution of these problems. For this reason, the membership of the conference was limited to about 200 people who have an active interest in the surface of the moon. A period of time after each speech was reserved for discussion, as was a half day at the end of the conference.

It was decided that the proceedings of the conference should be published as promptly as possible. The aim of this publication is to show the general trends of opinion about the surface of the moon with special reference to the Ranger photographs. It is felt that many physicists, chemists, geologists, astronomers, and mathematicians who could contribute to our understanding of the surface of the moon are inhibited by the feeling that their remarks might either be contradicted by facts well known to others or might be so obvious as to appear trivial. The reader of this book will be in a much better position to judge the state of present opinion on these subjects and to see the possible place of his contribution in the field.

The first half-day of the conference was devoted to the assessment of the Ranger photographs themselves by Urey, Kuiper, Shoemaker, and Gold. These papers show plainly that while scientific opinion is unanimous in regarding the lunar surface as composed of a porous rock, there is wide divergence about the mechanism chiefly responsible for the structure, whether external (Urey, Shoemaker, Gold) or internal (Kuiper).

The next two half-days were divided between discussion of other physical data and discussion of possible mechanisms by which the surface might have been produced. Hapke discussed the optical behavior of the moon in reflected light, including color, albedo, polarization, and phase function, with special reference to the influence of the solar wind in darkening the moon. Dollfus summarized French work in the same field, and Dzhapiashvili described very recent Soviet work on polarization. Kopal presented his evidence for extensive luminescence. The radar studies were described by Hagfors, and some critical thermal problems were discussed by Ingrao.

These reports on the physical data are not in any sense summaries of all that has been done. They are sufficient, however, to show that we presently have adequate information about the physical condition of the outermost layers of the moon down to a few centimeters below the surface and very little information about the chemistry. The outermost layer is, as mentioned, porous with characteristic sizes somewhere between 5 and 50 microns; it is a very poor conductor of heat and is a matte black without highlights. Its chemistry is uncertain except that it is probably a silicate; anything from an ultrabasic rock like a chondrite to a highly acid rock like a rhyolite remains possible, though sometimes only barely so.

Four speakers attacked the problem from the point of view of the internal mechanisms which might be responsible. Urey and Levin were especially concerned with the question whether the moon could have an interior as hot as would be implied by surface volcanism. Smith and O'Keefe were interested in tracing a resemblance to the ring-dike-and-ash flow terrains of the earth.

The recent successful soft landing of the Russian Luna IX vehicle on the moon has given us two essential pieces of information. The first concerns the flatness of the surface. This is clearly foreshadowed by the radar information which Hagfors and Gold have described here. It was Gold who first drew attention in 1955 to the fact that this flatness was greater than would be expected for a basaltic flow, and the new pictures bear him out.

Gold sought to explain the flatness in terms of a rather mobile material produced by erosion. The second piece of information gained from the Russian landing is that the mare surface is not as mobile as he suggested, at least where Luna IX landed. The alternative possibilities seem to be some kind of vacuum welding of eroded material; some kind of exceptionally flat superstructure on top of a basaltic flow, perhaps the foam which Kuiper has suggested; and an ash flow as discussed by Smith and O'Keefe.

After the scheduled papers, Öpik summarized the proceedings and a panel discussion was held, in which a deliberate attempt was made to bring out the most controversial points under the expert supervision of Whipple.

When a book such as this is published promptly after the conference, it is inevitable that there should be errors. It is hoped that the reader will overlook the errors for the sake being placed in the main stream of the developments of our times.

Thanks are due to Mr. C. P. Boyle and the staff of the Public Affairs Office at Goddard Space Flight Center, who took care of the administrative side of the meeting, to Mr. Robert Tanner and the staff of the Editorial Branch for their work in preparing the copy, to Miss Judith Holland for the indexing, and to Dr. H. J. Goett, Goddard's Director, for his support.

Wilmot N. Hess
Donald H. Menzel
John A. O'Keefe

Part I.

Interpretation of Ranger Photographs
and Related Topics

1.

Observations on the Ranger Photographs

H. C. Urey

University of California, San Diego, California

No one knows whether the moon escaped from the earth or whether it was captured by the earth. This latter idea is the one that has appealed to me, since it would mean that the moon would be an older object than the earth and might provide evidence of the history of the solar system before the earth was formed. It is also possible that it might reveal some of the early history of the earth, and particularly the evolution of life.

Of course, one thing that appears fairly certain is that there has been an enormous bombardment of the surface of the moon. It would seem to me unlikely that the most superficial bombardment, which is what we see, is the only bombardment that has occurred. I envision that the outer parts of the moon have been bombarded and bombarded and bombarded until even the preterrestrial history, if it is there, has been obscured by this effect. I can see no other possibility if the moon was captured by the earth.

The thermal history of the moon has been discussed, particularly by myself, by Gordon MacDonald, and, from the Russian school, by Levin and by Ruskol. My reason for discussing it is an observation that there are marked differences in elevation on the surface of the moon and that the moon apparently has a nonequilibrium shape. Watts, for example, has found that there are two points on the eastern limb that differ in elevation by about 10 km. Haskell, in the 1930's, studied the changing level of Canada and Fennoscandia and ascribed it to the recent melting of an ice sheet, and from this derived a figure for the so-called viscosity of the earth. Applying this to a difference

in level of 10 km on the moon and a distance between the two points of about 500 km leads to the conclusion that if the viscosity of the moon were the same as that of the earth, it would have leveled out in the course of about 250,000 years. This evidently has not happened and would lead to the conclusion that the moon is much colder than the earth and hence has a much higher viscosity. Regardless of whether viscosity is a good way to discuss this, it is nevertheless true that the earth's surface is rising in these two regions, Fennoscandia and Canada, in the course of some tens of thousands of years because of the melting of ice. A difference of 10 km of rocks in altitude on the moon is equivalent to about 2 km of ice on the earth.

There are other evidences of the irregular shape of the moon as determined from the astronomical data that have been discussed in the past. An elevation of about 1 km in the direction of the earth exists over that at the limbs, whereas some 60 m would normally be the equilibrium value. Different explanations for this have been offered. One holds that the moon is very strong, is at a low temperature, and has maintained this irregular shape for long periods of time. Another suggests that this bulge is caused by the variation in composition between the center of the lunar disk and the limbs of the moon. Even convection within the moon has been proposed as an explanation. I can give no decision at the present time as to which of these explanations is correct.

A discussion of the chemical composition of the moon and an attempt to calculate what might be the chemical composition as determined by its

TABLE 1–1: Comparison of the Density of the Moon with Calculated Densities of Meteoritic Matter

	Mean Observed	Mean Calculated
1. Density of low-iron group chondrites	3.51	3.574 g cm⁻³
2. Density of high-iron group chondrites	3.66	3.761 g cm⁻³
3. Low-iron group with albite converted to jadeite and SiO₂ in MgSiO₃		3.653 g cm⁻³
4. Density of Moon at low temperature and pressure (α and β for olivine)		3.361 g cm⁻³
5. Required iron content of (3) as FeO & FeS in order to have density (4)		9.65%
6. Cosmic Abundance, Si = 10⁶		2.24 × 10⁵
7. Suess-Urey Abundance		6 × 10⁵
8. Aller Solar Abundance		1.4 × 10⁵

NOTE: Previously published versions of this table are incorrect because of an error in the calculation of the density, ρ_0.
$\alpha = 3.3 \times 10^5$, $\beta = 7.9 \times 10^{-7}$. Mean temperature of the moon is 1100°C and the mean pressure 19,100 bars.

The values of α and β are those for forsterite and are taken from *Geol. Soc. Amer. Spec. Pap.*, 36, 33, and 36.

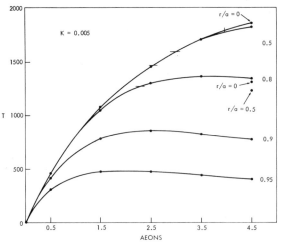

FIGURE 1–1: Temperature within a moon which was initially at 0°C throughout, calculated with K = 0.005. The horizontal lines across the curves for r/a equal to 0, 0.5, and 0.8 indicate the times of initial melting. The vertical line shows the time of complete melting.

density appear in Table 1–1; the assumption is made that it has the composition of the chondritic meteorites, except that iron is less. Using constants for enstatite, instead of those for olivine, results in a slightly lower concentration of iron. This differs considerably from what we would expect from solar abundance, as shown in the last lines of the table. Using the low iron group chondrites, we get about 6 × 10⁵ for the number of iron atoms, with silicon being equal to 10⁶, while the solar abundance is about 1.4 × 10⁵. That is, the moon seems to be low in iron, relative to the earth and meteorites, and to have more nearly the solar composition. Of course, if it escaped from the earth, we would expect the moon to have the composition of the mantle of the earth, which is one way to account for this discrepancy relative to the earth and meteorites.

There are other difficulties in regard to abundances of iron, as I have emphasized in recent years. Mercury appears to have too much iron, the meteorites vary substantially in the amount of iron in their composition, and all the terrestrial planets have higher densities than would be expected for solar non-volatile material.

Figure 1–1 shows a calculation for the temperature at various radii within the moon, using a thermal diffusivity of 0.005 cgs and calories. Gordon MacDonald used 0.006 for this constant. The time in billions of years is plotted to the right. The temperature elevation is plotted at the left. The cross marks are estimates I have made as to when melting will occur. The little horizontal bar shows when the beginning of melting should occur at the various levels, and the smaller vertical bar shows when complete melting should occur.

Deep within the moon, we expect that there will be some melting even extending out to eight-tenths of the radius, providing the moon accumulated cold. Gordon MacDonald's calculations are seen in Figure 1–2. He used slightly different constants but the values secured are very similar indeed. Figure 1–3 shows another calculation I made using slightly less potassium. There is an indication that potassium is less abundant on the moon than even in the meteorites, and no melting is to be expected on the basis of a cold moon if this potassium abundance is correct. The Russian calculations are very similar.

It seems to me that a cold moon, or a moon that is much colder than the earth, is entirely reasonable. We expect that an object much smaller than the earth should be, in general, much colder than the earth. Its ratio of surface to mass is 5.6, and surface to volume is about 4 times these ratios for

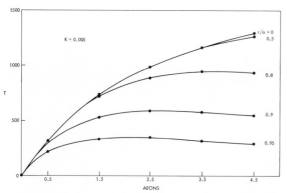

FIGURE 1-2: Temperature within a moon which was initially at 0°C throughout, calculated with K = 0.005.

the earth. This should lead us to expect that it will cool off much more effectively and lava flows should be less prominent on the moon than on the earth.

During the last 15 years I have approached a study of the moon from considerations of this sort. One of the great puzzles to me is how large differences in the level of the lunar surface could be maintained and how, at the same time, extensive lava flows could have occurred after these differences in level were established. I think the only explanation for this is that the lunar mountains have roots. It is very difficult, though, to see how such roots developed. Those who assume so easily that lava flows exist everywhere should try to answer these questions.

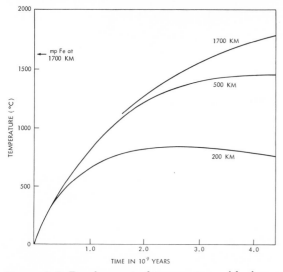

FIGURE 1-3: Development of temperature with time at various depths for a cold moon.

Now I wish to turn to the Ranger pictures. Figure 1-4 is one of the early Ranger VII photographs. Notice the ray area to the right and the absence of a ray area to the left. It is believed that many craters in this region are secondaries from Tycho. As pointed out, there is a flag effect in which the dust seems to have been blown away from the direction of Tycho; I believe this is a very reasonable explanation. I am going to assume that most of the secondary craters in this region are indeed secondaries from Tycho, 1,000 km away, and that the object producing these craters must have arrived at about 1 km/sec, as a minimum, and at perhaps greater velocities if they were nearly circum-surface trajectories to this point, as Gault maintains.

Among the group of these are the cluster of craters, which I think are secondaries, shown in Figure 1-5. If one uses Eugene Shoemaker's tables, one concludes that a crater of the size of one of these would be produced by an object having in the neighborhood of 10^{23} or 10^{24} ergs. If it came from Tycho, then we would conclude that the object was 10 or 100 million tons in mass, a rather surprising mass for an object accelerated at such high velocity, and thrown for such a long distance in a single piece. This surprises me, and I would like to see the estimate lowered. I have thought that perhaps the lower gravitational field of the moon, or perhaps the existence of less dense material than the materials of the earth, might be a possible way of reducing the calculation of mass.

I rather think that if objects of this size had been thrown about the moon, it is not surprising that others were thrown completely around the moon and perhaps produced those curious rays of Tycho that seem to come out of the western wall of that crater rather than out of its center, as I suggested some years ago.

Note the crater with rocks; there are others of similar size and appearance. One has a couple of small black spots in it and looks as though it might be another with rocks. There are long gouges near the secondaries and it seems doubtful that these are part of the same pattern of objects as the ones that produced the larger craters. It hardly seems probable that two sets of objects would travel at the same velocities from Tycho and arrive and produce these curiously different types of craters. I think Gault and Quaide have suggested that the

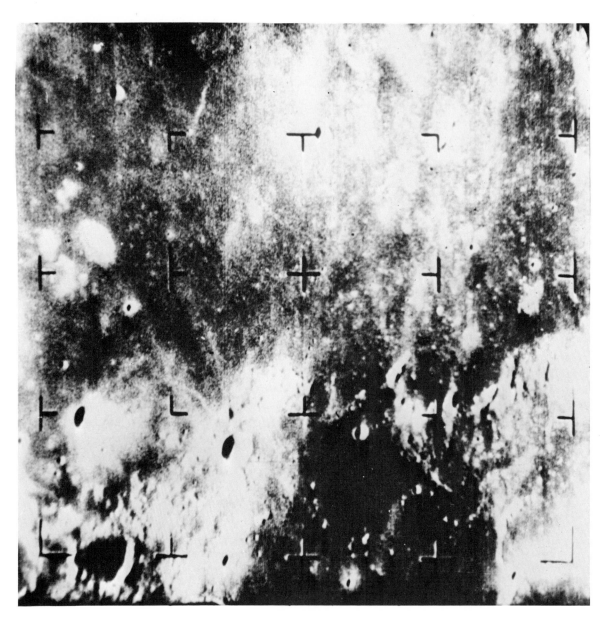

FIGURE 1–4: Early Ranger VII picture.

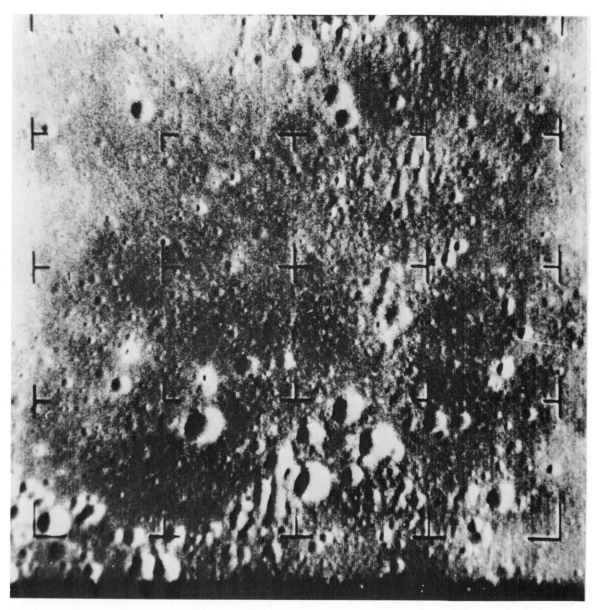

FIGURE 1–5: Cluster of secondary craters near Ranger VII impact.

longer shallow gouges came from Bullialdus, which is an older crater.

In Figure 1–6 there is some evidence of slumping in this region of the moon. Possibly the long gouges mentioned above are the result of slumping to crevasses below the surface—a suggestion I made quite sometime ago in the Experimenters' Meeting. The well-known dimples that were immediately obvious on the Ranger VII pictures perhaps can be explained also as due to draining to crevasses below the surface. Of course, materials on the surface of the moon would hardly flow easily like sand, and so one must expect that bouncing from one point to another occurs as a result of micro-meteorites and small macrometeorites. The dimple is then simply the absence of material coming back from the hole in the bottom and reflects the pattern of scattered material from the point on the moon's surface.

If this explanation proves to be correct (and I think Ranger IX photographs show many of these dimples) it would mean that there is a considerable layer of finely divided material on the surface of the moon, some 10 or 20 m, as indicated by this phenomenon, without any indication that this necessarily is the maximum thickness.

Figure 1–6 also shows one of these wrinkles which are so obvious from terrestrial photographs. I interpret it to follow through the crater near one wall. There has been some argument as to whether

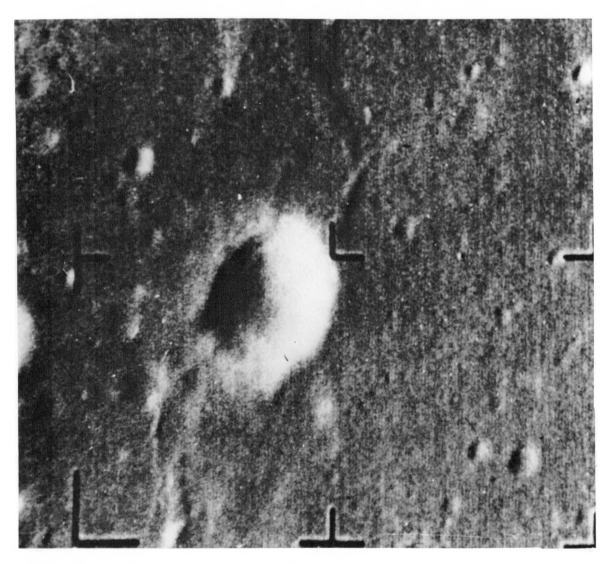

FIGURE 1–6: Ranger VII picture showing ridge intersecting crater.

it comes more directly across; I think there is no question about the fact that it goes down through the crater. I think this has significance; namely, that it is more recent than the surface of the maria floor. It has been suggested that the wrinkles have something to do with the lava flows postulated to have occurred in this region, but I think it is a little difficult to harmonize that idea with this wrinkle traversing the crater as it does. Also Eugene Shoemaker has noted that there are very sharp features on this wrinkle. If the erosion effect with which we are familiar occurs on the moon, I think it might have destroyed these sharp features of the wrinkle. This probably indicates that these wrinkles have something to do with material flowing up from below into these areas of the moon.

Figure 1–7 shows another wrinkle going down into a crater. A reasonable explanation for this is that the wrinkles are due to water coming from the interior along cracks of this kind. I conclude from the Ranger VII photographs that there has been considerable slumping to crevasses below the surface; there are indeed rather prominent cracks on the surface that seem to indicate this. I have outlined these more completely in the report on Ranger VII than I can do here.

The fact that large objects have been thrown from Tycho to Mare Cognitum encourages me to believe that material has been thrown off the moon, probably in sizes that would be recognized as meteorites when they arrived on the earth. This indicates it is quite probable that material has arrived on the earth from the moon, and much of our fear of contaminating the earth with exceedingly strange biological material is therefore not well founded.

With regard to the dimples, I have suggested two things: slumping into a hole and scattering from a hard object at the bottom of these dimples, ideas which I wrote in the Ranger VII report. I now believe that the idea of scattering from a hard object is wrong because we have found objects on the surface, not at the bottom of a dimple.

Jaffe made some models to demonstrate this idea and was able to duplicate a dimple by dropping finely powdered material on models of craters of the moon. It produced a very nice-looking dimple. Of course, this was done under atmospheric conditions, and his dust will move to some extent. I think maybe it has flowed in from the sides and has made this dimple. Whether a similar sort of flow could occur under the vacuum conditions of the moon is difficult to say. Perhaps his explanation of the dimples is better than my own.

Turning to Ranger VIII, we found at least one good dimple, curious collapse features, I am quite sure, and erosion, which Dr. Shoemaker will discuss in considerable detail. We also found projections from the surface in this case, that is, elevated areas that look like rocks sticking up above the surface. In general, all the features are very similar to Ranger VII.

Figure 1–8, from Ranger VIII, shows an easily recognizable little dimple; the rocks are at the lower right.

Figure 1–9 is a rectified picture of Mare Tranquillitatis that I made quite some time ago and have always thought looked much like a lava flow from Mare Serenitatis, into Mare Tranquillitatis, spreading to all the little bays and valleys. The liquid must have been at high temperatures, as pointed out some years ago by Dietz. Yet Mare Tranquillitatis and Mare Cognitum, in my opinion, look very much alike. I expected we might get some real evidence for lava flows that would be unique in this area, but I do not believe we did.

Now I want to turn to the Ranger IX photographs of Alphonsus. Figure 1–10 shows Alphonsus with the black halo craters. This is a terrestrial-based photograph, of course, and shows the general character. The central ridge is quite obvious and is aligned with other structures of the Imbrium collision. Mare Imbrium is at the north. This feature is either material thrown from Mare Imbrium or else material blasted from the crater wall which then fell across the surface.

In Figure 1–11 we have a picture of these walls and the so-called volcanos. These craters are surrounded by an area with fewer small craters, indicating that solid material has fallen in the neighborhood. What surprises me, and some of my friends, is that these craters are such bad examples of volcanos. It is difficult to believe that they are anything but collision craters. In fact, until these are compared to the Earth-based pictures they do not appear to be plutonic features at all. They have a steeper wall than the typical collision craters and have no extended rim above the surface. Debris seems to have filled the rilles to some extent. The density of small craters near them is less than it is in the rest of the crater, which I believe shows exceedingly mild volcanic activity. We picked this

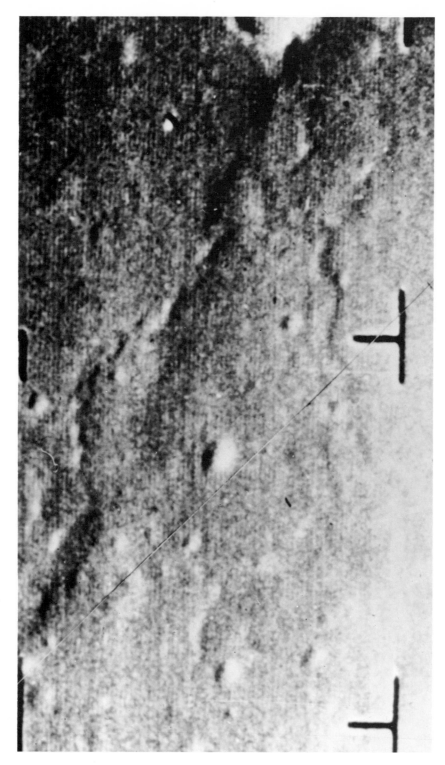

FIGURE 1–7: Ranger VII pictures showing wrinkle ridge entering crater (at top).

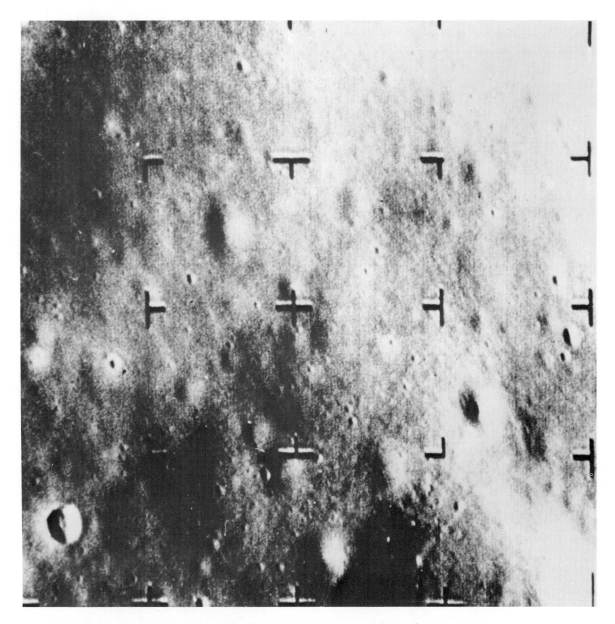

FIGURE 1–8: Ranger VIII picture showing a dimple and craters containing rocks.

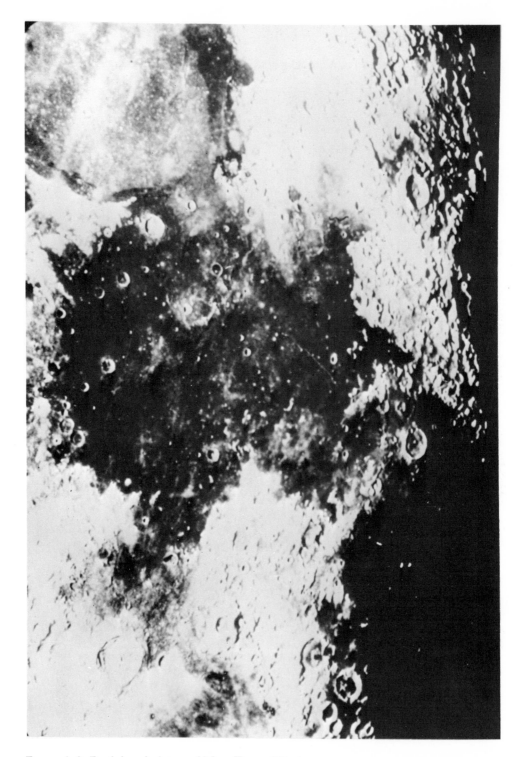

FIGURE 1-9: Earth-based picture of Mare Tranquillitatis.

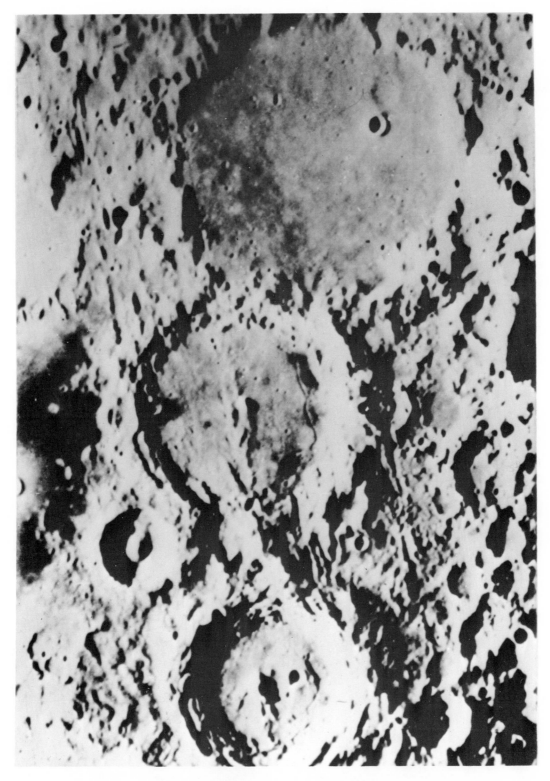

FIGURE 1–10: Earth-based picture of Alphonsus showing the black-halo craters.

crater for the Ranger IX landing area because it seemed to have more evidence of volcanic activity than any other place on the moon, yet the volcanic activity, in my opinion, is very slight.

The maria walls, the crater walls, are comparatively smooth. If you look at them closely, you will see that there is evidence of bombardment on the surface of these walls. I interpret this to mean that the crater wall material is different, chemically, physically, or both, from the crater floor; I do not know what the difference is.

We also see this beautiful so-called lake. As far as I can tell, it has the same level as the floor and is covered with craters just as the maria floor is.

There are other regions here that are comparatively smooth and do not seem to have as many craters per unit area on them as the crater floor. There are a couple of mountain peaks that I refer to as nickel-plated. They stick up above the surface and are somewhat bright. At least some nickel-iron

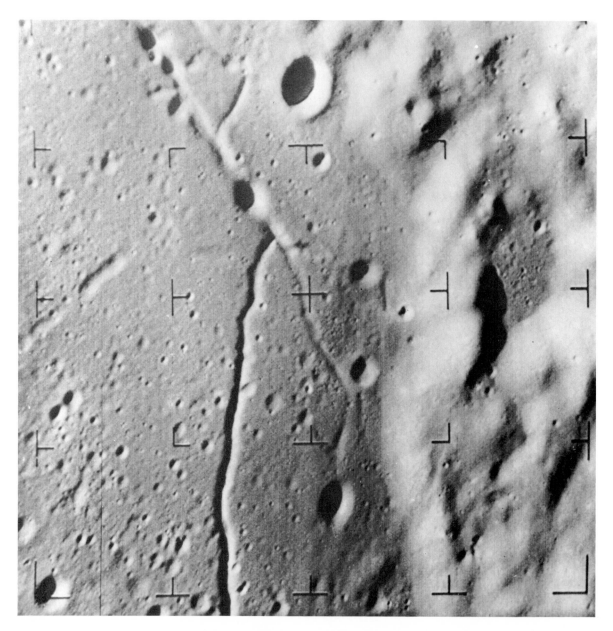

FIGURE 1-11: Ranger IX picture of Alphonsus wall, showing nearby rille and black-halo craters.

should be scattered around the moon. Whether this peculiar effect is due to metal or just white rocks, I cannot say.

The peculiar feature is that they do project somewhat above the rest of the terrain, as though they were rather stronger material on the average than the general surroundings.

Figure 1–12 shows evidence of slump features. The aligned region suggests to me that some features are not dissimilar from those seen in the maria floors.

There are also craters here that I believe can be classed as dimple craters, although they seem to vary from the dimpled crater to one that appears to have been made by sand flowing out of a funnel. This indicates that there are certain slump features in the mare floor.

Figure 1–13 is the final one of the B pictures,

FIGURE 1–12: Ranger IX picture of dimple craters (these are in the west end of the small rille on the left side of Figure 1–11).

which shows what is apparently a dimple crater. It has the rough, corrugated surface that was found in Mare Cognitum and has been particularly emphasized by Professor Kuiper. It is found again on the surface of another crater floor, and it is also present in the Ranger VIII pictures. Apparently all these smooth areas of the moon are covered with material of this kind.

Figure 1–14, one of the A pictures, shows the central region of the crater. A feature that interests me is the clifflike structure along the material from the Imbrium collision. The central region has fallen as a result of the Imbrium collision, and it therefore appears as though this central feature is later than the crater floor. This is contrary to what I had supposed—that these materials are not lava but dust and fragmented material that had fallen from the great collisions that produced the maria, such as Mare Imbrium, and that this material had come from Mare Imbrium. Of course, one cannot

FIGURE 1–13: Picture of dimple craters taken 5½ seconds before Ranger IX impact.

assume that it came from Mare Imbrium if the floor was deposited earlier than the central ridge material, which, I think, is very definitely a part of the Imbrium collision.

There are several volcanic craters that look as though they may have been produced by collisions followed by gas coming from the interior. But I think that such an origin is inconsistent with the crater that lies in a rille and is elongated; I do not expect that collision craters will be elongated. This indicates that the craters are due to explosive forces of some kind from the interior. They are surrounded by rather smooth material and a black halo; other similar ones can be seen. One that strikes me as being similar to the halo craters lies on a rille which comes right down in the shadow of the central peak. I have tried hard to judge whether Kozyrev's gases might not have come out of these craters, but of course one cannot tell what is under a shadow.

FIGURE 1–14: Ranger IX picture of the center of the crater Alphonsus, showing central peak and ridge.

I think Professor Öpik has presented very good reasons for believing that gases have not escaped and that Kozyrev's pictures are not evidence for gas at all. Some time ago I suggested that his C_2 was produced from acetylene coming from the interior of the moon and that the black areas might be black graphite. Of course, that suggestion depends on the accuracy of Kozyrev's work. If C_2 did not escape, there is no reason to expect that this black covering is graphite. It may be hydrogen sulphide or just plain black rock. Again, we see the smooth character, as Dr. Schurmeier has pointed

out, in this region. The material must be somewhat similar to the material of the crater floor.

Figure 1–15 shows the west side of the Alphonsus crater. The smooth area is Lake Titicaca, as named by Whitaker, a smooth area high on the wall. It seems unlikely that a lava flow would rise to this high level, something like 1.5 km above the crater floor, through highly fragmented material of the crater wall. Nature does some remarkable things, but the lava flow hypothesis doesn't look particularly convincing to me.

Figure 1–16 shows the crater without the halo

FIGURE 1–15: The west side of the Alphonsus crater, showing a lake-like feature in the wall.

that still has characteristics similar to those of the halo craters. In this case the surroundings are white. This area shows considerable evidence for slumplike features on the mare. For example, the rille continues up to a point and then appears as a group of craters. There is some evidence for slump features, as though great crevasses existed below the surface.

I think that the material in these pictures indicates very mild volcanic activity. It is not inconsistent with the idea that the moon has had very little volcanic activity. This is what I have expected because of the great differences in elevation on the surface of the moon. I wish to emphasize again that it is very difficult for me to believe that great differences in elevation can exist and, at the same time, that the moon can have had great, extensive volcanic activity.

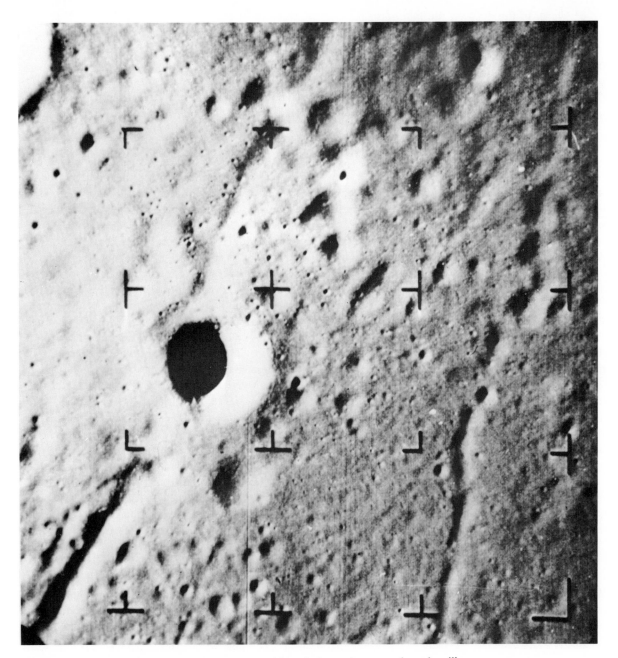

FIGURE 1-16: Ranger IX picture showing crater without halo, and nature of nearby rilles.

Some years ago Gold suggested that the maria were indeed filled with dust. Although I had certain criticisms of the idea, I found it a very interesting theory and suggested that possibly the large smooth areas of the moon were produced by fragmented material from the great collisions that produced the maria.

This, of course, is consistent with a cold origin for the moon and a relatively cold history since. It was necessary, however, to account for some way by which all of the dust left the mountain tops; I suggested that water was briefly present on the moon and that rains washed the dust from the mountains. I have not said much about it in the years since; we can make many suggestions about the moon, but we have rather great difficulty in proving that what we say is more than just possibilities.

I noticed yesterday, flying across Arizona, some of the mountain peaks sticking up above the surface of the desert, and next to them was a very smooth layer of material deposited by water, with a sharp soil line between the two. In fact, it was quite similar to the "lakes" on the eastern side of the moon among the peaks of the crater wall of Mare Alphonsus. I could see nothing of the kind, however, as I looked in another direction; it was not a universal phenomenon on the Arizona desert at all. And it is not a universal phenomenon on the moon, either, for that matter.

During the last ten years, I have made remarks of various kinds suggesting that water has been present on the moon temporarily. It certainly has not been there very long, for if it had, it should have produced extensive river valleys, which are not a characteristic of the moon.

It seems to me, however, that the black craters in the region of the moon which we picked out as probable places to observe volcanic activity indicate the theory of Gilbert, who wrote about the moon 72 years ago. Gilbert concluded "the moon is dead," and I think he was almost right. The moon is almost dead.

The smooth areas, however, have been discussed repeatedly and sometimes I think people are rather like birds—birds that are born with a map of the sky in their brains that enables them to fly from one part of the world to another. I think people are born with a great deal of knowledge of exactly what the moon is composed of. Among other things, they know definitely that on the moon the maria are

lava flows. This has been a popular idea for a great many years and some of our friends at the present time feel very certain about this. I am unhappy about it because of the thermal calculations and my expectation that a small object such as the moon would be much colder than the earth; hence the probability of extensive volcanism would be considerably less. As I said a moment ago, I have been very much intrigued by Gold's dust ideas. I have been critical of the first suggestion about the origin of the dust, although I believe that the effects are certainly there. However, I think they are more limited than he described in his first paper. Like all proposals of this kind that any of us make, they are likely to be only partly right, and we ought to be immensely pleased if they are only partly right. I think Gold has made a great contribution in calling our attention to the possibility of dust on the surface of the moon.

I feel much happier about my own suggestion, that is, that fragments from the great collisions supplied most of the materials of the maria, though I believe that much of Gold's discussion is still applicable. Gold and Kopal state that water is coming from the interior of the moon and they compare it to the earth. I am sympathetic to the idea of considerable water having been present on the moon briefly at some time, but I do not think that it will diffuse from the interior of the moon. My geological friends, at least, tell me that water has come to the surface of the earth through volcanism. The latest persons to emphatically state this to me were Francis Birch and Al Engel, whom I regard as reliable authorities in this field. If there has been water on the moon, it is much more likely to have come as a result of contamination from the earth at the time the moon escaped from the earth, or was captured by the earth.

I think there is evidence that some of the meteorites are coming from the moon, and I have been interested in the carbonaceous chondrites and the great controversy raging about this subject. Professor Nagy at California has continued work on this, with me as "Big Brother" looking over his shoulder part of the time. Some of these carbonaceous chondrites contain what I would expect to be a good residue for a primitive ocean of the earth. I believe that that ocean would contain ammonium salts. Ammonium salts were reported in these objects. I would also expect that that ocean would contain ferrous iron, and iron has been reported in these objects in the form of ferrous carbonate. There is considerable evidence in these meteorites for either materials of biogenic origin, or materials that look like the ones required for the first synthesis of life, either materials useful for the origin of life, or materials of primitive biotic origin. I am not sure that this is true but it would be most interesting if it were the case.

I am sure that what I say about the moon is heavily prejudiced—prejudiced by a desire that the moon shall be interesting. If the moon escaped from the earth, and if it is completely dry, it is a comparatively uninteresting object. You might think that a thrown-off portion of the earth would provide a record of the early history of the earth, but it would be so badly broken up by the events that occurred after the separation that it would take a large number of geologists many, many years to make a proximate unraveling of the story. If, however, when it did separate it carried along some of the primitive oceans of the earth, at a time when life was evolving, it would be fascinating beyond words. It would be wonderful to get a sample of material from the Apollo project that would enable biologists and micropaleontologists to study the properties of the moon in great detail.

If the moon was captured by the earth, however, it records some preterrestrial history and would be interesting from the standpoint of astronomy. It might tell us something about the origin of the solar system, and if during capture it became contaminated with material from the earth, perhaps we could learn the prehistory of the earth, the early history of the solar system, and possibly something about the origin of life on earth.

Levin argues that the moon accumulated from debris in the neighborhood of the earth. In this case the earth and moon should have the same composition, which definitely is not the case in my view or many geophysicists' views. He postulates a core of high-density silicates so that the percentage of iron in both earth and moon is the same as that of the sun. In this case the moon should contain about 6–10 per cent by weight of iron and the surface should contain this same low percentage. If my ideas are correct, the surface of the moon will contain amounts of iron similar to those of the chondritic meteorites, namely 20–35 per cent of iron, for I postulate a layer of earthlike material including the iron core on its surface.

This argument, in regard to the composition of meteorites, planets, the moon, and the sun has gone on for some years. The Russian school says everything consists of solar material except the meteorites and that the geophysicists are wrong about the iron core of the earth. Another group thinks that the astronomers are wrong about the composition of the sun and that their iron content of the sun is too low by a factor of about 6. I am only a simple country boy from Indiana who believes that the astronomers know the composition of the sun and that the geophysicists know the composition of the earth, and since they do not agree it is the duty of people who make theories and models to try to include all the data.

2.

Preliminary Analysis of the Fine Structure of the Lunar Surface in Mare Cognitum*

Eugene M. Shoemaker

United States Geological Survey, Flagstaff, Arizona

The principal new facts obtained from the Ranger VII photographs pertain to features of the moon's surface less than 300 m across and to the small details on larger features in the region of Mare Cognitum. These facts, when combined with prior knowledge about the moon obtained with Earth-based telescopes, permit the formulation of a comprehensive model of the fine structure of the surface of Mare Cognitum and the development of hypotheses about processes which have led to the formation of the observed small topographic features. The formulation of a model of the fine structure and the development of hypotheses about the surface processes are important for two reasons: (1) such a model and set of hypotheses may be used to predict the nature of the surface to be expected elsewhere on the moon and (2) formulation of a model consistent with the available facts is an essential first step in drawing inferences about the physical characteristics of the moon's surface which are not directly observable in the photographs. Coherence and bearing strength, properties important in the problem of landing spacecraft on the moon, are among the unknown physical characteristics of the lunar surface which we might attempt to infer. The predictions and inferences made from the model and working hypothesis that will be presented here are subject to test by further lunar exploration. I will start by reviewing the basic new facts that may be established from the Ranger VII photographs and by introducing pre-

viously known facts about the moon that are necessary to formulate a model of the fine structure of a mare surface.

I. MAJOR FACTS ESTABLISHED BY RANGER VII PHOTOGRAPHS

Two basic kinds of facts about a planetary surface can be obtained from photographs or television imagery: (1) the topographic configuration and (2) photometric characteristics. Information drawn from a photograph about other physical characteristics, the processes acting on the surface, the structure, and the history of the surface are necessarily inferences based on the first two basic kinds of information. In the Ranger VII pictures, the new information obtained concerns chiefly small topographic features. These may be classified, for purposes of discussion, into negative- and positive-relief features. A certain amount of information important to constructing a model of the fine structure may also be derived from detailed observations of the photometric heterogeneity of the surface.

Negative-Relief Features

The small topographic features revealed by the Ranger VII photographs of Mare Cognitum are almost all craters. Craters thus are not only the dominant large topographic features on the moon but also the dominant small topographic form revealed by Ranger VII. The new craters observed

* This paper originally appeared in *Technical Report No. 32–700*, published and copyrighted by the Jet Propulsion Laboratory, California Institute of Technology.

show a wide range of shapes. Some of them may be recognized as belonging to certain classes of craters observed through the telescope, but a large number of the craters less than 300 m across cannot be classified by shape alone in the categories previously recognized by telescopic observations. In my opinion, there are, with one minor exception, no unequivocally identifiable negative-relief features less than 300 m in length in the Ranger VII photographs that cannot be classified either as a form of crater, a group of craters, or a topographic detail within a crater.

Shapes of craters

Two basic types of craters well known from telescopic observations can be readily identified in the Ranger photographs: (1) primary craters and (2) secondary craters.* These types are defined on the basis of morphology and distribution. Small telescopically observable primary craters are typically very uniform in shape. They are nearly circular in plan and have a distinct smooth raised rim, generally of nearly uniform height. The inner walls are also smooth, uniform, and steep. The walls are slightly concave upward; typical slopes of the wall near the rim crest are greater than 35 degrees and, near the foot, about 25 degrees. The average slope of the crater walls is typically between 30 and 35 degrees. In primary craters larger than about 7 km in diameter, the steep wall terminates, at the foot, against a relatively smooth, level, circular floor. The diameter of the floor diminishes with size and is very small or absent in primary craters less than 7 km across; the walls thus extend to the center, or almost to the center, of these small craters. As has been remarked by Dr. G. P. Kuiper,[1] these remarkably uniform craters look as though they had been turned out on a lathe. They are scattered across the maria and other parts of the moon without apparent control by other surface features. Some of them are surrounded by a bright halo, or a system of rays; others are not.

With increased diameter, the primary craters are less regular in form. At 25 km diameter, some irregular hummocks are found in the rim, terraces

occur on the crater walls, and irregularities begin to appear on the primary crater floor. At 50- or 60-km diameter, the rims are typically rugged and hummocky, the rim crest is uneven and commonly roughly hexagonal in plan, the walls are terraced, and the floor is marked by scattered hills and one or more prominent irregular peaks near the center of the crater. A swarm of small craters, which are barely resolvable through the telescope, surrounds a primary crater of this size. The greatest areal density of small craters in the swarm occurs at a distance of about one crater diameter from the rim crest of the primary. At greater distances, the secondary craters are clustered within rays or within arcuate and loop-shaped bands extending outward, in some cases hundreds of miles, from the primary crater. These small craters, whose forms can be distinguished only in the swarms that occur around very large primaries, are called secondary craters. Several thousand of them can be observed in the swarms that surround large primary craters such as Tycho (Figure 2–1), Copernicus (Figure 2–2), Aristoteles (Figure 2–3), and Langrenus (Figure 2–4).

The secondary craters that can be photographed with earth-based telescopes (generally 1 km across or larger) are in most cases readily distinguishable in form from very small primary craters. The secondary craters are typically elongate, are generally shallower than primaries of corresponding size, and in most cases have low irregular rims, or in some cases no rims at all. Many elongate secondaries, when observed visually through the telescope under favorable conditions, are found to be composite, consisting of several small craters strung end to end or merged together. The secondary craters are found around primary craters with rays and also around those without rays. Around the ray craters, the secondaries occur within the rays and within the bright halo surrounding the ray crater. Secondary craters associated with primary ray craters, which are referred to the Copernican System of Shoemaker and Hackman,[2] generally exhibit raised rims and are commonly observed to be composite (Figure 2–5). In the swarms of secondaries around rayless

---

* Other classes of craters known from telescopic observations, such as chain craters and craters at the summits of domes, are apparently not represented in the high-resolution Ranger VII photographs and will not be discussed here.

[1] G. P. Kuiper, "The Exploration of the Moon," *Vistas in Astronautics*, Vol. 2, ed. by Alperin, Morton, and Gregory, New York: Pergamon Press (1959), pp. 273–313.

[2] E. M. Shoemaker and R. J. Hackman, "Stratigraphic Basis for a Lunar Time Scale," *The Moon—Symposium No. 14 of the International Astronomical Union*, ed. by Z. Kopal and Z. K. Michailov, London: Academic Press (1962), pp. 289–300.

FIGURE 2–1: Telescopic photograph of Tycho showing part of its associated swarm of secondary craters. The secondary craters appear as very small, closely spaced craters, many of which have diameters just above the limit of resolution and are scattered over much larger older craters surrounding Tycho. The diameter of Tycho is 88 km. (Photograph from Lick Observatory.)

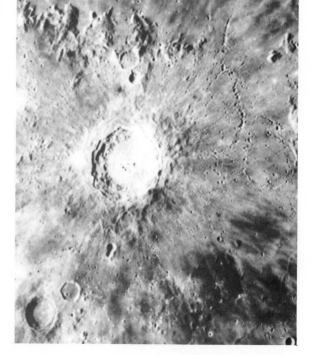

FIGURE 2–2: Telescopic photograph of Copernicus showing part of its associated swarm of secondary craters. The secondary craters are distributed over both mare surfaces and over the more rugged terrain surrounding Copernicus. Several hundred secondaries appear as very small craters which are just barely resolved. Distinct bands or loops of secondary craters are easily observed in the lower right-hand corner. The diameter of Copernicus is 91 km. (Photograph from Mt. Wilson Observatory.)

FIGURE 2–3: Telescopic photograph of Aristoteles showing part of its associated swarm of secondary craters. A well-defined loop of secondary craters extends to the upper right from Aristoteles on to the surface of Mare Frigoris. The diameter of Aristoteles is 91 km. (Photograph from Lick Observatory.)

primaries, the secondary craters are generally devoid of rims and are commonly long shallow trenches of simple form (Figure 2–6). Secondary craters on the maria associated with rayless primaries are referred to the Eratosthenian System of Shoemaker and Hackman.[3] Both Copernican and Eratosthenian secondaries are represented in the Ranger VII photographs.

Primary craters. Primary craters are well represented in the Ranger VII photographs. They range in size from craters previously known from telescopic observation—such as Bullialdus, 62 km in diameter, and Darney, 16 km in diameter—down to 2 m in diameter, the size of the smallest identifiable primary craters shown in the last photographs obtained from the P cameras. For craters smaller than 7 km in diameter, the shape is essentially the same, down to the smallest crater observed. On all photographs obtained from Ranger VII, from the smallest to the largest scale, the primary craters occupy a very small fraction of

[3] *Ibid.*

the mare surface, typically less than 1 per cent. At all scales, they appear to be randomly scattered over the field. Over the range of size from 30-m diameter to the largest primaries photographed, some primary craters are surrounded by bright halos or rays, and some are not (Figures 2–7 through 2–10). Discrete bright halos are associated exclusively with primary craters, but no demonstrable variations in albedo have been found to be localized around craters smaller than 30 m in diameter.

Secondary craters. Secondary craters are abundantly represented in the Ranger VII photographs. In Mare Cognitum, a majority of them may be identified with three major secondary crater swarms. Two of these swarms coincide with major ray systems—one with the ray system of Copernicus and one with that of Tycho. Only relatively large secondary craters of Copernicus are resolved in the Ranger VII photographs, as Copernicus rays are not present in the highest-resolution pictures. These large secondary craters of Copernicus are best

FIGURE 2–4: Telescopic photograph of Langrenus showing part of its associated swarm of secondary craters. The secondary craters extend outward from the hummocky rim of Langrenus as subradial bands of elongate craters on Mare Fecunditatis. Langrenus, 131 km in diameter, is the largest ray crater on the sub-Earth side of the moon. (Photograph from Lick Observatory.)

FIGURE 2–5: Diagrammatic sketch of a composite secondary crater of Copernican age. Copernican secondary craters that can be resolved with Earth-based telescopes generally have distinct raised rims and may show internal cusps. (Based on telescopic observation of M. H. Carr.)

FIGURE 2–6: Diagrammatic sketch of a secondary crater of Eratosthenian age. Eratosthenian secondary craters generally have no observable raised rims or internal cusps. (Based on telescopic observations of M. H. Carr.)

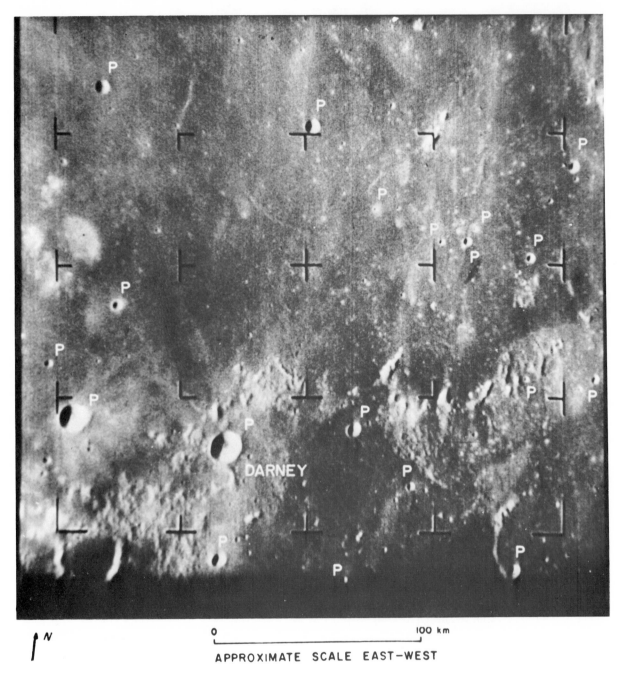

N

0   100 km

APPROXIMATE SCALE EAST—WEST

FIGURE 2–7: A-camera photograph 151 showing distribution of telescopically resolvable primary craters (designated P).

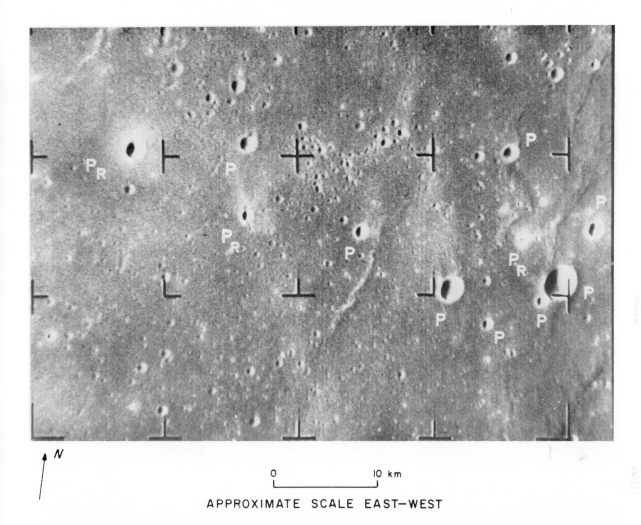

FIGURE 2–8: Part of A-camera photograph 189 showing group of primary craters of intermediate size. (Primary craters without rays are designated *P*, and primary craters with rays are designated *P_R*.)

portrayed near the northern margins of A-camera photographs 155 to 158. Most of the secondary craters in the highest-resolution photographs are in a ray belonging to the Tycho system. The third major swarm of secondary craters is associated with the Eratosthenian primary crater Bullialdus, and the differences in form between the larger Copernican and Eratosthenian secondaries are well illustrated in the A-camera photographs.

The swarm of secondary craters within the Tycho ray shown in the last four A-camera photographs (Figures 2–11 through 2–14) exhibit details of form and spacing too small to be seen through the telescope. The majority of readily identifiable secondary craters in this ray range from 200 to about 400 m across and from about 200 m to 1 km in

length. Most of the more elongate craters are found to be composite, composed of two, and in some cases three, craters merged together. Craters of this type are closely spaced, forming a rough crescent-shaped pattern opening to the northwest (astronautical co-ordinates). The convex boundary of the crescent is the approximate margin of the ray, and the largest secondary craters tend to occur along the southern part of the crescent, which is the part nearest Tycho. Along the northern arm of the crescent, the secondary craters are smaller and shallower in proportion to their width, and the rims are distinctly more rounded.

In the southern or proximal part of the crescent-shaped secondary crater cluster are a number of circular, relatively deep craters ranging in diam-

N

0   10 km

APPROXIMATE SCALE EAST–WEST

FIGURE 2–9: A-camera photograph 197 showing identifiable primary craters. (Primary craters without rays are designated P, and primary craters with rays are designated P_R. Most of the primary craters identified are too small to be resolved with Earth-based telescopes.)

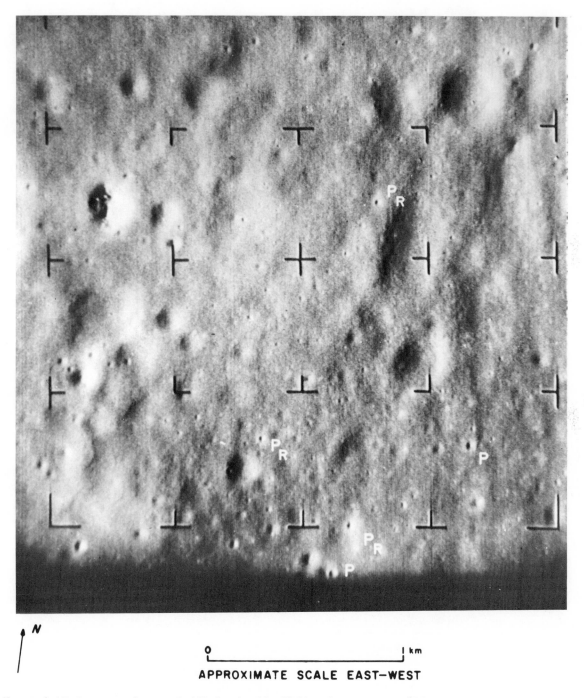

FIGURE 2–10: A-camera photograph 199 showing identifiable primary craters. (Primary craters without rays are designated P, and primary craters with rays are designated P_R. The smallest ray craters identified are about 30 m in diameter and are the smallest ray craters so far identified on the moon.)

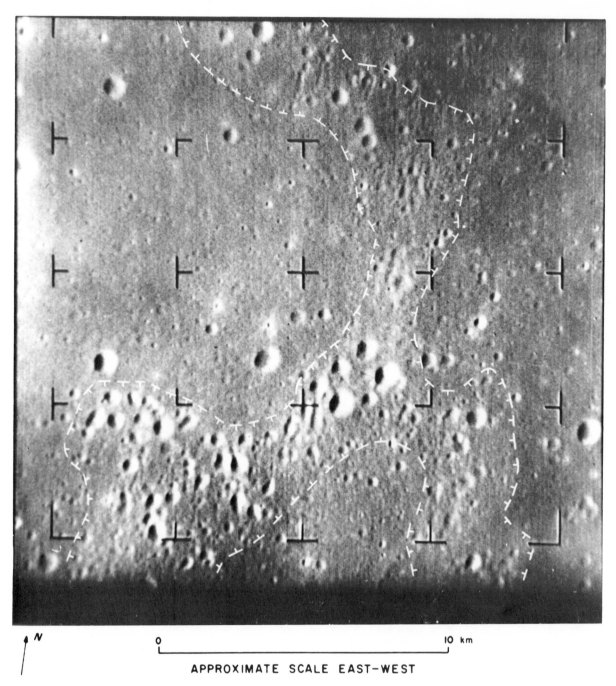

N

0   10 km

APPROXIMATE SCALE EAST–WEST

FIGURE 2–11: A-camera photograph 196 showing approximate boundaries of Tycho ray and distribution of secondary craters within the ray. (Dashed lines with tick marks are ray boundaries; tick marks are on ray. Note the abundance of elongate, composite craters in the ray. Nearly all of the elongate craters and most of the circular craters in the ray are interpreted as secondary craters of Tycho.)

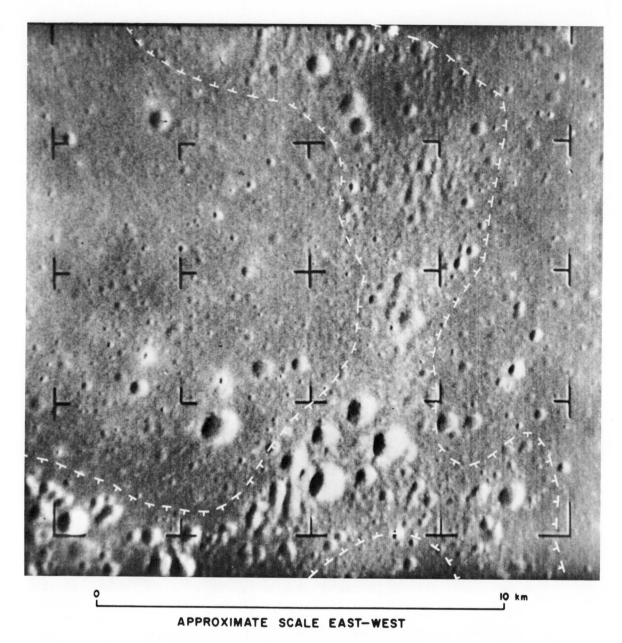

APPROXIMATE SCALE EAST–WEST

FIGURE 2–12: A-camera photograph 197 showing approximate boundaries of Tycho ray and distribution of secondary craters within the ray. (Dashed lines with tick marks are ray boundaries; tick marks are on ray. Note the abundance of elongate, composite craters in the ray. Nearly all of the elongate craters and most of the circular craters in the ray are interpreted as secondary craters of Tycho.)

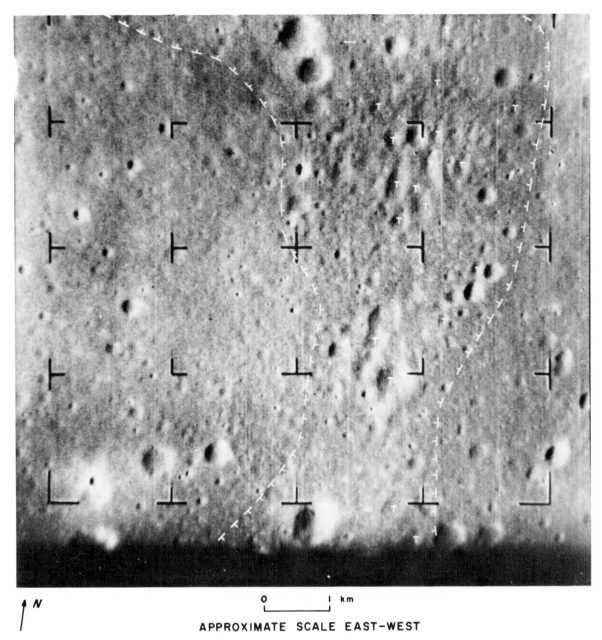

N

0 — 1 km

APPROXIMATE SCALE EAST-WEST

FIGURE 2–13: A-camera photograph 198 showing approximate boundaries of Tycho ray and elongate secondary craters of Tycho. (Dashed lines with tick marks are ray boundaries; tick marks are on ray. Elongate Tycho secondary craters are designated *T*. Only the largest elongate secondary craters that may be identified with reasonable confidence as belonging to the Tycho secondary crater swarm are marked.)

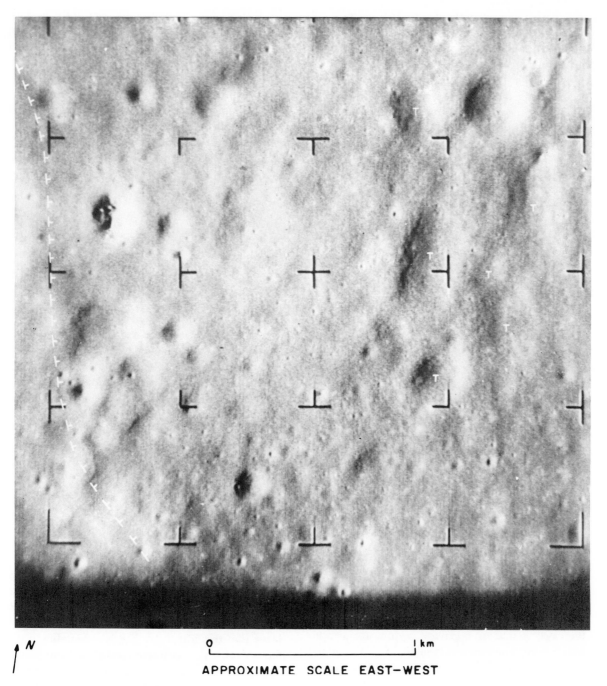

N

0 |_____| 1 km

APPROXIMATE SCALE EAST–WEST

FIGURE 2–14: A-camera photograph 199 showing approximate boundaries of Tycho ray and elongate secondary craters of Tycho (cf. Figure 2–13).

eter from about 400 to 900 m. Except for the lack of well-developed rims, some of these circular craters could be mistaken for primaries, were they not clearly members of the swarm. In another small Tycho ray northwest of the crescent-shaped cluster, all of the largest craters, which form a loose cluster on the end of the ray nearest Tycho, are circular in form. From the fact that they form a cluster and occur within the ray, it is inferred that these also must be secondary craters. It appears that the more open the cluster, the greater is the tendency for the secondaries to be circular.

A somewhat higher proportion of circular to elongate secondaries occurs in the rays of Tycho shown in the highest-resolution A-camera photographs than in the typical swarms of secondaries observed through the telescope. It appears that, in a given secondary swarm, the smaller and more widely spaced the craters become, the greater is the tendency for individual circular rather than elongate composite craters to be formed.

Large Eratosthenian secondary craters, most of which are probably related to Bullialdus, are widely scattered over the southern part of Mare Cognitum. A row or band of secondary craters may be traced north from Bullialdus crossing the southern margin of the mare (Figures 2–15 and 2–16). Near the north end of this row is a very prominent elongate crater over 3 km in length and about 1 km in maximum width (Figures 2–17 and 2–18). It is devoid of any discernible rim. The edges of the crater are rounded, and it is much shallower than primaries of 1-km diameter. The elongate crater has a vague internal structure and appears to be composed of about four barely recognizable smaller craters strung end to end. Three closely similar composite secondary craters of about the same size may be seen in this general region of the mare. The one which lies closest to the Ranger impact point is shown with highest resolution in photographs A191–193 (Figures 2–19 through 2–21). Between the rays, in this same general region, are a large number of smaller, less well-defined secondary craters, most of which are also probably a part of the Bullialdus swarm. They can be distinguished best in photographs A182–186 (Figures 2–22 through 2–25). In the northwest corner of photographs A187 and 188 (Figures 2–26 and 2–27) is a series of extremely shallow and very elongate faintly marked features which may belong to an unidentified Eratosthenian secondary swarm, possibly older than that of Bullialdus.

Elongate small secondaries can be recognized in the Ranger VII photographs over the range of length from 3 km down to 1 m. At least two sets of very small secondaries appear in the last part of the sequence of photographs obtained by the P cameras, one set with long axes tending north-northwest and the other set north-northeast (Figures 2–28 and 2–29). These small secondary craters are probably related to two or more nearby primary craters in Mare Cognitum, possibly two primaries about 1 km across or smaller with conspicuous bright halos that lie south of the Ranger impact point (Figure 2–8).

Craters less than 300 m in diameter. Nearly all craters in Mare Cognitum shown on the Ranger VII photographs that are greater than 1 km across can be classified as primary or as secondary craters. The majority of them are primary. In the diameter range from 300 m to 1 km, most of the craters can also be classified as primary or secondary, but craters of this size are predominantly secondary. Beginning at a diameter of about 500 m and extending to smaller sizes, a number of craters are observed in the Ranger photographs that are not clearly assignable either to the primary or to the secondary class. These craters are circular in plan but have low rounded rims and are shallower than typical primary craters (Figures 2–30 and 2–31). In the areas between the rays, these shallow circular craters do not clearly belong to any recognizable secondary crater swarm. They may be degraded primary craters or isolated round degraded secondary craters. At diameters less than 300 m, the crater population is dominated by this third indefinite type.

All gradations in form may be observed among circular craters less than 300 m across. They range from steep-walled primaries to craters with no discernible raised rim and with such gentle interior slopes that the craters are barely detectable in the photographs. For each size class from 2- to 300-m diameter, a complete sequence of crater forms with varying depth and slope of wall may be found between the primary and the shallowest, barely detectable craters. The edges of the shallow craters are invariably rounded and, in fact, are generally so smooth in the shallowest ones as to defy precise location.

FIGURE 2–15: Part of A-camera photograph 108 showing Bullialdus, position of band of Bullialdus secondary craters, and Mare Cognitum. (Secondary craters of Bullialdus are not resolved in this photograph.)

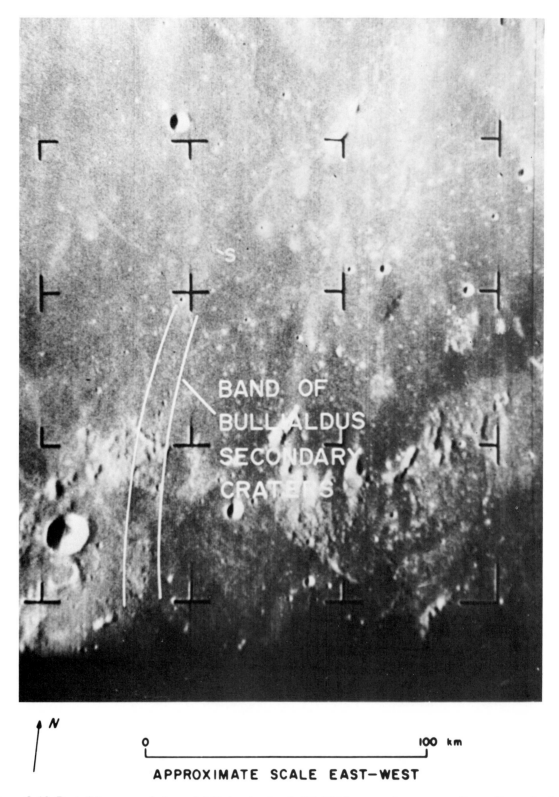

FIGURE 2–16: Part of A-camera photograph 156 showing band of Bullialdus secondary craters. (Large Eratosthenian secondary crater beyond band, designated S, probably also belongs to the Bullialdus secondary crater swarm.)

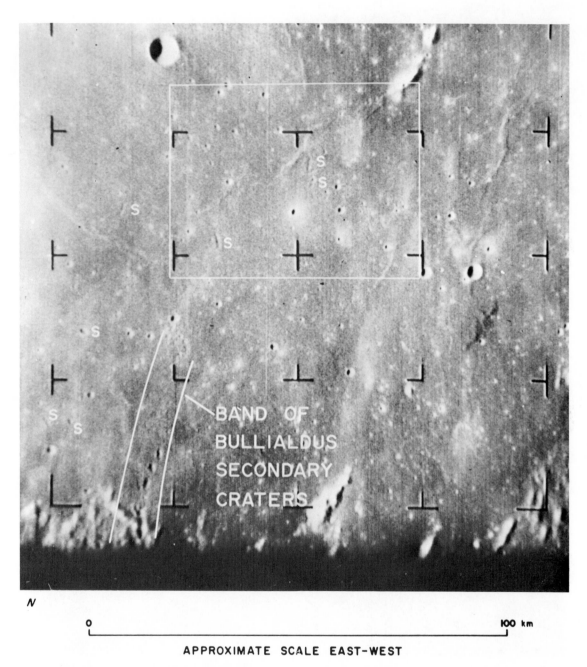

FIGURE 2–17: A-camera photograph 176 showing band of Bullialdus secondary craters and scattered large Eratosthenian secondary craters (designated *S*) that may belong to Bullialdus secondary crater swarm. (Inset shows position of Figure 2–18.)

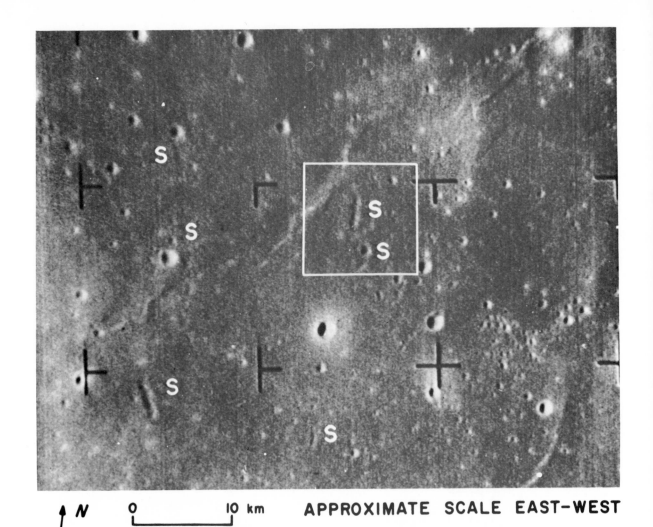

↑ N   0 |——————| 10 km   APPROXIMATE SCALE EAST-WEST

FIGURE 2–18: Part of A-camera photograph 185 showing distribution of large Eratosthenian secondary craters (designated *S*). (Inset shows position of Figures 2–19 and 2–20.)

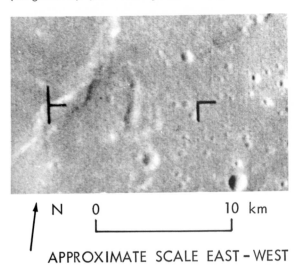

↑ N   0 |——————| 10 km

APPROXIMATE SCALE EAST-WEST

FIGURE 2–19: Part of A-camera photograph 191 showing small cluster of large Eratosthenian secondary craters. This cluster is probably part of Bullialdus secondary crater swarm.

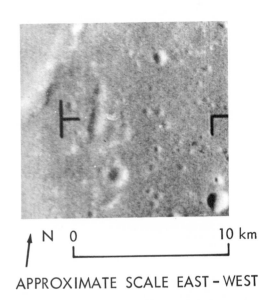

↑ N   0 |——————| 10 km

APPROXIMATE SCALE EAST-WEST

FIGURE 2–20: Part of A-camera photograph 192 showing small cluster of large Eratosthenian secondary craters.

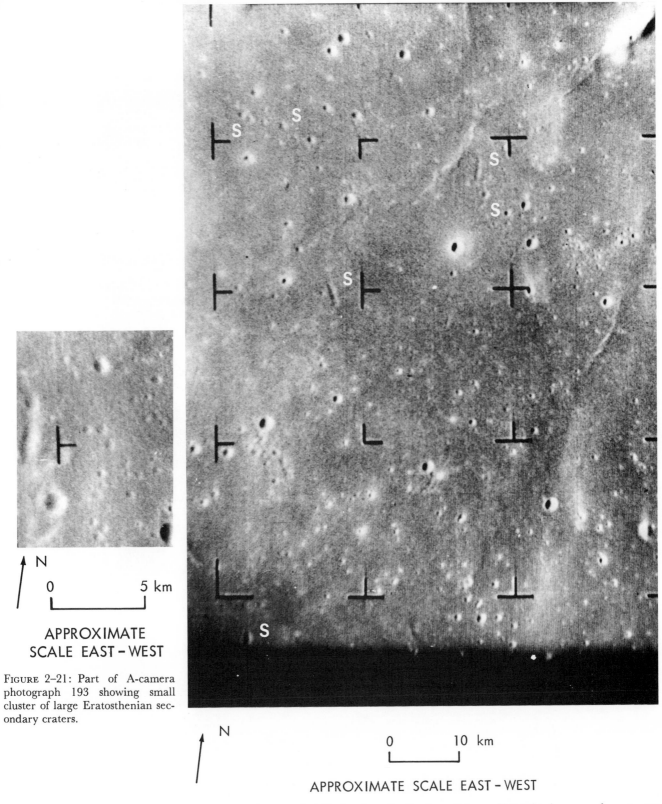

N

0   5 km

APPROXIMATE
SCALE EAST – WEST

FIGURE 2–21: Part of A-camera
photograph 193 showing small
cluster of large Eratosthenian secondary craters.

N

0   10 km

APPROXIMATE SCALE EAST – WEST

FIGURE 2–22: Part of A-camera photograph 182 showing distribution of large Eratosthenian secondary craters
(designated S).

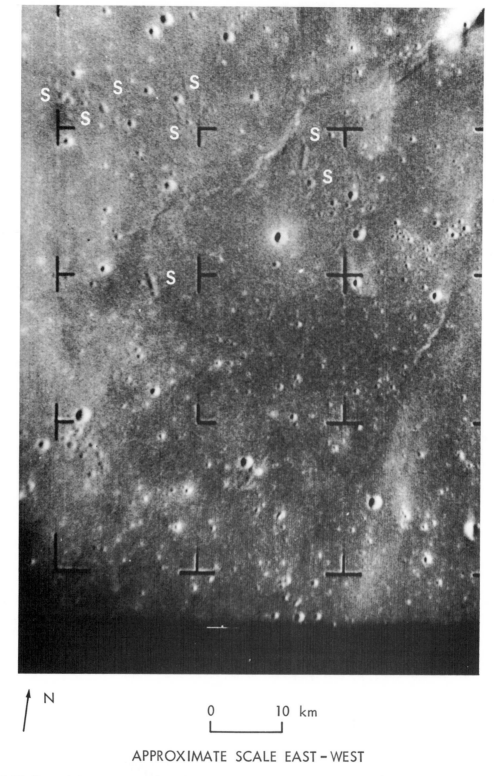

N

0   10 km

APPROXIMATE SCALE EAST - WEST

FIGURE 2–23: Part of A-camera photograph 183 showing distribution of large Eratosthenian secondary craters (designated *S*).

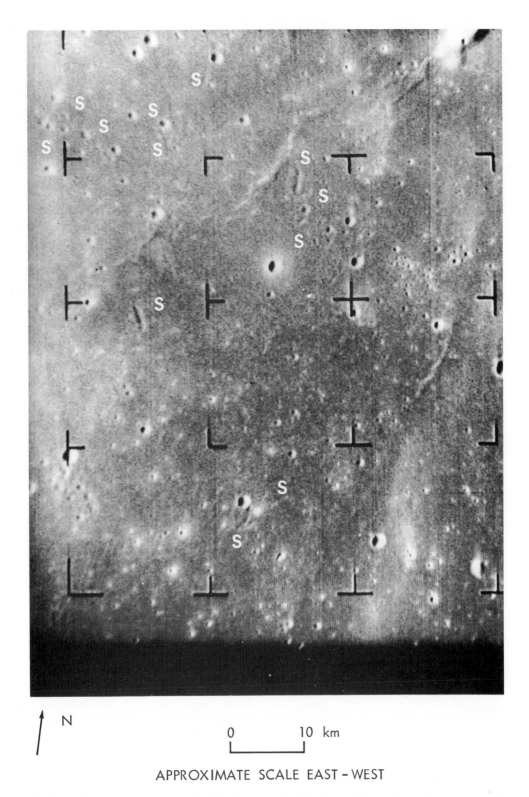

N

0   10 km

APPROXIMATE SCALE EAST – WEST

FIGURE 2–24: Part of A-camera photograph 184 showing distribution of large Eratosthenian secondary craters (designated S).

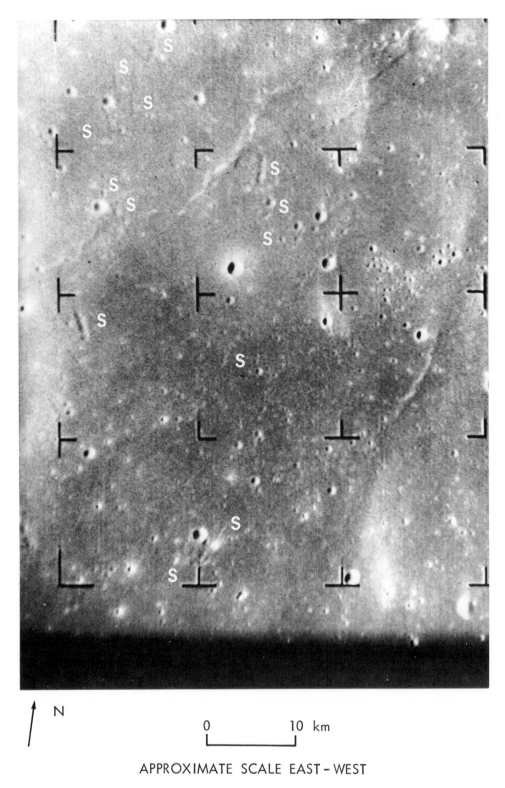

N

0   10 km

APPROXIMATE SCALE EAST – WEST

FIGURE 2–25: Part of A-camera photograph 186 showing distribution of large Eratosthenian secondary craters (designated *S*).

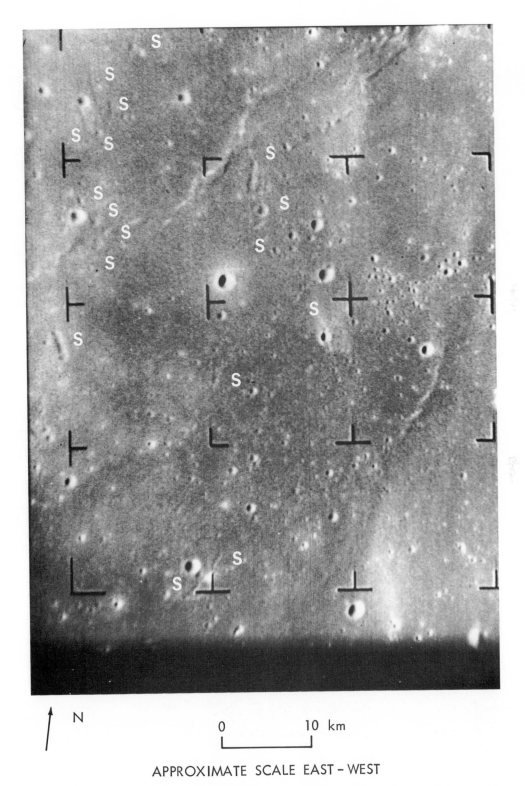

N

0   10 km

APPROXIMATE SCALE EAST – WEST

FIGURE 2–26: Part of A-camera photograph 187 showing distribution of large Eratosthenian secondary craters (designated *S*). Very shallow craters in upper right-hand corner may belong to a secondary crater swarm older than Bullialdus.

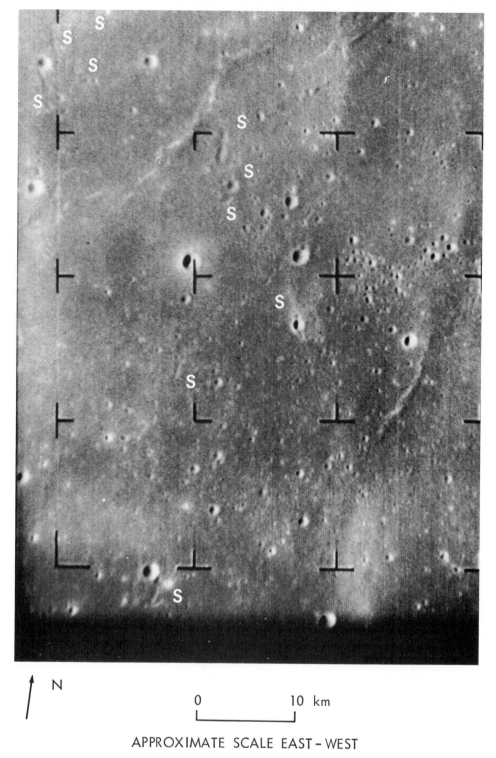

FIGURE 2–27: Part of A-camera photograph 188 showing distribution of Eratosthenian secondary craters. Very shallow craters in upper right-hand corner may belong to a secondary crater swarm older than Bullialdus.

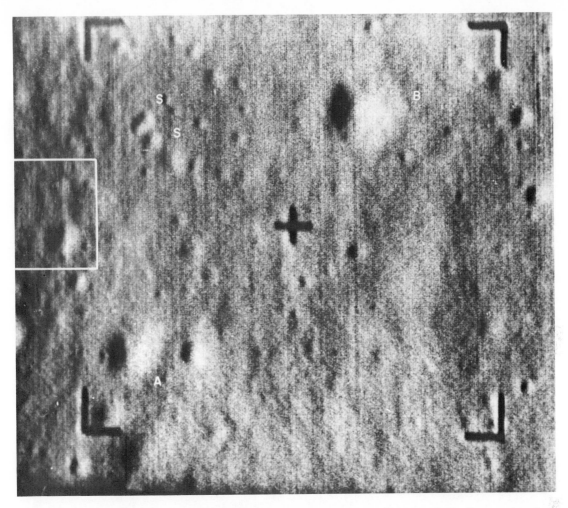

FIGURE 2–28: Last P₄-camera photograph showing small secondary craters (designated *S*). Craters *A* and *B* are identical with craters marked *A* and *B* in Figure 2–35. (Inset shows position of part of Figure 2–36.)

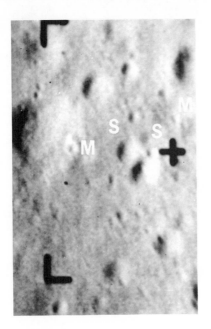

FIGURE 2–29: Last P₃-camera photograph showing small secondary craters (designated *S*) and mounts with summit craters (designated *M*). The length of the secondary craters is exaggerated by image motion, but the secondary craters are also elongated in the general direction of motion of the image as shown by the presence of small circular craters in the same part of the field.

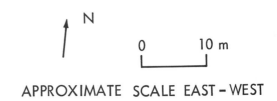

APPROXIMATE SCALE EAST–WEST

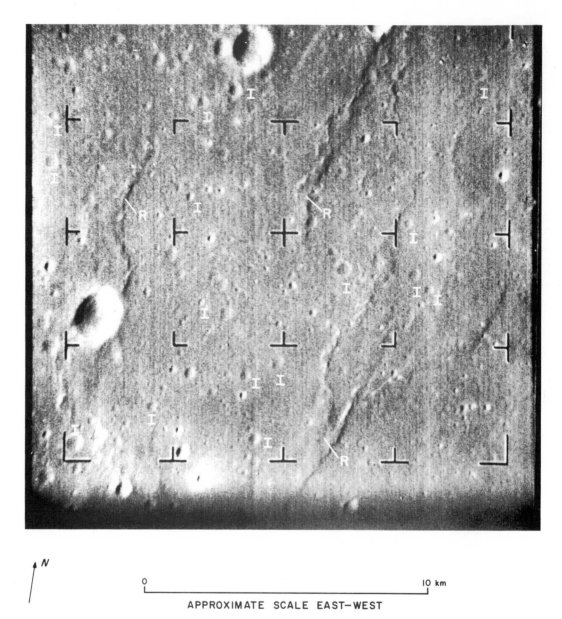

FIGURE 2–30: B-camera photograph 194 showing distribution of shallow circular craters with low rounded rims (designated *I*) of indefinite classification. Only the largest of the indefinite craters are marked. Major ridges are designated *R*. Note craters on ridge in upper right.

FIGURE 2–31: B-camera photograph 197 showing distribution of shallow circular craters with low rounded rims (designated *I*) of indefinite classification, and elongate and irregular craters (designated *E*). Only the largest craters in these two classes are marked. A prominent ridge in the lower right part of the photograph is designated *R*. Note the presence of sharply defined branches or spurs on the ridge.

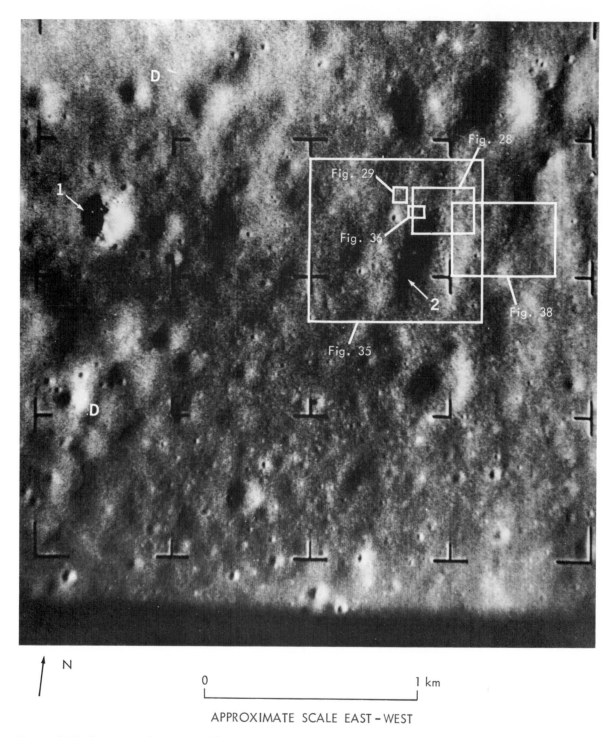

FIGURE 2–32: A-camera photograph 199 showing location of irregular features of craters, dimple-shaped craters (designated *D*), and other figures. Crater with prominent protuberances is designated *1*. Scarp in wall of elongate Tycho secondary crater is designated *2*. Subdued scarps in nearby secondary crater occur within inset showing position of Figure 2–38.

Many very shallow round-edged elongate or irregular craters less than 300 m in diameter are also represented in the high-resolution photographs (Figure 2–31). They occur both in and between the major rays.

Irregular features of craters. Certain unusual features of the small craters photographed by Ranger VII have attracted comment and speculation out of all proportion to the frequency of their occurrence. Foremost among these is the cluster of somewhat angular protuberances in the bottom of the deepest crater (the so-called rock) shown in A199 (Figure 2–32). There are perhaps as many as half a dozen other small craters, seen in the highest-resolution A and B photographs, that also have irregularities near the crater center (for example, in Figure 2–33), but none is shown with the same clarity of detail as the 200-m-diameter crater in A199. In this crater, there appear to be at least three discrete protuberances, one much larger than the other two. Slopes on the sides of these protuberances in places exceed the local sun elevation angle of 23 degrees. In addition to the protuber-

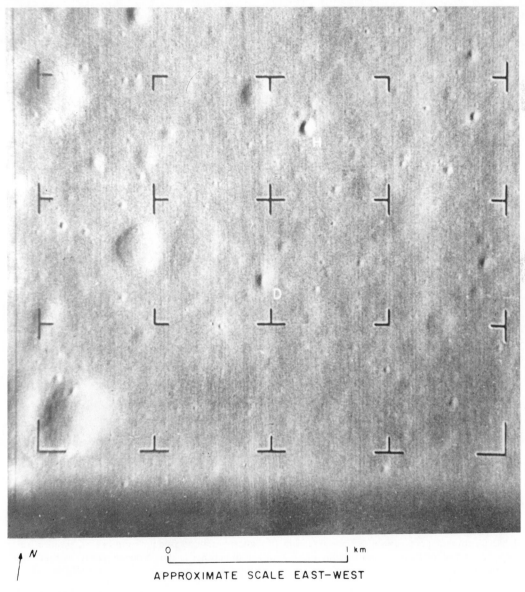

N

0   1 km

APPROXIMATE SCALE EAST–WEST

FIGURE 2–33: B-camera photograph 200 showing dimple-shaped crater (designated *D*) and small craters containing humps or protuberances (designated *H*).

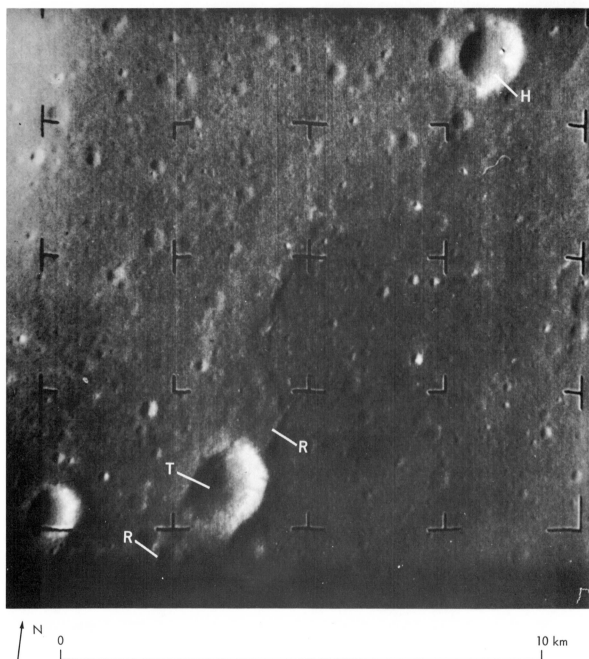

N   0   10 km

APPROXIMATE SCALE EAST – WEST

FIGURE 2–34: B-camera photograph 196 showing ridge (designated *R*) intersected by a crater, position of terrace within crater (designated *T*), and hump (designated *H*) in wall of another crater. Note streaks in wall of crater that intersects the ridge.

ances, there are two rows of smaller irregular features, consisting of very small bumps and troughs that extend up the walls of the crater. On the west wall, above the largest protuberance, is a distinct scallop or depression that resembles a landslide scar and is about the same width as the protruding mass below. The east wall is distinctly concave except near the floor of the crater, where it is sharply convex adjacent to the protruding body. A radial profile of this wall resembles the longitudinal profile of certain types of small landslides with rounded toes or snouts. The resemblance in form suggests that the east wall of the crater has slumped. Indeed, all of the features seen within the crater may have been formed by slumping, the largest protruding mass having been derived from the upslope scallop or scar.

A plausible alternative explanation of the relatively rare protuberances in small crater floors is that they represent remnants of impacting masses that formed the craters. To account for the survival of the impacting mass and the ratio of the diameter of the largest protuberance to the diameter of the crater, the impacting bodies would most probably be low-velocity lunar fragments ejected from nearby primary craters.

Other features observed in a few craters suggest that slumping has occurred, at least locally. These include scarps, terraces, and a few vague linear features within the craters which may be described as subdued scarps. One of the best examples of an abrupt scarp within a crater occurs in a 500-m-long rounded Tycho secondary crater in A199 (Figure 2–32). This is an irregular branching scarp along the western wall of the crater; it is steeper than 23 degrees over most of its length. The irregular trace and local branching of the scarp suggest that the material of the lower part of the western wall and crater floor has slipped downward a few meters. Very subdued scarps that are similarly irregular in trace appear in two smaller shallow irregular secondary craters nearby. I believe they are probably similar in origin to the abrupt-scarp.

A rounded terrace occurs midway up the western crater wall of a primary crater approximately 1.5 km in diameter in B196 (Figure 2–34). This feature has been interpreted by Dr. H. C. Urey as the extension of a mare ridge that intersects the rim crest of the crater, and as a laccolith by Dr. Kuiper; it may be related to similar terraces that have probably been formed by slumping in much larger primary craters. A hump in the wall of another crater of about the same size (Figure 2–34) resembles a landslide.

A few unusual craters about 100 m across are observed in A199 (Figure 2–32) and B200 (Figure 2–33), whose form may best be described as dimple-shaped. Each of these craters has a low, broadly rounded rim that is convex upward and a wall which appears to be nearly conical or locally convex upward; at the foot of the wall, in some of these craters, the slope seems to steepen slightly, forming a funnel-shaped pit at the center. It is likely that craters of this form have also been modified by slumping or mass movement, the upper parts of the walls having flowed toward the center, the converging toe of the flowing mass enclosing the pit.

Distribution of craters

The Ranger VII photographs contain a great deal of critically important new information on the size and spatial distribution of craters less than a kilometer in diameter in Mare Cognitum. The size-frequency distributions in Mare Cognitum of primary and secondary craters greater than a kilometer across are close to the average for other maria. The newly observed distributions of small craters represent extensions of three orders of magnitude of diameters below the previously known size distributions for larger craters on the maria. It is important to distinguish between the distribution of the craters in the major ray systems and the distribution in the areas between the rays.

Ray areas. With increasing resolution, the observed ray areas in Mare Cognitum begin to fill up very rapidly with craters less than a kilometer across, but their distribution within the rays is irregular. Some parts of the crescent-shaped ray shown in the last four A-camera photographs (Figures 2–11 through 2–14) are more than 50 per cent occupied by secondary craters 300 m across and larger. In the northern part of the same ray, however, only 20 per cent of the ray is occupied by craters of that size. Craters larger than 300 m are similarly sparse in the ray to the northwest of the crescent-shaped ray. For craters much smaller than 300 m in diameter, data are available only for the crescent-shaped ray; all of the highest-resolution photographs depict the northern branch of this ray, where the abundance of larger craters is lower than average for the entire crescent.

The last A-camera frame, 199 (Figure 2–14), illustrates what appears to be a fairly representative part of the northern part of the crescent-shaped ray. Craters ranging from 300 to 10 m in diameter can be detected in this photograph. Approximately 50 per cent of the field is covered by craters. The vast majority of these belong to the indefinite class of shallow craters, and many of them are extremely shallow depressions. Craters ranging in size from 10 to 50 m cover 25 to 30 per cent of the field and

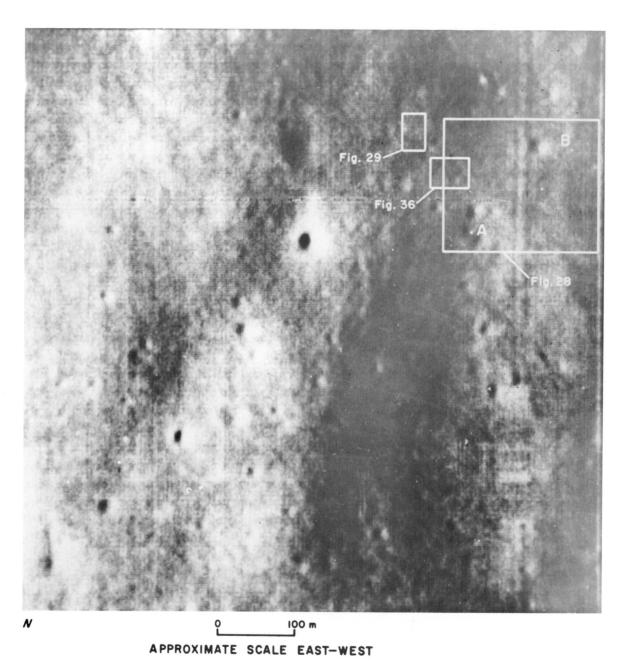

FIGURE 2–35: Part of A-camera photograph 199 reproduced from digitized magnetic tape record after removal of part of coherent noise by digital computer processing. Contrast has been enhanced to portray small craters more clearly. Computer processing and reproduction of photograph were carried out under the direction of Robert Nathan. Craters *A* and *B* are identical with craters marked *A* and *B* in Figure 2–28.

are scattered more or less uniformly over the larger craters and the areas between them. Many of the smallest craters are difficult to distinguish on the photographs reproduced in the A-Camera Atlas, but their occurrence and distribution can be observed on enhanced photographs from which some of the coherent noise has been removed by Robert Nathan of the Jet Propulsion Laboratory (Figure 2–35). The profusion of small craters on the larger craters is well illustrated in a segment of photograph A199 prepared by Nathan. The majority of the smallest craters are very shallow and have rounded rims.

The last photographs obtained from the P_1, P_3, and P_4 cameras (Figures 2–36, 2–29, 2–30), which appear to be fairly representative samples of the ray area covered in A199, show that between 25 and 50 per cent of the field is occupied by craters ranging from 1 to 10 m in diameter. Again, smaller craters are scattered over the larger ones.

The cumulative size-frequency distribution of craters in the ray area sampled by the Ranger photographs is illustrated in Figure 2–37. This figure shows the distribution for all types of craters, from the largest observed in Mare Cognitum down to craters 1 m in diameter. For craters larger than 1 km across, the curve represents the entire mare, but for smaller craters the data are necessarily restricted to areas covered by the Ranger photographs, and only a very minute area is sampled to

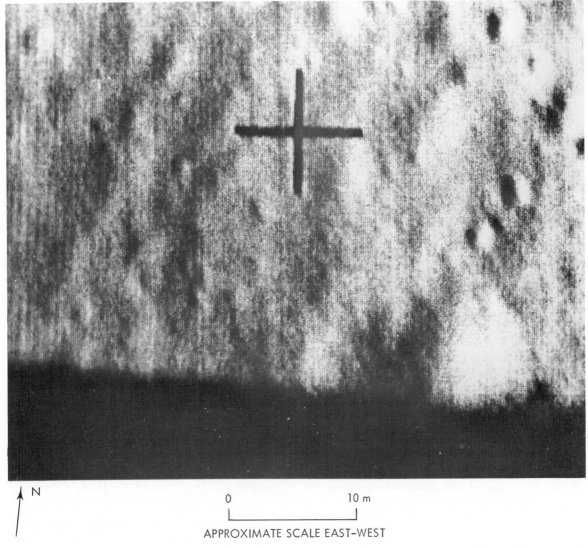

N

0   10 m

APPROXIMATE SCALE EAST-WEST

FIGURE 2–36: Last P_1-camera photograph. Note low ridges trending slightly east of north in left half of photograph.

obtain the estimated frequency of the smallest craters.

Areas between rays. Between the rays, the craters are more evenly distributed than in the rays. The spatial density of craters less than 1 km in diameter increases more slowly with increasing resolution than in the rays. In the sampled areas, covered chiefly by the B-camera photographs, craters greater than 300 m across are fairly widely spaced, occupying about 5 per cent of the total field. With further increase in resolution, the field fills rapidly with craters less than 300 m in diameter. Approximately 20 per cent of the field is covered with craters ranging from 50 to 300 m in diameter in the areas between the rays for which data are available (from highest-resolution B-camera photographs). Again, the large majority of the craters are of the indefinite shallow type, most of them very shallow with gently rounded rims. The distribution of craters much smaller than about 50 m in diameter in the areas between the rays cannot be obtained from the Ranger VII photographs because most of the pictures showing smaller craters are within the crescent-shaped ray. Thus, the range within which small craters would be expected to be found abundantly scattered over larger craters lies beyond the limits of best resolution for the Ranger coverage of the areas between the rays.

The cumulative size-frequency distribution for all types of craters between the rays is shown as a separate branch of the distribution curve in Figure 2–37.

Positive-Relief Features

One of the most striking facts revealed by the Ranger VII photographs is the rarity of small positive-relief features other than crater rims. New features of this type revealed in the Ranger VII pictures include small details on a few mare ridges in areas between major rays and a number of very small low mounds in the crescent-shaped Tycho ray observed in the highest-resolution P-camera photographs.

Mare ridges photographed with resolution much greater than that of previously available photographs are portrayed best in B-camera pictures 194 to 197 (Figures 2–30 and 2–31). The fine details revealed are similar to somewhat larger topographic forms observed on mare ridges through the telescope. Of particular interest are sharply defined

branches that extend short distances off the main ridges, generally where the main ridges change strike. Some of these branches and some small cross ridges as short as 300 m in length are 50 m or less in width and have local slopes on their flanks approaching 20 degrees. Similar small spines occur locally along the crest of a main ridge. Perhaps one of the most significant observations about these ridges is their local sharpness of form compared to the rounded rims of most of the small craters nearby. Parts of the ridges, however, are pitted with craters (Figure 2–30).

At least two kinds of mounds appear to be present in the highest-resolution photographs of the

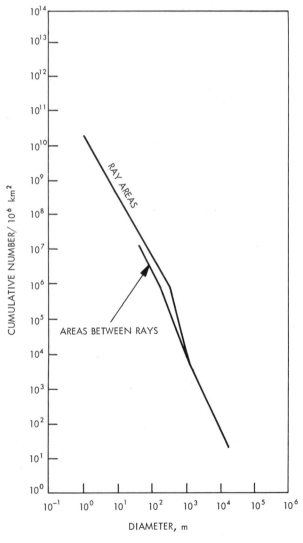

FIGURE 2–37: Cumulative size-frequency distribution of craters in Mare Cognitum (based on distribution of craters observed in Ranger VII photographs).

crescent-shaped Tycho ray. These are best portrayed in the last P_1- and P_3-camera photographs (Figures 2–36 and 2–29). Much of the field in the last P_1 photograph is occupied by gentle, irregular, elongate mounds trending north-northeast. Many small craters are superimposed on the mounds, which enhances their irregular appearance. The mounds are 5 to 15 m across and are separated only by narrow, winding, shallow depressions. Similar features are observed with poorer resolution on the next-to-last P_3- and P_4-camera photographs (see, for example, Figure 2–38). The mounds are locally well defined along a faint curved trough or subdued scarp, to which they give a very ropy appearance. Dr. Kuiper has compared the surface here to the bark of the Ponderosa pine. The bark-like appearance is heightened by the presence of many small craters superposed on and interspersed with the mounds. Relief on most of these mounds probably does not exceed a few meters, and the slopes are predominantly less than 10 degrees (estimated). I believe that these gentle mounds are probably an intrinsic feature of the ray and will discuss them in more detail in the interpretive section to follow. A few other gentle mounds, ranging from a meter to a few meters across, have been detected in the last P_3-camera photograph, chiefly by photometric

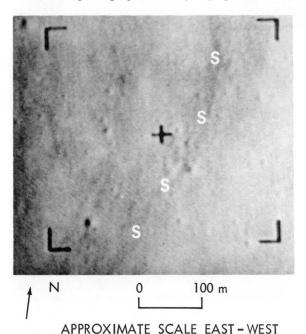

N   0   100 m

APPROXIMATE SCALE EAST – WEST

FIGURE 2–38: Second-to-last P_4-camera photograph showing ropy appearance of subdued scarp (designated S).

measurements. These features have such gentle slopes that they are not readily detected by visual examination.

One other unusual kind of positive-relief feature occurs in the last P_3-camera photograph (Figure 2–29). It consists of a mound with a crater in the summit; it may be regarded as a crater with a greatly exaggerated rim. There are at least two definite examples of this feature in the photograph. The largest is an elongate mound about 5 m in length and 3 m across, with a 1-m-diameter crater in its summit. This mound occurs just inside the rim crest of a 10-m-diameter crater. The height of the mound measured in a direction normal to the large crater wall is about ⅓ m.

Albedo Variations

Variations in surface brightness due to variations in albedo of the surface rather than to variations in slope occur at many different scales in Mare Cognitum. Albedo variations over broad areas of the mare surface and local bright halos around primary craters larger than 300 m in diameter are well known from telescopic observations. The major broad variations in albedo are due to ray systems of Tycho and Copernicus. Another well-known feature is a local, very dark streak in the southeastern part of Mare Cognitum that corresponds with a low, irregular ridge about 15 km in length. This ridge has been classified as part of the Procellarum group of Imbrian age by Eggleton,[4] who considered it to be closely related to domes observed elsewhere on the maria. It is well portrayed in about 100 of the A-camera photographs, but no critical new details of the ridge have been observed from these pictures.

Information of prime interest obtained from the Ranger VII photographs about the variations in albedo chiefly concerns the very small features and the sharpness of gradient of albedo variations. The margins of major rays observed through the telescope are diffuse, but the cause of this diffuseness could not be determined from the telescopic observations. One possibility was that the rays were composed of discrete, telescopically unresolved bright patches, and that the bright patches became more widely spaced toward the ray margins. The high-resolution photographs of the Tycho rays obtained by Ranger have revealed that the bright walls of some of the telescopically unresolvable craters in the rays contribute to the over-all ray brightness, but that this effect is superimposed on a background of more uniform albedo between the bright-walled craters. This background albedo of rays diminishes gradually toward the ray margins. The margin of the ray cannot be located more closely than about 1 km because of the gradual change in albedo, even in the high-resolution pictures. Over distances of a few tens of meters, areas in the crescent-shaped Tycho ray appear to be nearly uniform in albedo except near very small, superimposed ray craters.

The smallest observed ray craters are about 30 m across and have bright halos or rays of comparable width. The margins of these halos and rays have the steepest gradients of albedo variation observed.

Another albedo variation feature of considerable interest is the pattern of radial and irregular streaks in the walls of certain primary craters. The best example occurs in the crater intersecting the mare ridge shown in B196 (Figure 2–33). Alternate bright and dark streaks may be observed running down the illuminated crater wall. Well-defined streaks range in width from about 100 m to the limit of resolution, and some taper noticeably

downward. These streaks may be produced by sliding of weakly consolidated debris on the crater walls. A more detailed interpretation is given below. Similar but more irregular streaks occur on the wall of a primary crater, 3 km in diameter, shown in Figures 2–39 through 2–41. These streaks have widths up to several hundred meters.

II. CRATERING AS THE DOMINANT SURFACE PROCESS

So far as may be judged from the Ranger VII photographs, cratering is the dominant process responsible for the small relief features of Mare Cognitum. This conclusion was anticipated before the evidence from Ranger VII was at hand, because the moon is essentially devoid of an atmosphere and its surface is directly exposed to the bombardment of small and large meteoroids crossing the earth's orbit. Because of the lack of an atmosphere, furthermore, most of the familiar processes which sculpt and modify the surface of the earth would not be expected to occur on the moon. Thus, the bombardment by meteoroids and larger objects from interplanetary space is one of the few processes that, demonstrably, must occur on the moon, and it is the only process that can also be demonstrated to be capable of producing

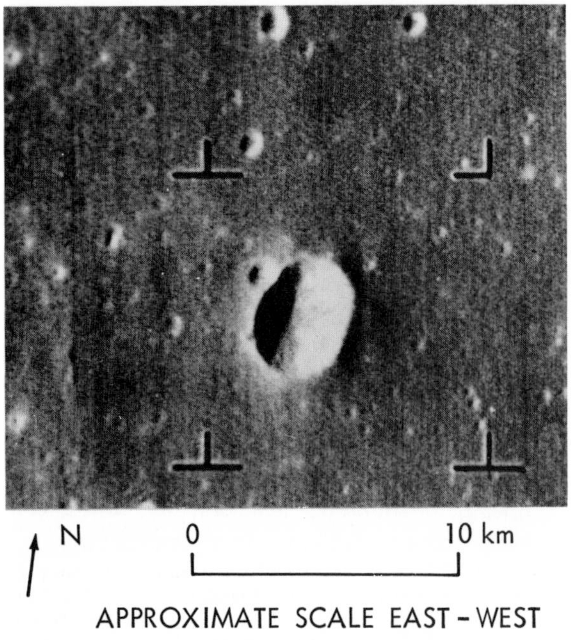

N   0   10 km

APPROXIMATE SCALE EAST – WEST

FIGURE 2–39: Part of B-camera photograph 182 showing crater with prominent streaks on crater wall.

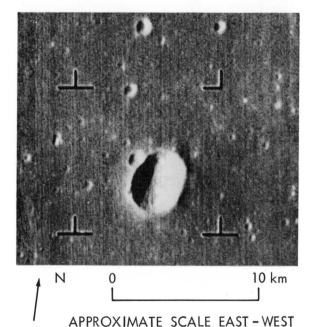

N   0   10 km

APPROXIMATE SCALE EAST – WEST

FIGURE 2–40: Part of B-camera photograph 183 showing crater with prominent streaks on crater wall.

most of the small surface features observed in Mare Cognitum.

All of the small relief features discovered in the Ranger VII photographs can, in my opinion, be explained as being a direct result of the formation of craters by impact. No other process is required, except for local mass movement (slumping), to account for the full range of small features (other than the sinuous mare ridges) and crater shapes observed. In accordance with the principle of simplicity (Occam's razor), impact cratering (with minor mass movement) will be adopted as the single process leading to the development of the present relief in the region of Mare Cognitum around the Ranger VII impact point. The adoption of this point of view does not mean that the possible occurrence of other processes is not recognized, but simply that no compelling evidence is known that other processes have had an observable effect on the topography of the features observed.

Darkening of materials near the moon's surface by high-energy solar radiation will be adopted as an auxiliary process leading, in conjunction with cratering, to the observed albedo variations in Mare Cognitum. It has been well established that high-energy solar radiation must reach the surface of the moon and that such radiation will produce color centers and darkening of many silicates in the laboratory.

N 0 10 km

APPROXIMATE SCALE EAST - WEST

FIGURE 2–41: Part of B-camera photograph 184 showing crater with prominent streaks on crater wall.

With these initial assumptions, we may proceed to examine how the features observed in Mare Cognitum are to be accounted for, in detail, and then to construct a model of the fine structure of the mare surface that is consistent with the observational facts and assumed processes. It will be shown that such a model can be developed that is also consistent with the presently known facts about the physical characteristics of the moon's surface obtained from telescopic observations.

Cratering Rate as a Function of Crater Size

The primary craters on the maria are here interpreted as having been formed by impact of meteoroids and larger interplanetary objects such as asteroids and comet nuclei. Detailed evidence for this interpretation has been presented elsewhere[5–12] and will not be reintroduced here. According to this interpretation, the observed primary craters were formed by solid objects, ranging in diameter from several centimeters to a few kilometers, striking the solid surface of the mare at high velocity and propagating a shock locally in the mare material.[13–14] This process has probably continued intermittently since the maria were formed, possibly as long as several billion years.

[5] Shoemaker, "Interpretation of Lunar Craters," *Physics and Astronomy of the Moon*, ed. by Z. Kopal, London:Academic Press (1960), pp. 283–359.

[6] Shoemaker, Hackman, and Eggleton, "Interplanetary Correlation of Geologic Time," *Advances in the Astronautical Sciences*, Vol. 8, New York: Plenum Press (1963), pp. 70–89.

[7] G. K. Gilbert, "The Moon's Face—A Study of the Origin of Its Features," *Philos. Soc. of Wash. Bull.*, Vol. 12 (1893), pp. 241–292.

[8] R. S. Dietz, "The Meteoritic Impact Origin of the Moon's Surface Features," *J. Geol.*, Vol. 54 (1946), pp. 359–375.

[9] R. B. Baldwin, *The Face of the Moon*, Chicago: University of Chicago Press (1949).

[10] H. C. Urey, "The Origin and Development of the Earth and Other Terrestrial Planets," *Geochimica et Cosmochimica Acta*, Vol. 1 (1951), pp. 209–277; correction, Vol. 2 (1952), pp. 263–268.

[11] Urey, "The Origin and Significance of the Moon's Surface," *Vistas in Astronomy*, Vol. 2, New York: Pergamon Press (1956), pp. 1667–1680.

[12] Kuiper, "On the Origin of the Lunar Surface Features," *Proc. Nat. Acad. Sci.*, Vol. 40 (1954), pp. 1096–1112; criticism by Urey, Vol. 41, pp. 423–428; further comments by Kuiper, pp. 820–823.

[13] Shoemaker, "Impact Mechanics at Meteor Crater, Arizona," *The Solar System, Vol. IV—The Moon, Meteorites, and Comets*, ed. by B. M. Middlehurst and G. P. Kuiper, Chicago: University of Chicago Press (1963), pp. 301–336.

[14] R. L. Bjork, "Analysis of the Formation of Meteor Crater, Arizona, a Preliminary Report," *J. Geophys. Res.*, Vol. 66 (1961), pp. 3379–3387.

The secondary craters are interpreted as having been formed by impact of comparatively low-velocity fragments of the moon ejected from primary craters located both on the maria and on other parts of the lunar surface. (For an explicit account of the interior ballistics of a primary crater and the exterior ballistics of secondary crater-forming fragments, see Shoemaker and Hackman [fn. 2], pp. 329–340.) The rays on Mare Cognitum, under this interpretation, consist primarily of mare material ejected from the secondary craters but contain a small admixture of material derived from the more distant primary craters with which the rays are associated.

Primary cratering rate

The rate of primary cratering on the lunar maria may be estimated from three independent sources of information: (1) the present observed fall of meteorites and micrometeorites on Earth, (2) the distribution and age of ancient impact craters and eroded structures of possible impact origin on Earth, and (3) the distribution of primary craters on the maria and estimates of the absolute ages of the maria. Shoemaker, Hackman, and Eggleton (fn. 6) showed that their interpretation of the terrestrial geologic record of impact cratering for the past several hundred million years is fairly consistent with a very large extrapolation in time and mass of the observed mass-frequency distribution and rate of fall of meteorites. They further demonstrated that the size-frequency distribution and average number of craters per unit area on the maria that may be predicted from the terrestrial data are consistent with the telescopically observed primary crater distribution on the maria (for all except the largest craters) if (1) the maria are assumed to be several billion years old and (2) the impact rate has remained relatively constant since they were formed. The observational uncertainties of the meteorite infall and of the geologic record of cratering, however, are rather large. Perhaps the greatest source of uncertainty in comparing the terrestrial data with the lunar data is the large extrapolation backward in time of the predicted cratering rate.

The expected or predicted size-frequency distribution of primary craters on the lunar maria, based on extrapolating the estimated size-frequency distribution of meteorites falling on the earth to

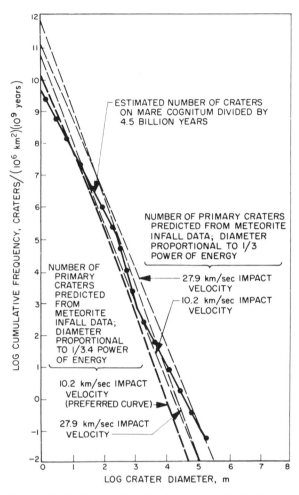

FIGURE 2–42: Cumulative size-frequency distribution of craters in Mare Cognitum compared with cumulative size-frequency distribution predicted for primary craters on the basis of meteorite infall data. Size-frequency distribution of craters in Mare Cognitum estimated from Ranger VII photographs is an average for ray areas and areas between rays.

very large sizes and on extrapolating the present estimated rate of infall backward in time, is illustrated in Figure 2–42. The meteorite infall data used are from Brown;[15] it is necessary to adopt likely impact velocities for the meteorites striking the moon and to employ empirical scaling laws in order to derive crater size-frequency distributions. An age of 4.5 billion years, the approximate age of

[15] H. Brown, "The Density and Mass Distribution of Meteoritic Bodies in the Neighborhood of the Earth's Orbit," J. Geophys. Res., Vol. 65 (1960), pp. 1679–1683; "The Density and Mass Distribution of Meteoritic Bodies in the Neighborhood of the Earth's Orbit—Addendum," J. Geophys. Res., Vol. 66 (1961), pp. 1316–1317.

the earth,[16] has been arbitrarily assumed for the maria in order to compare predicted primary crater distributions with the observed crater distribution on Mare Cognitum.

Four curves are shown in Figure 2–42 which correspond, respectively, to the probable upper and lower limits of average meteorite impact velocity on the moon and to two empirical scaling laws. A lower limit of average impact velocity of 10.2 km/sec and an upper limit of 27.9 km/sec are adopted. Objects striking the moon at these velocities would have entered the earth's atmosphere at 15 and 30 km/sec, respectively, had they encountered the earth instead. The actual modal velocity of entry of meteorites into the earth's atmosphere has been estimated as close to 15 km/sec.[17] The diameters of the craters formed by impact have been assumed to be proportional either to the $\frac{1}{3}$ power or to the $\frac{1}{3.4}$ power of the kinetic energy of the meteorites. Diameters of very small hypervelocity impact craters in rock are approximately proportional to the $\frac{1}{3}$ power of the projectile kinetic energy,[18] whereas the diameters of large craters formed by shallow subsurface explosion in rock and alluvium are approximately proportional to the $\frac{1}{3.4}$ power of the energy released.[19] Crater sizes are scaled from the Jangle U nuclear explosion crater, 82 m in diameter, which is intermediate in size between the largest and smallest craters observed on Mare Cognitum. Small corrections for the differences in size and capture cross-section of the earth and moon have been made in estimating the lunar impact rate from the terrestrial meteorite infall. (For a discussion of these corrections and the crater scaling relationships, see Shoemaker, Hackman, and Eggleton [fn. 6].) The four curves define an envelope of the predicted size-frequency distribution of primary craters formed on the lunar maria in 4.5 billion years.

[16] C. C. Patterson, G. R. Tilton, and M. G. Inghram, "Age of the Earth," *Science*, Vol. 121 (1955), pp. 69–75.

[17] F. L. Whipple and R. F. Hughes, "On the Velocities and Orbits of Meteors, Fireballs, and Meteorites," *J. Atmos. and Terres. Phys.*, Suppl. 2 (1955), pp. 149–156.

[18] H. J. Moore, D. E. Gault, and E. D. Heitowit, "Change in Effective Target Strength With Increasing Size of Hypervelocity Impact Craters," *U.S. Geol. Sur., Astrogeol. Studies Ann. Prog. Rep., Aug. 25, 1962, to July 1, 1963*, Pt. D (1964), pp. 52–63.

[19] G. W. Johnson, *Note on Estimating the Energies of the Arizona and Ungava Meteorite Craters*, UCRL-6227, United States Atomic Energy Commission, 1963.

The cumulative size distribution of all craters observed in the Ranger VII photographs (data for ray areas and nonray areas combined) falls within the envelope of predicted distributions of primary craters for all except the smallest and largest craters (Figure 2–42). This rough agreement might be taken to indicate that the assumptions adopted in deriving the predicted distributions are approximately correct and that most of the observed craters are of primary impact origin. There is a serious discrepancy, however, between the slopes of the observed crater distribution and the predicted distribution curves. Except for a short segment of the observed crater distribution curve, for craters ranging in size from 250 to 1,000 m, that is, steeper than the predicted distribution curves, the observed crater distribution curve is less steep than the predicted distributions. For this reason, there is no way to improve significantly the match between the predicted and observed distributions simply by assuming a different age for the maria.

The predicted crater distribution, based on the best estimate of the modal impact velocity of meteorites on the moon and the preferred scaling law for craters between 1 and 1,000 m in diameter, is shown in Figure 2–42 by a heavy line. The discrepancy between this curve and the observed crater distribution is large. At a diameter of 1 m, the observed cumulative number of craters is an order of magnitude less than predicted. If the age of the maria were reduced an order of magnitude to bring the predicted and observed number of craters into agreement at 1-m diameter, the number of all craters larger than 10 m in diameter would be more than an order of magnitude too great.

The deficiency of observed small craters that can be identified as primary is actually much greater than an order of magnitude. In Figure 2–43, the size-frequency distributions of identifiable primary craters and of the total number of craters observed in the Ranger VII photographs are plotted as separate curves. The curves join at about 1-km crater diameter, as all craters (in Mare Cognitum) much larger than this that are observed in the Ranger photographs are identifiable primaries. At 100-m diameter, the total number of craters is an order of magnitude larger than the number of identifiable primary craters, and at a few meters diameter, the total number of craters is more than an order of magnitude larger than the number of

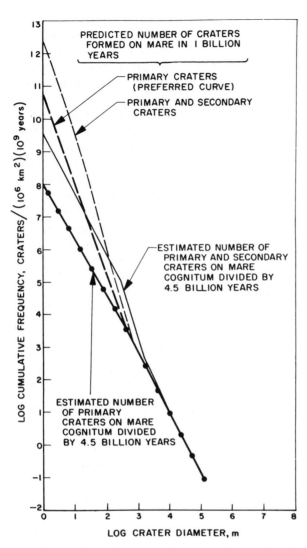

FIGURE 2–43: Cumulative size-frequency distribution of primary craters and of all craters on Mare Cognitum compared with predicted size-frequency distribution of primary craters and of all craters.

identifiable primaries. The predicted number of primary craters (preferred curve) is more than two orders of magnitude greater than the observed number at a few meters crater diameter. This difference between predicted and observed numbers of primary craters decreases rapidly with increasing crater size, and the two curves intersect at about 400-m diameter.

Secondary cratering rate

The rate of formation of secondary craters on the moon can be estimated from the primary cratering rate if the size distribution of secondary craters formed by individual primary craters is known. A predicted size distribution of secondary craters on the maria can be derived from the observed distribution of large primary craters, moreover, without any assumptions about the age of the maria or about the cratering rate. To do this, we require empirical data on the size-frequency distribution of secondary craters produced by primaries of different sizes. Around very large craters on the moon, part of the size distribution of secondary craters can be observed directly. To obtain the expected size distribution of secondaries around small primary craters, such as those near the limit of telescopic resolution, it is necessary to examine experimental craters on the earth.

Craters formed by large charges of high explosives or by nuclear devices detonated at shallow depth beneath the earth's surface have the form and structure of primary meteorite impact craters.[20] Most such explosion craters, if they are larger than about 50 m in diameter, are surrounded by a swarm of secondary impact craters. Secondary craters are also formed around primary impact craters produced by ballistic missiles,[21] even in cases in which the primary crater is as small as 10 m in diameter. Part of the secondary crater field around the nuclear explosion crater Sedan is illustrated in Figure 2–44. This explosion crater is about 400 m in diameter, and is surrounded by more than 5,000 secondaries that may be resolved in the aerial photograph. The secondary craters, which were formed by impact of large masses of rock and alluvium ejected on ballistic trajectories from the main crater, range in diameter or minor axis from 2 to 32 m (Figure 2–45). Smaller secondary impact craters were also formed but cannot be distinguished on the aerial photograph. The cumulative number of the larger secondaries is approximately a simple power function of the crater diameters. A power function with an exponent of −4 is plotted in Figure 2–45 for purposes of comparison. The cumulative number for the smallest secondary craters observed, however, lies significantly below the comparison curve. This is due in part to the fact that only a few of the craters with diameters near the resolution of the photograph were recognizable and counted.

[20] Shoemaker, "Impact Mechanics."
[21] H. J. Moore, R. Kachadoorian, and H. G. Wilshire, "A Preliminary Study of Craters Produced by Missile Impacts," U.S. Geol. Sur., Astrogeol. Studies Ann. Prog. Rep., July 1, 1963, to July 1, 1964, Pt. B (1964), pp. 58–92.

0   500 m

APPROXIMATE SCALE

FIGURE 2–44: Aerial photograph of Sedan nuclear-explosion crater (Nevada) showing associated swarm of secondary craters. Several thousand secondary craters are resolved in the original photograph.

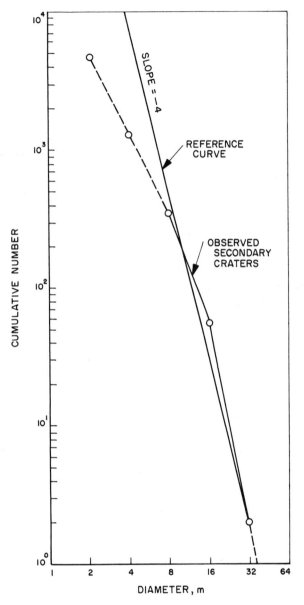

FIGURE 2–45: Cumulative size-frequency distribution of Sedan secondary craters.

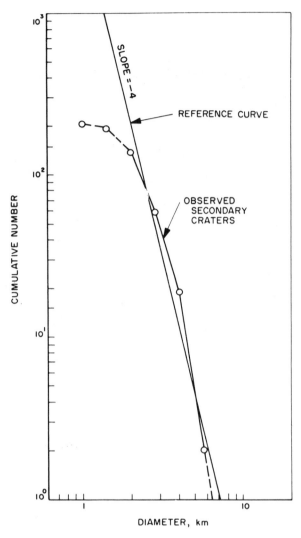

FIGURE 2–46: Cumulative size-frequency distribution of Langrenus secondary craters.

The size-frequency distribution in each of the large secondary crater swarms on the moon is similar in form to the distribution observed around Sedan. As an example, the size-frequency distribution of a part of the secondary craters in the swarm around Langrenus is shown in Figure 2–46. Only the craters formed on Mare Fecunditatis (Figure 2–4) have been measured and counted. Again a power function with an exponent of −4 has been plotted for comparison, and the cumulative number for the smallest observed secondary craters falls significantly below this comparison curve. As in the case of the Sedan secondaries, only a few of the craters with diameters or minor axes near the limit of resolution of the photographic plate were detected and counted. In order to obtain the true form of the size-frequency distribution of secondaries down to the limit of the smallest formed, it would be necessary to have much higher-resolution photographs, such as those acquired by Ranger VII.

If it is assumed that the number and size distribution of secondary craters observed in the crescent-shaped cluster in the highest-resolution Ranger VII A-camera photographs is representative of the number and size distribution of small

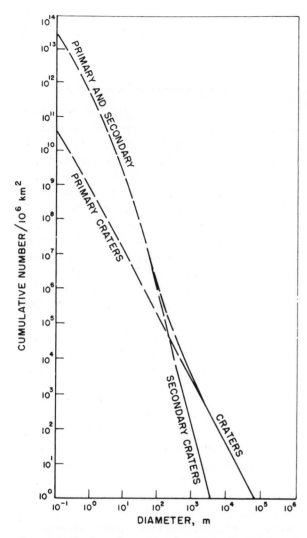

FIGURE 2–47: Cumulative size-frequency distribution of primary craters on lunar maria and calculated distribution of primary and secondary craters. (Solid lines indicate telescopically observed distributions; dashed lines are extrapolated or calculated.)

secondary craters in the rest of the Tycho ray system, then we may estimate the cumulative size-frequency distribution of the Tycho secondary craters from the largest secondaries formed around Tycho down to 200 m in diameter. It is found, on this basis, that the cumulative number of secondary craters is very closely a simple power function of the diameters up to a cumulative number of 100,000 craters. The exponent of the function that best fits the estimated size distribution is slightly greater than −4. (The absolute value of the exponent is slightly less than 4.) The largest secondary crater formed by Sedan is about 1/12

the diameter of the primary crater, whereas the largest secondary in the swarm around Langrenus is about 1/20 the diameter of the primary. If we assume that this ratio varies smoothly with increasing diameter for primary craters from the size of the Sedan to the largest (Langrenus), it is possible to derive the expected size-frequency distribution of secondaries for any primary crater of intermediate size. Each of these expected distributions will be a simple exponential function with an exponent which may be taken to be the same as that estimated for the Tycho secondary crater swarm. It will be assumed that these distributions extend to a cumulative number of 100,000 craters, as in the case of Tycho.

The observed size-frequency distribution of primary craters on the lunar maria is illustrated by the solid segment of the line labeled "primary craters" in Figure 2–47; the dashed segment of this line represents an extrapolation of the telescopically observed distribution of primary craters. The size-frequency distribution of secondary craters, obtained by integrating all of the secondary craters expected for each primary on the lunar maria, is illustrated in Figure 2–47 by the line labeled "secondary craters". The solid segment of the line represents that part of the predicted distribution which is observable through the telescope. It may be seen that the total number of predicted secondary craters becomes equal to the number of primary craters (extrapolated) at a diameter a little above 200 m, which is a size just beyond the limit of visual resolution under the best seeing conditions at a large telescope. The actual distribution of secondary craters on the lunar maria observed through the telescope follows the predicted distribution very closely. With decreasing crater size below 200 m, the predicted number of secondary craters rapidly becomes much greater than the number of primaries expected on the basis of extrapolation. At a diameter of 10 m, the predicted number of secondary craters so greatly exceeds the number of primaries that the curve for the distribution of both primary and secondary craters is not distinguishable from the secondary crater curve in the illustration. For diameters smaller than 10 m, several hundred secondary craters are expected for each primary.

A majority of the telescopically resolvable secondary craters are clustered within the major swarms of secondaries around large primary craters.

At the outer limits of these swarms and beyond, however, are regions in which a significant number of secondary craters can be found that cannot definitely be assigned to any given primary. These unassigned secondaries constitute a background field of craters with much more uniform distribution than is observed within the swarms. Most of the large number of secondary craters too small to be resolved telescopically will probably not be assignable to individual primary craters because there will be extensive overlap of the swarms associated with small primaries. Thus, the small secondaries will contribute to the background field of craters. The mare surfaces, therefore, not only will contain small secondary crater swarms but also are expected to be more or less uniformly peppered with such craters.

An improved estimate of the number of small secondary craters that have been formed on the maria may be made utilizing the predicted number of small primary craters based on the terrestrial meteorite infall data (Figure 2–43). For this estimate, an age of 4.5 billion years is again adopted for the maria. The primary crater distribution curve adopted in making the estimate has two branches: (1) the preferred curve for primary craters based on meteorite infall data, which extends from diameters of 1 m to about 400 m, and (2) the observed crater frequency distribution of the lunar maria for diameters above 400 m. The cumulative number of secondary craters larger than 1 m in diameter, predicted on this basis, is more than 10 times greater than the number based on the observed primary crater distribution extrapolated to small sizes. The predicted cumulative number of all craters larger than 1 m in diameter, however, is only about 50 times greater than the predicted cumulative number of primary craters of that size.

The predicted cumulative number of all craters 1 m in diameter and larger (improved estimate) is more than two orders of magnitude greater than the estimated cumulative number of all craters based on the Ranger VII photographs. At diameters of a few meters, the ratio of the cumulative number predicted to the number observed is about the same for all craters (primary and secondary combined) as it is for the primary craters. As in the case of the primaries, this ratio decreases rapidly with increasing crater size, and the predicted and observed curves cross at a diameter between 100 and 200 m. For areas between the rays, the predicted and observed cumulative curves for all craters join at about 300 m.

Effects of Interaction of Small Craters

The predicted number of small craters, both primary and secondary, is so great that craters are expected to have been formed everywhere on the lunar maria. On the basis of the improved prediction for the cumulative number of all craters larger than 1 m across, the mare surface would have been cratered repeatedly at any one place. It is impossible, in other words, to observe the predicted number of craters larger than 1 m across because their total area is more than 10 times as great as the area of the mare surface. The effect of repeated cratering is to reduce the total number of craters that are observable. Not only are small craters lost by the superposition of others of the same size or larger, but the repeated formation of small craters will also finally obliterate larger ones.

The effects of repeated formation of small craters on larger ones are complex and can only be worked out in terms of a statistical model of cratering frequency, the mechanics of cratering, and distribution of crater ejecta. For each cratering event, material is removed at the site of the crater and is widely distributed in the surrounding area. On the basis of the velocity distribution of fragments observed in impact experiments with rock targets, about half of the fragments ejected from a small crater on the moon will land within a kilometer of the crater. The other pieces will be thrown greater distances, and, in the case of primary impact craters, a small percentage of the material thrown out will be widely scattered over the moon's surface. Some material is even ejected at the velocity of escape from the moon. For each cratering event that removes material from a small spot and scatters it widely over the surrounding area, there will be many events in the surrounding area that will throw material back.

If no material were lost from the moon by ejection at escape velocity, the ultimate effect of repeated formation of craters of one size would be the development of a layer of fragments with an average depth approximately equal to the depth of the individual craters. At any one point, this layer of fragments would be removed or reduced in thickness from time to time by the formation of a new crater and would then be gradually covered again by fragments ejected from a large number of craters formed elsewhere. The average time of turnover or renewal of the layer at any one point would be just equal to the time required to form the number of craters whose total area is equal to the surface of the moon.

If a frequency distribution of crater sizes like the cumulative size distribution predicted for the lunar maria is next considered, the effects of repeated cratering are more complicated. The smaller the crater diameter, the higher the turnover rate of the surface layer. The average depth of the fragmental layer on the maria will be given by the product of the average depth of the craters larger than the size at which the total area of the craters is equal to the area of the maria and the ratio of the specific volume of the crater ejecta to the specific volume of the original material. The layer will have been stirred to this depth just once on the average. At successively shallower depths, the material is stirred or the layer is rejuvenated more frequently, and, in its uppermost part, the layer is stirred by the repeated formation of very small craters a very large number of times. The turnover rate of the uppermost millimeter of the moon's surface, on the basis of estimates of the micrometeorite influx[22] and the secondary particles generated by this influx,[23] may be greater than once every 100 years.

On a surface on which a fragmental layer is continuously being formed and reformed, the shapes of craters are modified by two processes. The first process is the actual destruction or removal of a crater by the formation of a new crater the same size or larger. In the second process, a crater is gradually covered over by fragments and peppered with smaller craters in the time intervals between larger cratering events. The latter process, if it continues sufficiently long, will so modify the shape of the crater that it is no longer recognizable. With a given flux of impacting objects, the distribution of observed craters below some limiting size will tend toward a steady size-frequency distribution. Within the size range in which the distribution remains steady, new craters are formed as fast as

old ones disappear from the combined effects of the two crater-modification processes. There will be a steady number of craters of any given size, no matter how long the cratering continues, and craters of a given size will exhibit a complete range of shapes from fresh unmodified forms to forms so modified that the crater is barely discernible. Above this limiting size, craters may undergo considerable modification in shape, but those formed do not disappear. With a given flux and size distribution of impacting objects, the limiting size of the steady crater size distribution gradually increases with time.

Craters only slightly larger than the limit of the steady size-frequency distribution will generally be significantly modified by covering with fragmental material and peppering by smaller craters. The precise effects of this process vary with position on the crater. As the majority of smaller craters are expected to be secondary impact craters, formed by fragments traveling generally along relatively low-angle trajectories,[24] the rim of the large crater is exposed to a greater flux of impacting objects than the floor. The rim is also exposed to a greater flux than the more level areas surrounding the crater, and the floor is exposed to a lower flux than the surrounding terrain. Thus, more material tends to be ejected from the rim than is thrown back from the surrounding region, and the floor of the crater tends to receive more material from the surrounding region, including the rim, than is thrown out. In the crater floor, the fragmental layer tends gradually to become thicker, whereas the crater rim, or any protruding or positive-relief feature on a mare surface, will tend to be covered by only a thin layer of fragments relative to the surrounding region and to be continuously worn away. As the crater floor slowly fills up and the rim is ground down, the differences in the solid angles of space subtended by points on the surface of the rim and on the floor become less pronounced, the differences in rate of cratering diminish, and the rates of filling of the crater and degradation of the rim decrease. The ultimate effect of repeated formation of smaller craters on a large one, however, would be to reduce the rim and raise the floor to the same

[22] C. W. McCracken and M. Dubin, *Dust Bombardment on the Lunar Surface*, United States National Aeronautics and Space Administration Technical Note D-2100, 1963.

[23] D. E. Gault, E. M. Shoemaker, and H. J. Moore, *Spray Ejected from the Lunar Surface by Meteoroid Impact*, United States National Aeronautics and Space Administration Technical Note D-1767, 1963.

[24] See Shoemaker and Hackman, "Stratigraphic Basis," and M. H. Carr, "Trajectories of Objects Producing Copernicus Ray Material on the Crater Eratosthenes," *U.S. Geol. Sur., Astrogeol. Studies Ann. Prog. Rep., July 1, 1963, to July 1, 1964,* Pt. A (1964), pp. 33–41.

level as the surrounding terrain, thus completely eliminating the topography of the crater. A deposit of fragments would be present over the floor of the original crater, with a maximum depth equal to the depth of the original crater below the level of the surrounding terrain.

Let us turn now to the interpretation of the difference between the predicted number of craters and the estimates of the actual size-frequency distributions of craters in Mare Cognitum obtained from the Ranger VII photographs, as illustrated in Figure 2–43. The predicted curves represent the numbers of craters expected to have been formed in each million square kilometers of the mare surface per billion years, whereas the curves based on the Ranger VII data represent the observed distribution of craters per million square kilometers divided by 4.5 to give the equivalent rate of cratering per billion years. If the predicted curves for the number of craters formed are valid, they should join the observed curves, both for primary craters and for primary and secondary craters combined, at the limiting size for steady frequency distribution. This occurs at approximately 400 m for the primaries and at approximately 300 m for all craters in the areas between the rays. Inasmuch as the locations of the points at which the observed and predicted curves join are very sensitive to small variations in the predicted curves, this difference is not judged to be significant. In the ray areas, the observed number of all craters is actually above the predicted curve between 150-m and about 1-km diameter because of the local concentration of Tycho secondary craters in the rays, but is below the predicted curve for diameters less than 150 m. Except for this local concentration of Tycho secondaries, the difference between the observed and the predicted curves may be interpreted as representing the numbers of craters lost by the effects of crater interaction. At diameters of a few meters, several hundred times as many craters have been formed and subsequently destroyed as are now present on the mare surface. The ratio of craters formed on the maria to those now present is essentially the same for primary and secondary craters.

Interpretation of Areas between Rays

In the areas between the rays, the distribution and shapes of craters observed in the Ranger VII photographs are here interpreted as being the result of the cumulative effects of primary and secondary cratering since Mare Cognitum was formed. Many primary craters smaller than 1 km in diameter that are seen in the high-resolution A and B pictures show some rounding of rims and somewhat shallower depths than would be expected for typical primary craters. The rounding of rims is probably due to the eroding effects of the interaction of small unresolved craters, and the anomalously shallow depths are probably the result of partial filling by ejecta from surrounding craters and by landslides and talus derived locally from the crater walls. The almost complete lack of raised rims on the large elongate secondary craters in the Bullialdus secondary swarm and on other secondaries, labeled S in Figures 2–17 through 2–27, is also interpreted as due to erosion by repeated formation of small craters. If these craters originally had rims like the larger secondaries of Tycho and Copernicus, 10 to 20 m of rim material has been worn away. The floors of these ancient craters are probably underlain by a fairly thick layer of slowly accumulated fragmental material.

With craters of progressively smaller diameter, it becomes increasingly difficult to distinguish modified primary from secondary craters. Erosion and filling due to small crater interaction apparently have modified the shapes of primaries less than 1 km across so that they approach secondary craters in form. Craters smaller than 300 m in diameter exhibit a complete range of shapes from typical primaries to craters so shallow that they are scarcely detectable in the Ranger photographs. In this size range, it is impossible to distinguish eroded primary from circular secondary craters. On the basis of the predicted numbers of primary and secondary craters, the circular craters of the indefinite class larger than about 200 m in diameter (labeled I on Figures 2–30 and 2–31) are predominantly eroded primaries but probably include some secondaries. The elongate to irregular craters (labeled E on Figure 2–31) are probably chiefly eroded secondaries. The depth of fill of most of the nearly flat circular craters larger than 200 m across may be nearly equal to the depth of uneroded primaries of the same size, between one-fourth and one-third the crater diameter. The average depth of fill in the eroded secondaries is probably significantly less, as the depth of unmodified secondary craters is typically as low as one-tenth of the crater diameter.

Most of the indefinite craters less than 200 m in diameter are believed to be secondary craters, many with forms modified by small-crater interaction. All except the youngest probably have been partially filled with fragmental material. The average depth of the fragmental layer in these craters should be less than 10 m.

Between the resolvable craters, the surface of the mare is probably covered nearly everywhere with a fragmental layer of crater ejecta which varies irregularly in thickness from place to place. On the basis of the number of small craters, chiefly secondaries, predicted to have been formed, the average thickness of this fragmental layer in the areas between the rays should be of the order of 1 m. This estimate has a large uncertainty owing primarily to the unknown magnitude of two effects on the small crater ejecta: (1) change in specific volume of the lunar surface material due to fragmentation or to shock compression and (2) loss of material ejected at escape velocity from the moon. The first effect depends on the initial porosity of the lunar surface material, specifically, the mare material. If it is originally rock with little pore space, a large expansion of specific volume can result from fragmentation and loose stacking of the fragments. This expansion is typically of the order of 30 per cent for fragmentation of dense rocks on the earth, but it may be much greater for very open stacking of small fragments on the lunar surface. If the mare surface material is initially very porous lava, on the other hand, as Dr. Kuiper has suggested, then shock compression of the ejecta may be greater than the volume expansion normally expected from fragmentation and loose stacking, and the net volume of ejected material may be less than its original volume.

Loss of material ejected at escape velocity from the moon also depends on the porosity at the lunar surface as well as upon the total flux and velocity distribution of impacting primary particles. Where dense rocks are exposed, as much as an order of magnitude more mass may be ejected from the moon than strikes the lunar surface, at impact velocities typical for comets and meteor streams.[25] The amount lost is much less for low-velocity impacts and from porous materials. The total mass of material lost from the mare surface due to high-velocity impact of small solid particles and to

energetic solar protons may be roughly of the same order as the mass of fragments now present on the mare surface.

Original small surface features of Mare Cognitum have probably been strongly modified, and the smallest features completely removed, by repeated cratering. The only small features I have observed in the Ranger VII photographs which might be original features of the mare surface are the sinuous ridges and a few low scarps that may be related to the ridges. Some parts of these ridges are pitted with craters, but other parts are surprisingly sharply formed. From telescopic studies, it is known that the ridges cross color and albedo boundaries elsewhere on the maria; they appear to be younger than the mare material. It is possible that some of the ridges have been actively rising late in the history of the moon, although most of them are probably ancient features of the maria.

Most of the small linear features I have found in Mare Cognitum, and most of the lineaments illustrated by Strom (data presented by Dr. Kuiper) I believe are either secondary craters or rows of secondary craters. The azimuths of the major axes of composite secondary impact craters and of rows of secondary craters are related to the azimuths of major lineaments on the lunar crust through transformations given by the ballistic equations for secondary fragment trajectories and the equations describing particle acceleration during cratering.[26] Relatively few of the lineaments, in my opinion, are related directly to subjacent structures that extend into the mare material beneath the debris layer.

Interpretation of Ray Areas and the Highest-Resolution Pictures

The highest-resolution photographs obtained with Ranger VII show small parts of the mare surface that lie entirely within a ray of Tycho. The surface features shown in the last few P-camera pictures provide an insight into the fine structure of this ray, possibly a typical ray, but are probably not representative of the areas of the mare surface between the rays.

One of the most significant facts revealed by the last few A-camera pictures is that the ray, which is identifiable on both low- and high-resolution

[25] Gault, Shoemaker, and Moore, *Spray Ejected*.

[26] Shoemaker and Hackman, "Stratigraphic Basis," pp. 329, 334.

photographs as a bright streak, is coincident with an area of closely spaced secondary craters. Although the spacing and size of craters varies within the ray, the areal density of craters more than 200 m across, particularly the elongated secondary craters, is nearly everywhere greater inside the ray than in the areas adjacent to the ray. The gradational photometric boundary of the ray lies within a kilometer of the limit of the closely spaced secondary crater field.

If the craters are of secondary impact origin, as I believe, the extent of the bright material of the ray is coincident with an area which may be expected to be covered by coalescing and overlapping rim deposits of material ejected from the craters. By far the largest proportion of this material at any one place is probably derived from the nearest large secondary crater, although fragments are probably also present which have come from other secondaries within the ray and from the primary crater.

The ray is thus viewed here simply as an area nearly continuously covered by a layer composed of the collective throwout of the cluster of Tycho secondary impact craters and by the craters themselves. This thrown-out material is brighter than most of the material exposed elsewhere on the mare because of several factors discussed in detail below. The shape and extent of the ray are controlled strictly by the shape and extent of the secondary impact crater cluster.

In the ballistic theory,[27] the pattern of a secondary crater swarm and the shape of individual crater clusters are controlled by the breakup pattern of the lunar material as it is ejected from the primary crater. The size distribution of secondaries within each ray is related directly to the size distribution of fragments within each ray-forming cluster of fragments ejected from the primary crater. These fragments traveled through space on closely similar but slightly divergent ballistic trajectories as an expanding cluster. Fragments which were on the end of the cluster nearest the center of the primary crater as they left the primary received slightly greater accelerations, were ejected at somewhat higher angles, and traveled farther than the fragments originally farthest from the center of the primary. The radial sequence of points at which the individual fragments struck the lunar surface

is, therefore, the reverse of the radial sequence of the positions from which they were ejected from the primary crater. Fragments derived from positions farthest from the center of the primary tend to be larger than fragments from positions nearer the center, and there is thus a general tendency for the largest craters to occur near the end closest to the primary in each secondary crater cluster.

Circular secondary craters are probably formed by impact of individual large secondary fragments, whereas elongate and composite secondaries are formed by impact of two or more fragments that traveled very close together. Both theory and the evidence from "ballistic shadows" in the ray pattern of Copernicus[28] indicate that the angles between the horizontal plane at the impact points and the trajectories of the secondary fragments are low. Composite craters, therefore, tend to be elongate roughly in the azimuths of the trajectories, but their major axes will not, in general, be precisely radial from the primary crater. The fact that many of the elongate secondary craters are distinguishably composite and are generally not aligned precisely along radials from the primary shows that they are not just grooves plowed by individual fragments striking the lunar surface at low angles.

The elongate secondary craters in the distal or "downrange" part of the Tycho ray seen in the last Ranger VII pictures seem to have more rounded rims than craters of similar size in the proximal or "uprange" end of the ray. This apparent rounding may be due to the smoothing effects of throwout from the craters in the uprange end of the ray. It is possible that many of the shallow craters 10 to 50 m across that are revealed by enhancement of the last A-camera photograph (Figure 2–35) are "tertiary" craters—that is, they are secondary craters of the large uprange Tycho secondaries. These small shallow craters are strewn over the larger elongate Tycho secondaries shown on the last A-camera photograph, and their combined effect would be to reduce the relief of the larger craters.

The thickness of thrown-out material on the rims of the secondary craters probably varies with position on the rim. It may be expected to be thickest near the crests of the secondary crater rims and to become progressively thinner at greater distances

[27] *Ibid.*, pp. 289–300.

[28] Carr, "Trajectories of Objects."

from the rim crests. A substantial thickness of thrown-out material should be present in the immediate vicinity of the closely spaced Tycho secondary craters shown in the last A-camera photograph. This includes the area shown in the last P-camera pictures. The thickness of thrown-out material on the rim of the 500-m-long secondary crater shown in the last P_1-camera photograph (Figure 2–36) should be of the order of 10 m. Material at this locality on the rim probably has been expelled from depths greater than 10 m within the secondary crater. If the craters I have identified as Tycho secondaries in Figures 2–13 and 2–14 are truly of secondary impact origin, I do not believe that the original surface of the mare can possibly be exposed in the area shown in the last half-dozen P-camera photographs. The low mounds of the "tree-bark" structure of the surface revealed in these photographs are probably original features of the throw-out deposits of the large secondary craters. The structure may be the result of flowing motion of the thrown-out material as it skids along the surface after ejection from the secondary crater. I have observed similar topography on material thrown out of explosion craters. Although the tree-bark structure is most easily observed along scarps sloping away from the sun, it is probably present over most of the area covered by the last P-camera photographs.

The presence of small, possibly original, features of the throw-out deposits suggests that the ray was formed very late in lunar history. Tycho can be shown to be one of the youngest large ray craters on the moon because its rays are superimposed on most other types of lunar terrain, and no telescopically resolvable younger features have been found to be superimposed on Tycho. Its age may be on the order of one-hundredth the age of the maria—50 million years or less. The smallest craters observed in the highest-resolution P-camera photographs are superimposed on and still younger than the Tycho secondary craters and the ray. Even the smallest craters resolved in the last P-camera pictures may have diameters greater than the upper limit of the steady size-frequency distribution for craters formed since the ray. If this is the case, the ratio between the number of observed and predicted small craters on the ray may give the ratio of the age of the ray to the age of the mare (assuming the cratering rate has remained constant).

The absence of small blocks or pronounced bumps in the ray may indicate an upper limit to the size of fragments in the material thrown out of the secondary craters. If the fragments had dimensions as large as the width of the mounds in the tree-bark structure, one might expect them to stick out here and there. Fragments this size could not have been worn away by repeated cratering, or the mounds would have disappeared, too, although small blocks with dimensions near the limit of highest resolution of the P-camera photographs could have been worn away by still smaller craters. It is tentatively concluded that most of the fragments in the throwout from the 500-m-long secondary shown in the last A-camera photograph are less than 1 m across.

The fine structure of the ray, as interpreted here, is fairly complex. Within the ray, a layer of varying thickness of thrown-out material, derived from varying depths beneath the surface at the sites of the secondary craters, overlies an older mare surface. Immediately beneath the throw-out layer, there is probably a layer of fragmental debris in most places that is nearly identical with the debris layer exposed in the areas between the rays. The surface of this older debris layer was probably pitted with craters with distributions of size and shape similar to that observed between the rays. Where the layer of ray material is relatively thin, away from the rim crests of the secondary craters, most of the larger craters observed are probably thinly veneered pre-ray craters. A very thin debris layer produced by repeated formation of very small craters is expected to be present on top of the ray material.

Interpretation of Photometric Variations

Variations of brightness of the lunar surface recorded in the Ranger VII photographs are due to variations in albedo of the lunar surface materials and to variations in slope of the surface. First, the causes of albedo variations on Mare Cognitum will be examined in terms of the cratering processes. A method will then be outlined for estimating the distribution of slopes on the mare surface in areas with uniform albedo.

Interpretation of albedo variations

The principal variations of albedo on Mare Cognitum occur across the rays and along steep crater walls. Certain other features in the albedo pattern such as the unusual dark ridge in the mare shown in the A-camera pictures and the relatively high albedo of the highlands surrounding Mare Cognitum and of certain ridges, peaks, and broad high areas within the mare, are observable in the Ranger VII photographs and are well known from telescopic observations. Some of the observed differences in albedo are probably related to differences in mineral or chemical composition of the materials exposed, but it is instructive to examine whether most of the albedo variations on the mare can be explained as being a result of physical processes operating on materials of essentially uniform composition.

If it is assumed that the material at the mare surface is susceptible to darkening by high-energy solar radiation, then most of the materials making up the fragmental layer produced by repeated cratering have been affected by the darkening processes, as most fragments in the layer have been at the surface at one time or another. The darkening is assumed to take place as a result of a number of possible solid-state changes produced both by ultraviolet radiation and by solar-proton bombardment.[29] Laboratory studies of these effects suggest that most of the darkening takes place within a millimeter or a few millimeters of the exposed surfaces of common rock-forming minerals and depends upon the radiation dose, up to the point at which the albedo of the mineral or material in question reaches a limiting minimum value. If this is the case for the materials of the maria, the precise effects of irradiation on a repeatedly stirred layer of fragments on the order of a meter or many meters thick are complex.

The average rate of darkening of the surface of an initially unirradiated fragmental layer would depend on the relative rates of darkening of an undisturbed surface and of turnover and exposure of fresh material. If the time required for the albedo of an undisturbed surface to reach a lower limiting value under solar irradiation is much shorter than

the average time of turnover or renewal of the fragmental layer to the depth significantly affected by radiation darkening, then nearly all parts of the surface will be more or less uniformly darkened after a period of time not much greater than the period required to darken an undisturbed surface completely. If, on the other hand, the turnover time of the upper part of the fragmental layer is much shorter than the time required for the albedo of an undisturbed surface to reach a lower limiting value, then the radiation darkening effect will be spread through the material of the fragmental layer during early stages of darkening, and the upper surface of the layer will gradually become darker after a period of time much longer than the period required to darken an undisturbed surface completely. The latter case appears to offer a good model for the lunar maria.

In the case in which the darkening rate is relatively slow, material initially at and near the surface will be partially darkened first, and this partially darkened material will gradually be dispersed downward in the fragmental layer with repeated formation of craters of varying size. As the fragments that are initially at the surface are, on the average, exposed longer than fragments initially near the base of the fragmental layer, fragments nearest the surface will tend to receive a greater radiation dose in a given period of time than fragments nearest the base of the layer. A vertical gradient of darkening therefore tends to develop in the fragmental layer, the fragments in the lower part of the layer being darkened the least and the fragments in the upper part darkened the most. With increasing time and radiation dosage, this gradient of darkening, scaled to the thickness of the fragmental layer, tends toward a steady vertical distribution of partially darkened and completely darkened material. At any one place, the vertical gradient of darkening will generally be irregular and will depend on the specific history of cratering at that place. The albedo of the material at the surface, if measured on a scale that is coarse with respect to the mean fragment size, tends toward a lower limiting value that is somewhat higher than the lower limiting albedo of an undisturbed surface, because a certain proportion of fresh and partially darkened fragments will always be present.

Fresh craters that penetrate the base of the fragmental layer may be expected to have fresh or relatively unirradiated material in the deposits of

---

[29] Since the preparation of the manuscript of this report, Hapke has described the primary cause of darkening of fine rock powders by experimental irradiation of 2-kev H and He ions as due to the formation of coatings of non-stoichiometric compounds on the undersurfaces of the particles. The general process of darkening of the lunar surface by this mechanism would not be greatly different than outlined in this section (B. W. Hapke, *Laboratory Photometric Studies Relevant to the Lunar Surface*, paper presented at 118th American Astronomical Society meeting, Lexington, Ky., 1965).

throwout on their rims. These deposits form the rays and the bright halos around primary craters on the maria. If the craters are much deeper than the fragmental layer, relatively fresh material may also be expected to be exposed in the crater walls.

Most telescopically resolvable crater walls and other slopes on the moon that are steeper than 30 degrees have relatively high albedos. This is probably attributable to intermittent sliding and rolling of fragments down slopes which are at or near the angle of repose. A steady gradient of darkening will not develop on such slopes because any thoroughly irradiated material exposed at the surface tends to migrate down the slope and accumulate at the foot, and fresh unirradiated or partially irradiated material is continually or intermittently exposed. Crater walls and other steep slopes will tend to become thoroughly darkened only after the slopes are sufficiently reduced by sliding, slumping, and other degradation processes to become stabilized. The slopes with high albedo are thus interpreted as continually or intermittently active talus deposits. The streaks in the steep crater walls revealed in the Ranger VII photographs (Figures 2–34 and 2–39 through 2–41) probably correspond to individual rock slides. Bright streaks correspond to parts of the talus that have slid or been exposed most recently and the intervening dark streaks to the material that has remained stable the longest.

Interpretation of the albedo of a large ray with many secondary craters is complicated by the presence of high-albedo material on steep secondary crater walls. Variations of albedo across a large ray measured at the telescope are demonstrably due in part to the presence of unresolved bright crater walls. It might be argued that this effect occurs at all scales at which the ray is observed and that the relatively high albedo of a ray is due entirely to the presence of small crater walls and other steep slopes in the ray. Both the observed distribution of slopes in the highest-resolution Ranger VII pictures of the Tycho ray and the model of the surface processes presented above, however, suggest that the contribution of high-albedo steep slopes to the integrated albedo of the ray diminishes rapidly with decreasing length of slopes. In areas of the Tycho ray between the recognizable secondary craters, the surface of the ray is probably not significantly rougher, on the average, than areas outside the ray. Yet, at all scales, the albedo of the ray is generally signifi-

cantly brighter than the areas outside the ray. The high albedo of the ray material should probably be attributed chiefly to the excavation of fresh material from the secondary craters late in lunar history. The period of time elapsed since the ray was formed has been insufficient for the development of a steady gradient of darkening in the debris layer at the top of the layer of ray material.

The rays have gradational boundaries, probably as a result both of the initial distribution of fresh material and of the later processes of darkening of this material. The ray is conceived here as formed by a discrete, nearly continuous layer of ejecta. At the margins of the layer, however, the ejecta are probably discontinuously scattered as individual fragments across the pre-existing terrain. The original boundaries of the ray, therefore, tend to be diffuse. In addition, under the process of darkening suggested for mare material, the thinnest and marginal parts of the ray will become dark more quickly than thicker central parts, because already darkened material surrounding and beneath the ray will be mixed more rapidly with the thin marginal parts of the ray by repeated cratering.

The occurrence and distribution on the mare of very small primary craters with bright halos and rays provide a means of estimating the maximum depth to unirradiated or relatively weakly irradiated material. As the smallest primary ray craters observed in the Ranger VII photographs are about 30 m in diameter, the indicated maximum depth to relatively undarkened material at the sides of these craters is about 6 to 8 m, the approximate depth of the craters. The actual depth to undarkened material at these localities in the Tycho ray is probably much less. Small primary craters are so rare in the highest-resolution photographs that the smallest new ray craters present in the area are not likely to have been photographed.

Photometric measurement of slopes

The brightness of any element of the lunar surface recorded by Ranger VII depended on the albedo of the surface material, the angle of incidence of the sunlight on the surface, and the angle of emission of the light scattered toward the spacecraft cameras. The relationship between these variables is known as the photometric function of the surface. From photometric measurements of resolvable small areas on the moon, made through the telescope, it may be shown that the photometric function for small areas of the lunar surface exhibits a characteristic that is highly unusual for scattering surfaces. If the direction of illumination and the angle between the incident ray and the scattered ray (the phase angle) are held constant, then the brightness of a surface element of a given albedo depends only on the component of the slope of the surface in the plane of the phase angle. As the direction of solar illumination is constant and known for any Ranger VII photograph and the phase angle may be determined from the orientation of the cameras and position of image elements within the photographs, it is possible to measure the component of slope in the phase-angle plane for each image element from the response of the television camera, if certain assumptions are made about the albedo of the surface. Preflight photometric calibrations of the cameras may be employed to calculate the brightness of the image element either directly from the transmitted video signal, as recorded on magnetic tape, or from measurements of the film records of the pictures obtained with a microdensitometer.

An initial investigation by photometric analysis has been made of the distribution of slopes in the northern half of the area covered by the last Ranger VII P$_3$ photograph.[30] The small area studied, about 40 m across, lies within the crescent-shaped Tycho ray, a short distance from the rim crest of a relatively large Tycho secondary crater. No demonstrable variations in albedo are present within the area photographed, and it was assumed that the albedo was uniform for purposes of calculating the slopes. A photometric function was adopted for this part of the lunar surface based on numerous telescopic measurements of the lunar maria and adjusted to the estimated albedo of the ray material. The partial photometric function used, for the phase angle of 38 degrees, is illustrated in Figure 2–48. Here, the brightness of the lunar surface, measured in foot-lamberts, is shown as a function of the component of slope in the phase-angle plane, measured with reference to the local horizontal plane. All phase angles for the part of the photograph analyzed were within about 1½ degrees of 38 degrees.

[30] The work has been carried out for me by a group of my colleagues in the U.S. Geological Survey, among whom Howard A. Pohn and Laurence C. Rowan should receive special mention.

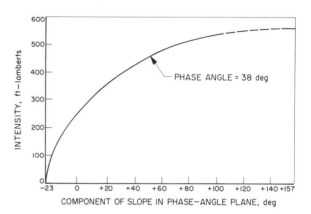

FIGURE 2–48: Partial photometric function adopted for estimation of slopes from highest-resolution Ranger VII photographs. (Graph shows relation between brightness of lunar surface and component of slope in the phase-angle plane for a phase angle of 38 degrees. Positive angles indicate slopes dipping toward the sun; negative angles indicate slopes dipping away from the sun.)

The brightness of each image element was calculated from microdensitometer measurements of a duplicate negative prepared from one of the prime negatives of the Goldstone station. Approximately 12,000 measurements were employed. Photometric corrections were made for shading and vignetting in the television cameras. Slope components were calculated from the corrected brightness values using the adopted photometric function. Individual slope components were then connected serially along the approximate traces of the phase-angle planes to construct a sequence of topographic profiles of the surface in the phase-angle planes. A topographic map was prepared from the array of profiles by adopting a series of control points to which the profiles were arbitrarily adjusted and by making simplifying assumptions about the orientation of the phase-angle planes.

A three-dimensional physical model of the area mapped was next prepared from the map by Howard A. Pohn and Walter C. Black, at an approximate scale of 1/20, and fine topographic details of the surface which had become smoothed out or lost by the measurement and calculation procedures were restored with the aid of light projected on the model from an angle appropriate to simulate the original solar illumination of the lunar surface. A cast of the model was then prepared and photographed, and a new topographic map was prepared by special photogrammetric techniques by Richard H. Lugn.

The cumulative frequency distribution of the components of slope in east-west vertical planes, measured from the improved topographic map, is illustrated in Figure 2–49. This distribution is for slope components with a base length of 1 m. The median slope is about 5 degrees. About 90 per cent of the slopes are less than 16 degrees and 10 per cent less than 1 degree. The form of the frequency distribution is similar to that observed on a number of different terrains on earth.

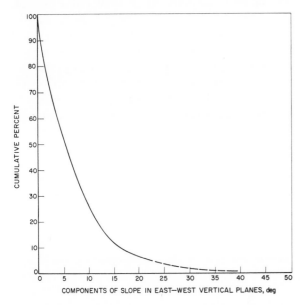

FIGURE 2–49: Estimated cumulative frequency distribution of slopes in half of area shown in last P_3-camera photograph.

III. PRELIMINARY MODEL OF FINE STRUCTURE

Information on the fine structure of the surface of Mare Cognitum derived from study of the Ranger VII photographs is consistent with the evidence about the fine structure that had been obtained previously from physical measurements made at the telescope. The telescopic data that bear on the fine structure may be classified in four categories: (1) photometric properties at the visible wavelengths, (2) polarization of scattered radiation at visible wavelengths, (3) thermal emission at both infrared and microwave wavelengths, and (4) reflection characteristics for microwave radiation. No unique set of physical interpretations can be made from these measurements, but limits can be set on the porosity and probable grain size of the material at and very close to the lunar surface.

The photometric function of small parts of the moon's surface for visible wavelengths indicates that at some scale appreciably coarser than the wavelengths of light, the texture of the moon's surface is extremely rough and porous.[31] The shapes of the pores are such that open holes extend into the surface in all directions. This configuration can be achieved by an open fibrous network, loose stacking of needles or fibres, or extremely open stacking of angular grains. Measurements of the polarization of scattered sunlight with variation in phase show that the scattering surface is covered with very fine, nearly opaque grains.[32]

Variations of the thermal emission with phase and during eclipses, measured at infrared wavelengths, indicate that the average thermal inertia of the uppermost few millimeters of the lunar surface material is between one and two orders of magnitude less than the thermal inertia of ordinary dense rocks.[33] The low thermal inertia of the lunar surface material can be readily explained if the density of this material is about an order of magnitude less than ordinary silicates, which suggests that the bulk of the volume of the uppermost few millimeters is open pore space. Thermal emission measured at microwave wavelengths shows that the diurnal thermal wave propagated into the moon's surface by insolation damps out very rapidly with depth. The lunar surface material is probably highly porous to depths of at least several centimeters.

Interpretation of the physical properties of the surface from measurements of reflected microwave radiation is, unfortunately, ambiguous. The principal reflecting horizon, however, appears to have a dielectric constant close to the value for dry sand.[34] Probably, the reflection takes place at some shallow depth beneath the surface.

[31] Hapke, "Theoretical Photometric Function for the Lunar Surface," *J. Geophys. Res.*, Vol. 68 (1963), pp. 45–86.

[32] A. Dollfus, "The Polarization of Moonlight," *Physics and Astronomy of the Moon*, ed. by Z. Kopal, New York and London: Academic Press (1962), pp. 131–159.

[33] W. M. Sinton, "Temperature on the Lunar Surface," *Physics and Astronomy of the Moon*, ed. by Z. Kopal, New York and London: Academic Press (1962), pp. 407–427.

[34] J. V. Evans and G. H. Pettengill, "The Scattering Properties of the Lunar Surface at Radio Wave Lengths," *The Solar System, Vol. IV—The Moon, Meteorites, and Comets*, pp. 129–159.

Employing the data available from the Ranger VII photographs, together with the information obtained at the telescope, a reasonable model of a typical local area between the rays on Mare Cognitum can be described in the following way:

1. A layer of shattered and pulverized rock covers more than 95 per cent of the mare. It is of variable thickness and rests with irregular contact on the underlying substance of the mare.

2. The fragments in this layer or blanket of shattered rock have been derived by ejection from craters, most of them nearby, but some lying great distances away. Probably, about 50 per cent of the fragments have come from within 1 km of the site at which they are now found, but there is a finite probability, decreasing with the distance to the source, of finding a rock fragment in the debris layer which has been derived from any place on the moon. Except along the margins of the mare, the pieces of debris will be composed predominantly of mare material.

3. Fragments occurring at the base of the debris layer will, on the average, have been transported on ballistic trajectories a smaller number of times than pieces near the middle or top of the layer. Progressing upward from the base, the layer has been stirred or reconstituted an increasing number of times by smaller and smaller and more and more numerous cratering events. The uppermost millimeter of the debris layer is probably completely reorganized once every 10 to 100 years by the formation and filling of minute craters.

4. The average grain size of the debris layer tends to decrease from base to top because fragments in the upper part have been shocked and broken a greater number of times and have been ejected, on the average, from smaller craters. The exact distribution of mean grain size in the layer is controlled closely by the mass-frequency distribution of impacting interplanetary objects. Unfortunately, this distribution is not well known for objects with masses less than about 10^{-6} g. If the mass distribution of interplanetary particles given by McCracken and Dubin is adopted, the size distribution of fragments in the debris layer may be roughly described as follows: Near the base, fragments as large as several centimeters in diameter will be common, whereas most of the material in the uppermost few millimeters will be finely pulverized. Throughout the debris layer, the bulk of the fragments will probably average less than a millimeter in grain size, but heaps of coarse, blocky rocks may be expected to surround the larger craters. At any one locality, the vertical variation of grain size may be expected to be extremely irregular.

5. Beneath the blanket of shattered rock, the mare substance, if originally solid, will in many places show evidence of having been broken to greater depths by shocks of varying strength produced during development of the larger craters. If the original mare substance at a given locality was initially relatively porous, it will have been strongly compressed by shock.

6. The contact between the underlying mare substance and the debris layer has considerable local relief, consisting of the intersecting segments of the original floors of numerous craters that ranged from a meter to a few tens of meters across. Most of these old craters are no longer observable, and their remnants are now buried beneath younger impact debris.

7. The upper surface of the debris layer is pockmarked by craters ranging from less than a millimeter to several tens of meters in diameter (or larger, depending on the local area). Craters larger than 1 m in diameter occupy about 50 per cent of the surface; smaller craters occupy the rest of the surface and are also superimposed on the large craters. Minute craters, with dimensions of the order of a millimeter or less, probably cover nearly all of the surface and are superimposed on nearly all other features.

8. The debris layer typically varies in thickness from a few tens of meters to less than a millimeter. It is thickest where it covers the floors of some of the oldest and largest craters present and is thin, or possibly even absent, along the walls of very young craters that cut through the debris layer into the underlying mare material. Between craters less than 100 m in diameter, the average thickness of the debris layer is probably between $\frac{1}{2}$ and 1 m.

9. The porosity of the debris layer is expected to be of the order of 90 per cent at the surface. It

decreases rapidly with depth in most places, probably to less than 50 per cent at depths of a few tens of centimeters. Beneath these depths, the debris will have been compacted by shocks propagated from numerous small impact events.

10. The uppermost few millimeters of the debris layer is conceived as a fragile open network of loosely stacked, very fine grains. It is probably compressible under loads of the order of a few tens of grams per square centimeter. The bearing strength and shear strength of the material increase rapidly with depth, and at depths of a few tens of centimeters are probably similar to the bearing and shear strength of moderately consolidated dry alluvium on Earth.

11. The average albedo of the material in the debris layer ranges from that measured at the telescope, about 0.07 at the surface, to intermediate values at the base, probably between 0.10 and 0.15. The immediately underlying mare materials, when pulverized, have an albedo comparable to that of the brightest rays on the mare, about 0.25 to 0.30.

In the ray areas, a layer of coarsely crushed rock overlies the debris layer described here. The fragments in the layer of ray material probably have ordinary rock densities, and the bulk density of the ray material is not much lower than the density of the individual fragments. Depending on the age of the ray, a new porous layer of finely pulverized debris may be expected to form a thin veneer on the ray material. Judging from the shapes of the smallest craters shown in the last Ranger VII P-camera photographs, the ray material is essentially unconsolidated. It is probably similar in its gross physical characteristics to a mixture of dry sand and angular gravel.

3.

The Surface of the Moon

E. A. Whitaker

Lunar and Planetary Laboratory, University of Arizona, Tucson, Arizona

The main conclusions, which I consider to be still somewhat tentative, that have been reached at the Lunar and Planetary Laboratory in the last few months can be roughly summarized by the following eight points.

1. The maria and other dark areas appear to be best explained by fluid flows rather than ash flows, debris deposits, or dust aggregations. There is no reason to think the flows are not of lava.

2. The maria are not totally covered with a layer of cosmic dust, and the mixing of the surface layer appears to be local only, say up to about 3 or 4 km.

3. The so-called grid system of linear features that is seen particularly well in the highlands is continued in the maria as isolated bright mountain ridges, elements of mare ridges, chains of soft-edged craters, shallow valleys, and Dr. Kuiper's so-called tree-bark structure.

4. The mare ridges appear to be caused by intrusions through fissures; these intrusions failed to break out at the surface but instead formed sills at lower levels.

5. Isolated mare peaks and mountain ridges appear to be actual extrusions of this sill-forming material.

6. The soft-edged craters seen in the higher resolution Ranger photographs appear to be better explained as subsidence features, somewhat analogous to terrestrial karsts, rather than secondary or tertiary impact features.

7. The bright crater rays appear to consist of a thin layer of evenly distributed bright material rather than discrete clumps of the same.

8. The average depth of finely divided debris on the lunar surface may be of the order of 1 meter, although there does not appear to be any evidence for a sharp line of demarcation. Dr. Kuiper has computed a bearing strength of the order of 1 ton per square foot for the floor of Alphonsus, based on certain assumptions in the interpretation of the final Ranger IX frames.

Let us now deal with these points one by one. Firstly, why do we consider the maria to be lava flows rather than ash or dust deposits?

Figure 3–1 was made almost a year ago during a short and mainly cloudy duty at the McDonald Observatory in Texas with the 82-inch telescope. What I had hoped to do was to photograph the full moon through ultraviolet and infrared filters, at about 3,800 and 7,800 angstroms respectively. The weather was quite bad, however, and all I could obtain was a few rather mediocre photographs at later phases through gaps in the clouds. The results should thus be considered as only preliminary.

I combined a positive transparency of the infra-red photographs with an ultraviolet negative, the contrast of the positive being adjusted to eliminate as far as possible the differences of albedo. This technique produces a photograph such as this wherein the apparent reflectivity differences are actually differences in color; the lighter are the bluer, and the darker are the redder regions.

This portion of the moon is the Mare Serenitatis and Mare Tranquillitatis area. It has long been known that Serenitatis has a somewhat brownish or warm color, which contrasts quite strongly with the steely grey color of Tranquillitatis, although we know from measurements that even Tranquillitatis is on the red side of neutral. This color difference is emphasized in Figure 3–1 and is readily visible in low-power binoculars or telescopes. There is quite a strong line of demarcation between the areas of different color on the eastern side of Serenitatis, but this fades out on the western side.

Figure 3–2 is a similar composite photograph of the Mare Imbrium region. If you look carefully you can see Sinus Iridum, to the west of which is an area of quite red highlands. On the SE boundary of this area can be seen two very red features, which correspond to two unusual domelike mountains.

There are other lines of demarcation which run across the Mare such as those marked as *a* and *b*. Note the little river of redder material, *c*, which connects the red island surrounding the crater Le Verrier with a larger red area to the east. Aristarchus appears bluer than average; this has been confirmed by some work of Tom Gehrels at our laboratory a few months ago. He measured the absolute colors of some of the points contained in these photographs and verified the reality of the color variations shown in the photograph. The well-known, yellow, diamond-shaped area NW of Aristarchus appears dark, contrasting strongly with the crater itself.

We see that sharp boundaries exist between areas of redder and bluer material. Just how sharp these really are we don't know, since the seeing was rather poor when the original plates were taken, and the very slight blurring of the boundaries may be either because of the seeing or

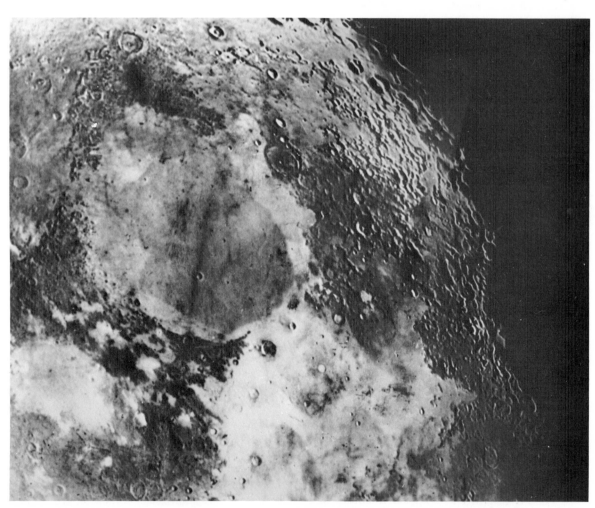

FIGURE 3–1: Composite UV-IR photograph of Mare Serenitatis and Mare Tranquillitatis region. Darker is redder. North is up in this and all succeeding figures.

actual mixing on the lunar surface. Do these boundaries correspond with any topographical features seen on the lunar surface? The answer is yes. Figure 3–3 shows a portion of Mare Imbrium at sunrise; the winding line *d* clearly indicates a low plateau, quite unlike mare ridges in appearance, and *e* traces the sunlit slope of a lower plateau. Figure 3–4 depicts the same area under sunset conditions and emphasizes the likeness between these features and terrestrial lava flows. The slope *e* corresponds exactly with part of the color boundary *b* in Figure 3–2, which strengthens this theory even further. The terminal slopes appear to be of the order of 30–100 m high, with angles of slope of approximately 2 or 3 degrees. The small "red river," of Figure 3–2 is at the positions marked similarly in Figures 3–3 and 3–4 but is not detectable in the photographs.

It can be seen that mare ridges cut right across the color boundaries without effect; these boundaries correspond only with the edges of the plateaus. Therefore we conjecture that these plateaus represent flows of material of a different color from the underlying mare surface. Could these be ash flows? Dr. O'Keefe is very favorably disposed toward ash flows, and indeed this is a distinctly possible hypothesis. However, terrestrial ash flows appear to run with great speed and tend to pass right over small obstacles in their path in the manner of a highly mobile fluid. In Figures 3–3 and 3–4 we find that the "flows" as exemplified by the plateaus tend to be stopped by small eminences, such as at *f*. For this reason, and also because of the terminal slopes of a few degrees, we consider that the plateaus represent flows of a fluid material rather than a gas-supported ash.

FIGURE 3–2: Composite UV-IR photograph of Mare Imbrium region. Darker is redder.

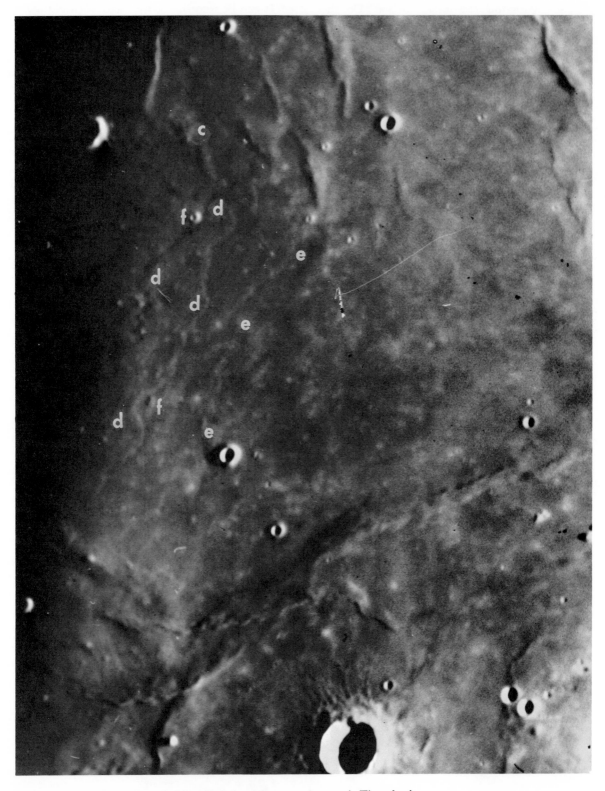

FIGURE 3–3: Sunrise on flows in Mare Imbrium. Crater at bottom is Timocharis.

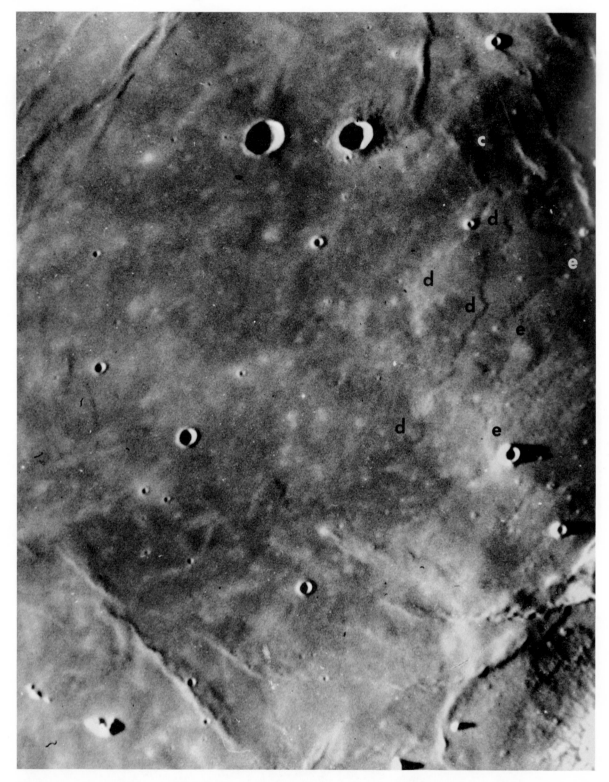

FIGURE 3–4: Sunset on flows in Mare Imbrium. Twin craters are Helicon and Le Verrier.

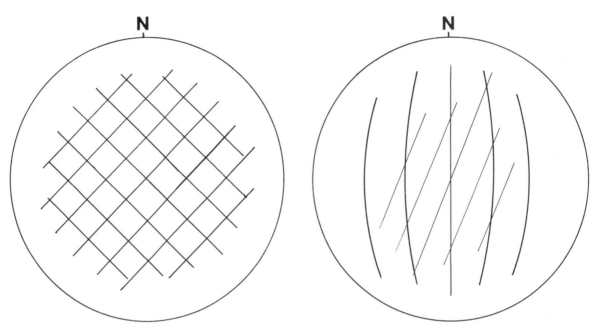

FIGURE 3–5: Diagrammatic representation of NE-SW and NW-SE lineament systems.

FIGURE 3–6: Diagrammatic representation of N-S and NNE-SSW lineament systems.

FIGURE 3–7: Clavius and neighborhood.

FIGURE 3–8: Clavius and neighborhood with a few of the lineaments emphasized.

The existence of the well-defined color boundaries on the moon indicates at once that the original lunar surface is not totally covered with a layer of cosmic dust, although the actual degree of contamination is indeterminable from this evidence.

Another research aspect that we have been pursuing for the last two years or so is in connection with the so-called grid system of linear features on the moon. Bob Strom, a geologist on our staff, has picked out all the linear features and alignments that are readily detectable on our best Earth-based lunar photographs, corrected their apparent position angles as measured on librated photographs to true position angles as would be seen in orthographic projection, and plotted them on a large chart. This procedure immediately makes it obvious that the lineaments fall into certain preferred directions, a result already obtained by Fielder over two years ago. By eliminating lineaments which fall into the various preferred directions one by one, we find that there are parallel, great-circle, and radial systems. The two parallel systems trend in NW-SE and NE-SW directions, as shown diagrammatically in Figure 3–5, and thus form systems of small circles. The N-S great-circle system at first appears to coincide with the lunar meridians, but closer analysis shows that it actually forms the bisectors, on the lunar surface, of the north- and south-facing angles of the intersections of the previous systems. A minor small circle system trends in a NNE-SSW direction, shown diagrammatically in Figure 3–6. The radial systems center on small areas near the centers of certain maria, viz. Imbrium, Nectaris, Crisium, Humorum, and Orientale.

Those various systems account for over 95 per cent of the lunar lineaments. From the very close analogy between the nonradial lunar systems and terrestrial lineament systems, particularly in regard to the preferred orientations, we attribute the lunar systems to fracturing of the lunar crust. The symmetrical disposition of the small- and great-circle systems with respect to the lunar axis and the mean direction of the earth suggest that terrestrial tidal forces may have played a role in the formation of these systems, and that the systems indeed represent crustal fracturing. Figure 3–7 depicts a lunar highland area in the region of Clavius; Figure 3–8 is of the same region with some of the lineaments emphasized.

Since the lineaments are seen in the highlands, the maria, and both inside and outside many craters, we consider that most of the lunar crust is

fractured to some depth in these preferred directions. Figure 3–9 is a Ranger VII photograph giving a view of the southern part of Mare Cognitum, in which parallel lineaments in both the mare and the surrounding "highlands" are well shown. Thus the features lettered g and h are closely parallel and belong to the NE-SW system; those lettered g are bright, elongated hills or mountain ridges, while those marked h are linear sections of mare ridges. The dark feature i, recently commented upon by Dr. O'Keefe, may be an extrusion of darker material and is also parallel to the other features. Such alignments are certainly not fortuitous. The pattern of alignments extends

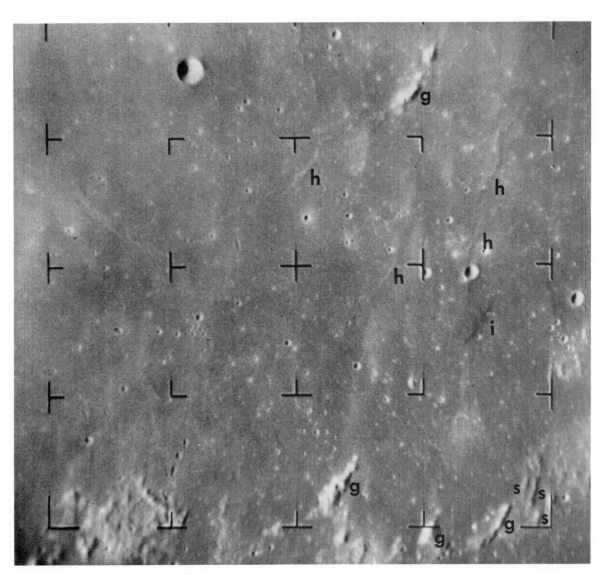

FIGURE 3–9: Portion of Mare Cognitum and its southern boundary (Ranger VII).

even to the resolution limits of all three Rangers, and Dr. Shoemaker has already drawn attention to the approximately orthogonal network of shallow features in some of the later Ranger IX photographs. Figure 3–10 shows four of the later Ranger VII P frames; the feature marked *j* is Dr. Kuiper's "boat," which is approximately diamond-shaped. Feature *k* is of a similar shape, although somewhat weaker and more irregular. The straight edges of these diamonds are aligned exactly with the N-S and NNE-SSW lineament systems. The tree-bark structure at *l* also trends in the latter direction, as it also does in the last two P frames (Figure 3–11).

Figure 3–12 is a Ranger VIII photograph of a portion of Mare Tranquillitatis, which displays several linear segments of mare ridges aligned precisely along NW-SE and NE-SW directions. Lineaments in the highland areas at the bottom of earlier photographs of this series are seen to be parallel to the mare ridges, indicating once again that both mare and highland topography are governed by the moon-wide fracture systems.

The fact that these systems manifest themselves as shallow, troughlike depressions less than a meter wide leads one to believe that the lunar surface is not completely covered by a thick layer of loose dust or debris, since such a layer would not be expected to duplicate, even weakly, the fine-scale

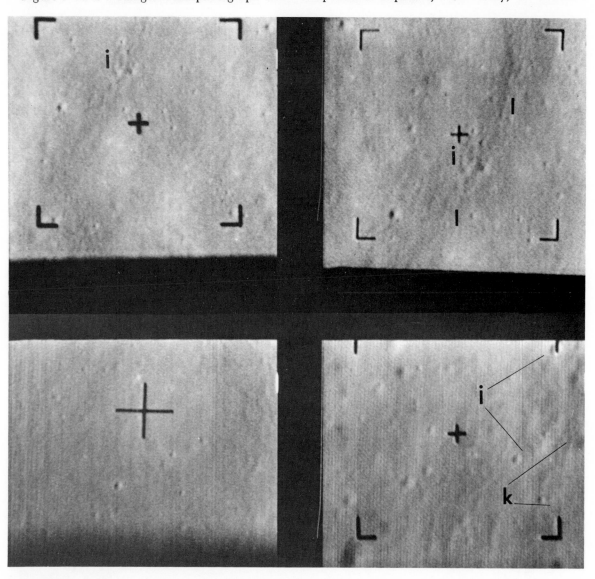

FIGURE 3–10: Boatlike structure in Mare Cognitum (Ranger VII).

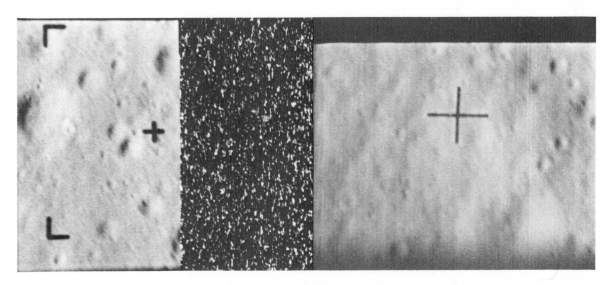

FIGURE 3–11: Highest-resolution photographs of surface of Mare Cognitum (Ranger VII).

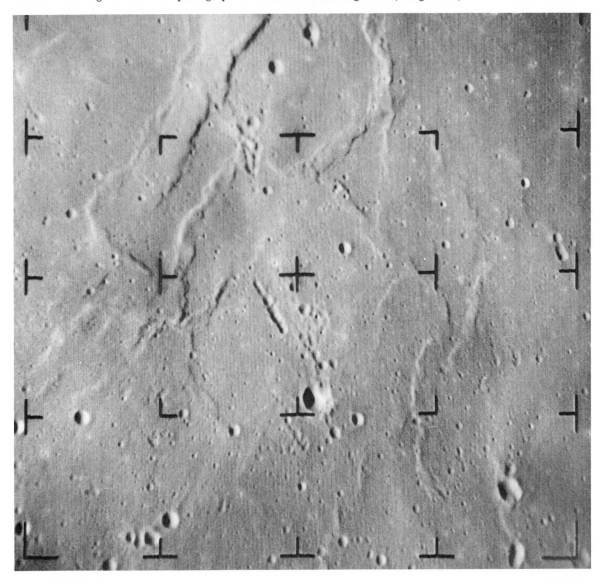

FIGURE 3–12: Lineament patterns in ridges in Mare Tranquillitatis (Ranger VIII).

topography of the harder, fractured, deeper layers. Similarly, the lunar surface appears not to have been subjected to anything approaching 100 per cent cratering, since this would also be expected to obliterate the fine lineaments.

Turning now to the subject of mare ridges, I think that there is almost unanimous agreement at our laboratory with Dr. Kuiper's explanation for them: that they have been caused by intrusions, presumably of lava, from below. The intrusive material is presumed to have risen through fissures, but failing to reach the surface, it spread out horizontally, instead, at some specific depth or depths, thereby raising the surface into the forms that we observe. Figure 3–13 shows a typical mare ridge in Mare Cognitum, but the solar elevation of 23 degrees is too high for the form to be displayed well. It is dissected by a crater 0.8 miles in diameter, which shows a soft, ledgelike feature near the base of the west inner slope. Dr. Urey interprets this ledge as representing the intersection of the ridge with the crater, implying a very gently dipping vein, but we consider the faint diametral line to represent the feeder fissure and the ledge possibly to represent the horizontal sill of intrusive material. The south wall of the crater appears to

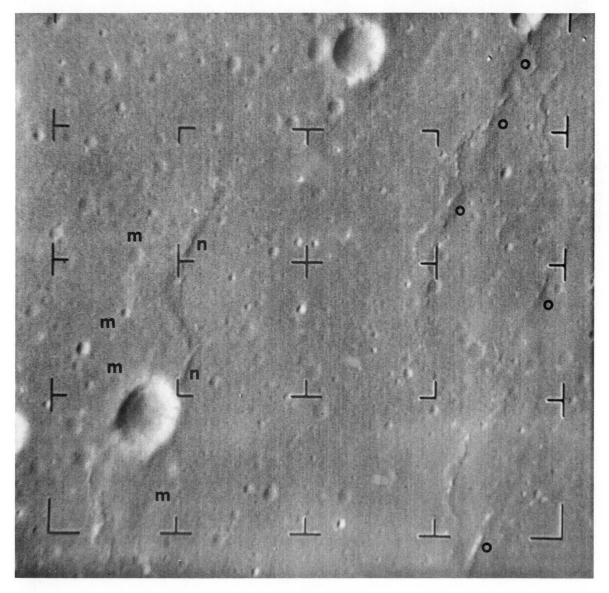

FIGURE 3–13: Dissected ridge in Mare Cognitum (Ranger VII).

show the feeder fanning out into separate dykes, but this may or may not be real. The ridge is about a mile wide in the vicinity of the crater, extending from *m* to *n*. The steepest portions of the ridge, at points marked *n*, are again seen to be linear and aligned NE-SW, as are also the sections *o*.

Although Figure 3–9 does not show it because of the high angle of illumination, low-illumination photographs of Mare Cognitum show that in all cases the bright mountain ridges *g* continue in the mare as mare ridges. This fact, coupled with the observation that many of the bright mountain ridges contain summit craters, lends support to the idea that these ridges are actual extrusions of the material which fails to reach the surface in the case of the mare ridges.

Let us now turn to the problem of the subtelescopic craters revealed by the Ranger series and their relationship to ray elements. Figure 3–14 shows the well-known group of secondary craters

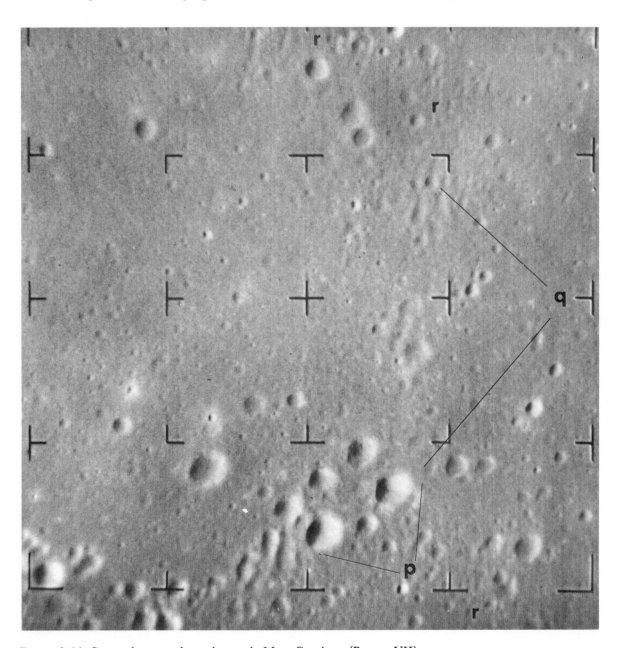

FIGURE 3–14: Crater clusters and ray element in Mare Cognitum (Ranger VII).

situated near the impact point of Ranger VII. Although most people consider groups like these to have been caused by the impacts of large blocks ejected during the formation of craters such as Tycho and Copernicus, my own personal view is that this may not be the case, since the depth and regular form of many of the craters of this type make them virtually indistinguishable from primary craters. For this and other reasons I have suggested that bright secondaries such as the group p in Figure 3–14 have been caused by the impacts of detached fragments of cometary nuclei. In this case, the main nucleus supposedly formed the crater Tycho, and the dust accompanying the fragments was responsible for the ray element q. The radial disposition of q from the group p with respect to Tycho is assumed to have been caused by the expansion of the gas cloud of vaporized cometary ices produced at Tycho, which deviated the dust but not the fragments. The group of

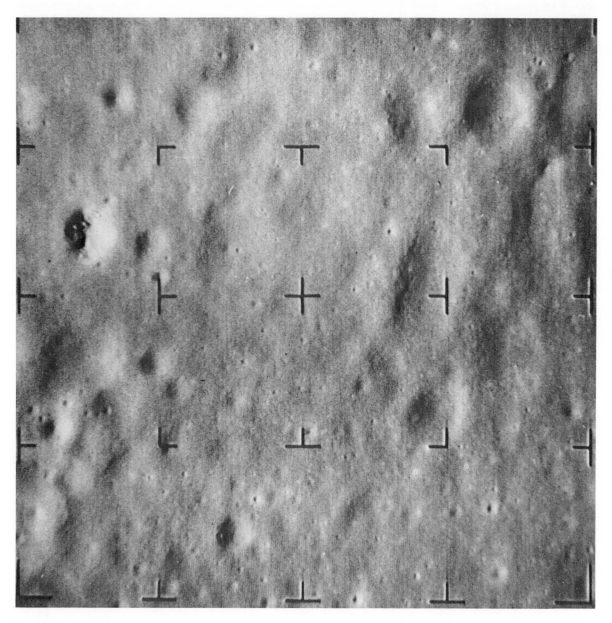

FIGURE 3–15: Elongated and circular soft-edged depressions in Mare Cognitum (Ranger VII).

shallow depressions situated in q was at first considered to be secondary or even tertiary craters and to be closely allied with the ray element. However, closer observation shows that several ray elements do not display shallow depressions and, conversely, that some groups of shallow depressions are not accompanied by rays, for example as at r. For this reason, I consider that the coincidence of ray q with the group of shallow depressions is fortuitous.

You will notice that these depressions show strong NNE-SSW alignment, a feature that is shared by some of the depressions at r. Dr. Shoemaker has attributed this parallelism to the method of ejection of elongated clots of material, the ballistic equations for which predict a certain degree of rotation of the clots, thus producing approximately parallel furrows. However, Bob Strom has compiled azimuth-frequency diagrams of the elongated craters surrounding Copernicus and finds that although the average direction is approximately radial, there is some deviation each side of this mean, even for adjacent craters. Since the larger Copernicus secondaries do not form groups displaying parallelism, we do not consider the parallelism of depressions q and r to be any criterion for an impact origin, but rather that it points to an internal origin, especially since the direction of alignment exactly matches the widely distributed NNE-SSW lineament system. Note that

the mountain ridges s in Figure 3–9 share this direction. We consider, therefore, that the depressions q are some type of collapse feature. Figure 3–15 shows these depressions with good resolution, and there appears to be no basic difference between the elongated and circular members of the group. Each type is virtually rimless and displays a roughly circumferential pattern of troughlike depressions which are indistinguishable from the tree-bark structure as seen in Figure 3–11. Since the randomly scattered soft-edged depressions appear to be identical with those in group q, one may hypothesize that all soft-edged depressions are collapse features.

Figure 3–16 shows a region in southern Indiana that we had taken earlier this year. The depressions are caused by the collapse of subterranean cavities formed by the dissolution of limestone by weak acids present in rainwater. The sun is at an elevation of about 15 degrees, and the resemblance between these features and the shallow depressions in Figure 3–15 is striking. The scene is snow covered, and unfortunately, the wind has produced an over-all wave pattern on the snow which we might liken to the grid pattern on the moon.

I will make only brief comments on a few of the Ranger VIII and IX photographs, since they were shown earlier today. Figure 3–17 is part of the floor of Alphonsus, and the density of craters is clearly

FIGURE 3–16: Snow-covered karst country, S. Indiana (courtesy of Dr. M. S. Burkhead).

greater than in M. Cognitum or M. Tranquilli- tatis. If we assume the maria have reached a steady state with regard to impacting bodies, how do we account for all the craters here? The rille near the right edge clearly degrades into a series of craters, and the higher resolution pictures show that these craters are indistinguishable from the other soft- edged depressions. The fact that rilles are appa- rently some type of collapse features lends further weight to our view that the soft-edged craters are collapse features.

Figure 3–18 is of the eastern floor of Alphonsus. The craters marked *t* are those surrounded by dark haloes, and they are seen to be absolutely rimless. From their relationship with the rilles and the dark patches, we assume they are of internal origin, but they do not seem to bear very close resemblance to terrestrial volcanic craters. The southernmost crater displays a line of bright spots some distance below the rim, each of which has a talus deposit descending from it.

The walls are on the right of the photograph, and

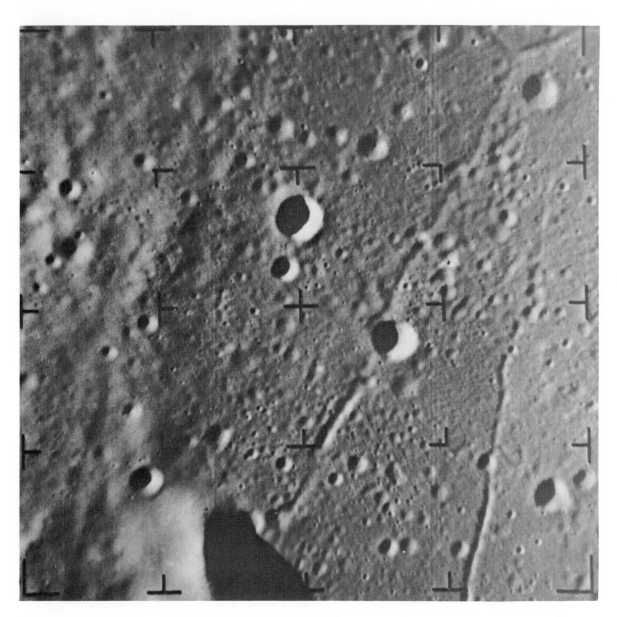

FIGURE 3–17: Floor and central peak of Alphonsus (Ranger IX).

their very smooth appearance rather bewildered us at first because we were not expecting this. The paucity of craters is difficult to explain if an impact origin is assumed for the majority of depressions we see on the level areas. If the soft-edged depressions are assumed to result from collapse, however, the lack of craters on the walls means that there have been few or no collapses there. This is borne out by the fact that nearly all small craters in the walls are sharp, deep, and circular, and therefore probably of impact origin.

The explanation that most of the impact craters in the walls have been erased by landslides does not seem to hold because we see sharp craters on fairly steep slopes, which suggests the material is not just a pile of unconsolidated rubble; and level areas in the walls do not contain more craters than the slopes. The peculiar, toasted-marshmallow appearance of the slopes may be caused by local slumping but this is conjectural. Later frames with higher resolution show some degree of striation, resembling the tree-bark structure of the level

FIGURE 3–18: Wall and east floor of Alphonsus (Ranger IX).

FIGURE 3–19: Mosaic of high-resolution photographs of floor of Alphonsus; arrows indicate positive relief features (Ranger IX).

regions. It is important to note that the walls have precisely the same albedo as the floor, except for the dark patches. The central peak, in its general smoothness and lack of craters, resembles the walls, but its albedo is greater than floor or walls. It is not easy to explain how a mountain like that is formed, particularly with such a sharp ridge; it may be somewhat similar to mountain ridges in the maria, but clearly there are points of difference.

Figure 3–19 is a mosaic of the last high-resolution photographs taken by Ranger IX; the arrows point to a number of positive features. Dr. Kuiper thinks these may be rocks that have been ejected from the only impact crater situated in the vicinity, which is the one marked u. He assumed that the ground and the rock were crushed equally on impact, and then, from measurements of typical heights and diameters and a valid assumption on the density of

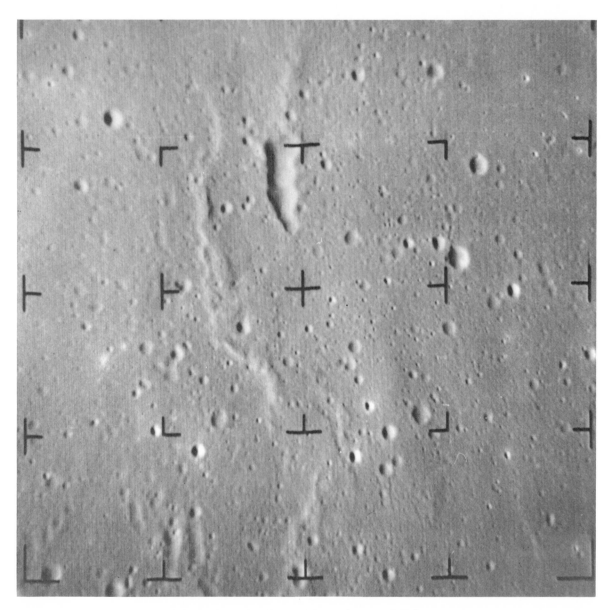

FIGURE 3–20: Related ridge and depression, Mare Tranquillitatis (Ranger VIII).

FIGURE 3–21: Field of volcanic dust and ash at Laimana crater, Hawaii; the crater *w* near center is 5 ft. in diameter (courtesy of Lyman Nichols, Hilo).

the material, calculated that the strength of the lunar surface could be as much as one ton per square foot. The distances from the crater enter into the calculation as well, of course. The smallest pits and rocks are less than 2 meters across. One ton per square foot is one kilogram per square centimeter, and I think it tends to represent an upper limit rather than a mean strength.

Figure 3–20 is a Ranger VIII photograph showing how a mare ridge suddenly becomes a depression, which fits in well with our ideas of lava intrusion and collapse following subsequent drainage.

Finally, Figures 3–21 and 3–22 are aerial photographs of a field of volcanic ashes and dust near Laimana crater in Hawaii caused by the 1960 eruption. We were given these and other photographs the day prior to a field trip to the area. We could see impact craters, *v*, caused by falling rocks, some having flat floors, some conical, some with central "peaks," and others without. Each had

walls of higher albedo than the surroundings. The craters, marked *w*, were clearly different and resembled somewhat the lunar dimple craters. The feature *x* looked exactly like a lunar rille, *y* was a dome, and the elongated depressions *z* looked like telescopic secondary craters. We thought that we might answer a number of lunar problems with one trip, but this did not materialize, of course. The flat floors were caused by wind blowing loose ashes in some craters; the white walls were caused by leaching of salts by rain; the rimless craters *w* were caused by ash draining through large holes in the lava bed underneath, as were the rille and elongated craters. The dome was merely an ash-covered excrescence of lava. Despite this, the trip served as an extremely valuable object lesson in what can and cannot be interpreted from a study of aerial photographs, and we are under no illusions concerning the accuracy or completeness of our interpretations of the Ranger records.

FIGURE 3–22: Aerial view of fields of lava and ash at Laimana crater, Hawaii (courtesy of Lyman Nichols, Hilo).

4.

The Surface Structure of the Moon

Gerard P. Kuiper

Lunar and Planetary Laboratory, University of Arizona, Tucson, Arizona

With two of the three Ranger VII Atlases and the Experimenters' Report of Ranger VII published,[1] and with the Ranger VIII and IX missions successfully accomplished, it is opportune to summarize some of the scientific results. This summary is based largely on analyses made at the University of Arizona laboratory, with participation by E. A. Whitaker, R. G. Strom, W. K. Hartmann, R. Le Poole, and others, and is taken in part from the author's Ranger VII Experimenters' Report.

It has been our first concern to assist in the full publication of the magnificent Ranger records, the only limitations being to eliminate redundancy and to ensure that the published records reproduce the full image content. Much time and thought has been given to this publication problem, and Mr. Whitaker has applied his full skill in the production and supervision of the photographic editions.

I. TEXTURE OF THE MARIA; DEPRESSIONS

The interpretive work has relied heavily on our collection of plates built up since 1959 at the McDonald 82-inch and the Yerkes 40-inch telescopes. This work includes a high-resolution study of the color distribution over the lunar surface by Mr. Whitaker, with photoelectric calibrations by Dr. Gehrels. The color study has shown that the

moon is not covered with a layer of cosmic dust, which would have obliterated the color differences in the maria, and that any migration of a hypothetical layer of lunar dust is restricted to subtelescopic dimensions because the color provinces observed have extremely sharp boundaries. Instead, the lunar surface materials such as crater rays, and structures such as the prominent flows observed on Mare Imbrium and Mare Serenitatis, each have characteristic colors, often quite uniform over a single feature, though adjacent structural features may differ markedly. The Earth-based photographs make plain that at least two of the lunar maria were built up as a succession of flows (presumably lava), with each flow having distinct terminal walls and therefore having had a very viscous or solid outer surface during deposition; yet, they are very extended up to 200 km in length on surfaces sloping about 1 degree or less, having a highly fluid interior at the time of deposition. The observed flows are 50–200 m thick but thinner flows may exist. On the basis of experiments with liquid silicates in vacuums and on certain rock formations found on Hawaii (materials thrown out of Laimana Crater in 1960 in liquid form and solidified in free fall) one can conclude that the flows observed on the lunar maria were originally covered by a layer roughly 10 m thick of extremely vesicular rock, with a bulk density of 0.1–0.3 and a bearing strength of the order of 1–5 kg/cm². Below this surface layer a denser rock is expected to exist, having resulted from solidification of the very fluid magma responsible for the bulk of the flow. A succession of flows, as observed on Mare Imbrium, will have led to an alternation of vesicular and denser layers.

A comparison of the last A and B frames of

---

[1] G. P. Kuiper *et al.*, *Ranger VII Photographs of the Moon. Part I*: *Camera "A" Series*, Pasadena, Calif.: Jet Propulsion Laboratory, NASA-SP-61 (1964); *Ranger VII Photographs of the Moon. Part II*: *Camera "B" Series*, Pasadena, Calif.: Jet Propulsion Laboratory, NASA-SP-62 (1965); R. L. Heacock *et al.*, *Ranger VII. Part II*: *Experimenters' Analyses and Interpretations*, Pasadena, Calif.: Jet Propulsion Laboratory, JPL-TL-32-700; NASA-CR-62347 (1965).

Ranger VIII, released on the day they were obtained, February 20, 1965, with the last A and B frames of Ranger VII, released on the day they were obtained, July 31, 1964, shows a remarkable similarity of the two maria, in spite of their appreciable color differences. Both show an abundance of very shallow dips of a variety of dimensions and depth-to-diameter ratios. Diameters of 50–100 m and depths of 1–10 m are typical. These shallow dips show no sign of explosive violence; the slopes are extremely smooth, somewhat like an inverted Gaussian curve rotated around its vertical axis of symmetry. The deeper depressions show signs of cracks and faults in the walls, and some depressions end not in a flat round bottom but in a point or a short horizontal line. These pointed depressions look somewhat like dimples and they might suggest drainage to subsurface cavities, but this does not explain their round shoulders, which are indistinguishable from those of depressions without the funnel.

The fact that both maria observed show these depressions, and that the lower sun angle of the Ranger VIII records (15 degrees vs 23 degrees) allows one to see them over the entire surface, suggests that a general property of a mare surface is involved. Basically the same structure is also found on the floor of crater Alphonsus observed on Ranger IX at a still lower sun angle, 10 degrees. This floor has other distinct mare-type properties.

One might at first suppose that the depressions could be due to a surface heavily cratered by post-mare impacts and subsequently buried by dust. But this is most unlikely on several grounds: (1) no single thickness can explain both the large and small dimple craters of similar proportion, and the shallow and deep depressions; (2) fine fractures (scale ∼ 1–2 meters) are observed that are structurally related to the depressions in a manner suggesting a brittle rock surface; (3) lineaments are observed over the entire mare surface to the smallest resolved dimension, apparently part of a global lineament pattern (see below); (4) there is no appreciable cosmic dust layer on the lunar surface (see above) and the post-mare impact rate of particles below 10^4 grams is about 1 gram/cm^2/ 4×10^9 years.[2]

Instead, the depressions resemble the gentle sinks often observed in karst-type formations.[3] There are four causes that may have led to sinks on the lunar maria: (1) local drainage of magmas below the surface rock-froth layer; (2) collapse of vapor bubbles as the temperature drops; (3) volume decreases resulting from solidification and cooling; (4) escape of gases. These mechanisms cause dips in the surface not unlike those due to solution of subsurface limestone in the karst formations. Cave-ins of the ceilings of the cavities lead to dimple craters. On the basis of the above identification, the sinks described are early post-mare in age. The absence of appreciable straight conical sections on the slopes of any observed dimple craters indicates that drainage of surface materials into the holes has been small. Recently this writer observed a basaltic lava flow, some 25 miles SSW of Grants, New Mexico, that even at high sun (50 degrees) showed numerous circular sinks (Figure 4–1). A ground study of these features is being planned.

II. CRATER RAYS

A second phenomenon discovered and interpreted on the basis of Ranger records is the fine structure of crater rays. While major rays were intentionally avoided in both missions because of their known disruption to the lunar surface, Ranger VII did impact in a small and weak Tycho ray, with several Tycho and Copernicus ray elements close enough to be partly resolved also. Generalizing from these data and after a re-examination of the best Earth-based records of the Tycho rays, one may say that each major crater ray is composed of many elements, each of which normally starts with an impact crater or a cluster of such craters, from which issues a separate ray element in the direction away from the central crater. Mr. Whitaker has pointed out that the orientation of the Tycho ray elements is most readily explained if the central Tycho explosion caused a blast over the lunar surface that blew the fine debris resulting from the secondary impacts downwind, away from the central crater. In order that this blast persist for the flight time of 10–30 minutes (Mare Cognitum is 1000 km away, v ≅ 1 km/sec) it is required that low-temperature volatiles be involved; sublimated

[2] C. W. McCracken and M. Dubin, "Dust Bombardment on the Lunar Surface," *The Lunar Surface Layer: Materials and Characteristics,* ed. by J. W. Salisbury and P. E. Glaser, New York: Academic Press (1964), pp. 179–214.

[3] W. D. Thornbury, *Principles of Geomorphology,* New York: John Wiley and Sons (1954), pp. 321 ff.

FIGURE 4–1: Aerial photo, by the author, of lava field 25 miles SSW of Grants, New Mexico, showing Moon-like circular sinks.

particles would not be effective in deflecting the ray elements. This leads to the hypothesis that ray craters, such as Tycho, were formed by the impact of a comet. Again, as Whitaker has pointed out, a comet is also a suitable source of the materials causing the bright impact craters or clusters of such craters at the head of each ray element because these resemble primary craters and are bright at full moon, unlike the secondaries caused by lunar ejecta much nearer to Tycho, which have no ray elements associated with them. The latter resemble, in distribution and numbers, the very numerous secondaries found around all other major post-mare craters, regardless of the presence of rays.

The ray elements appear to be a mere sprinkling of light-colored material over the affected areas of the lunar surface. They cast no shadows and seem to leave the pre-existing surface features essentially undisturbed. Their brightness is not due to unresolved bright craters except at the head of each ray element where, as stated, a small bright crater or a cluster of such craters is found, evidently the provenance of the bright ray material. These conclusions are substantiated in the Ranger VII Report (see fn. 1).

III. CLASSES OF CRATERS

The craters on the maria appear to be mostly of impact origin and this is presumably, though not necessarily, true for the post-mare craters on the terrae. The primary craters are by definition those due to impacts of cosmic masses. Both cometary and asteroidal (meteoritic) masses are involved. The first lead to craters accompanied by large complex rays, such as Copernicus and Tycho; the latter to craters without extensive rays though they may show small, symmetrical, bright halos.

Primary craters are surrounded by a swarm of much smaller secondary craters, apparently formed by lunar debris tossed out of the primary crater following the impact. They are strongly concentrated around the central crater, occur rather uniformly in all azimuths, and thin out rapidly beyond about three crater diameters. Often a substantial part of the lunar rock causing a secondary crater is still present on the crater floor. The dimensions of these masses relative to the craters they formed are consistent with their origin from primaries between 100–1,000 km away. The fact

that the masses remained reasonably intact after the primary explosion and the secondary crater formation shows that at least parts of the lunar rock are well consolidated.

Secondary craters occur around primaries of both cometary and asteroidal origin. Tycho and Copernicus are examples of the former; Bullialdus, Herschel, and the 46-meter crater observed on the last Ranger IX P frames are examples of the latter. The secondaries are apparently caused by lunar debris, impacting at $v \leq 1$ km/sec. They are invisible at full moon.

Primaries of cometary origin have a second class of craters associated with them, the single or clusters of smaller bright craters at the heads of the ray elements. These are not true secondaries, their impacts probably having occurred with velocities nearly identical with that of the central primary impact, about 50–55 km/sec for a parabolic comet on a nearly radial orbit near the earth. These craters, bright at full moon, are therefore associated primaries, presumably caused by the outlying parts of the complex body that constitutes a comet nucleus.

The sinks and dimple craters referred to earlier are not regarded as due to impacts but to collapse, for the reasons stated before. Chain craters as observed on the floor of Alphonsus by Ranger IX are also regarded as being of internal origin, due to collapse and slumping following fissure formation. The well-known dark-halo craters along rilles following the periphery of the Alphonsus floor are shown very well on the Ranger IX records. They are clearly of internal origin, having covered their immediate surroundings with a material that is presumably dark ash and cinders, not lava.

The Ranger records also show several craters or calderas, squarely placed on the summits of roundish mountains or mountain ridges. The assumption of an impact origin of these features is not credible on probability grounds. Instead, a volcanic origin is indicated for both the mounds and the craters or calderas.

The problems of pre-mare crater formation are tremendous, as are the craters themselves, particularly if the impact maria are included. This extension is logical on several grounds, and the magnitude of the impacts and their high frequency of occurrence during a comparatively short period in lunar history (for the smaller crater dimensions, about 10^4 times greater than at present) indicate that major problems of lunar origin and evolution are involved. It is expected that the Ranger VIII and IX data will yield many clues to this central problem of pre-mare history.

The Ranger results have defined the frequency of lunar craters below the 1-km limit set by the best telescopic surveys. Crater counts have been made under Dr. Shoemaker's direction at the Astrogeology Branch of the Geological Survey, and by Hartmann and Le Poole at this laboratory. There is good evidence that the slope defined in a log N vs log D plot of about -2.1, found for the telescopic (primary) craters on the maria, continues for the smaller primary craters down to the 120-m size. This is in agreement with theoretical predictions from data for asteroids and meteorites, assumed to be mostly responsible for the smaller primaries. And the steeper slope observed below D = 120 m (-3.5, approximately) is probably due to an actual increase in the number of meteorites below 10^8 grams mass. According to counts made by Mr. Le Poole, the frequency curves for sharp craters on Rangers VII, VIII, and IX agree very well in the region 1–100 m, indicating that secondary craters are not numerous among the sharp craters.

IV. MARE RIDGES, RILLES, AND LINEAMENTS

The mare ridges were, of course, known before but with the Ranger VII records we have made much progress in their interpretation. Through a fortunate accident one ridge in Mare Cognitum, which had been dissected by an impact crater, was observed well. It became apparent that the following model represents the essential observations. Because the surface color and albedo are not disturbed by the ridges, they are regarded as strips of the lunar mare crust that are uplifted. This uplift is caused by dikes that formed in subsurface fissures. The fissures resulted from dynamical causes, either encompassing the entire moon (related to the global grid system and perhaps due to changing tidal effects), or limited to a basin (radial and peripheral ridges in maria, etc.). Occasionally the ridges have broken through the mare surface; then the dikes are observable as light-colored walls, some 50–100 m wide, protruding as much as 10–20 m above the surroundings. The ridges and dikes frequently exhibit *en échelon* structure, apparently caused by subsurface branching of the fissures, with sometimes as many as four parallel branches approaching the surface. This mechanism accounts for the braided appearance of many of the ridges and also for the curious effects observed at ridge intersections.

Yet another phenomenon, also known before, has been clarified by the Ranger records: that of the rilles, often found just within the shorelines of a mare. From Earth-based observations these had been identified as graben.[4] The Ranger VIII photographs show the correctness of this description, including the *en échelon* of these rilles and their shorelines.

A further, most important type of structure in the lunar surface is the system of lineaments. Mr. Strom has made an analysis of this and has found the fine linear structures, so well portrayed in the five ACIC maps of the Ranger VII area (scales 1 : 1,000,000; 500,000; 100,000; 10,000; 1,000; 350), mostly belong to the global system he has recently investigated.[5] From the visibility of fine lineaments Strom has estimated that any layer of displaced lunar debris scattered on the original igneous surface will be less than 2 m thick.

V. BEARING STRENGTH OF A MARE FLOOR

The absence of loose rocks widely scattered over the lunar surface, first noted on the Ranger VII records, caused wide comment. It seemed in marked contrast, if not conflict, with the presence of both primary and secondary impact craters.

It was soon realized, however, that this is precisely what is to be expected if the mare surface is composed of brittle and vesicular material of low bulk density, as expected from lava flows exposed to a vacuum. The probable limiting bearing strength of 1–5 kg/cm² or 1–5 tons per square foot was so derived (see fn. 1); and it was concluded that rocks of about 1 m³ and larger would be found only within 100–200 m from a primary crater.

[4] Kuiper, "The Exploration of the Moon," *Vistas in Astronautics*, Vol. 2, ed. by M. Alperin and H. F. Gregory, New York: Pergamon Press Books (1959), pp. 273–313.

[5] R. G. Strom, "Analysis of Lunar Lineaments, I: Tectonic Maps of the Moon," *Commun. Lunar and Planet. Lab.*, Vol. 2 (1964), pp. 205–216.

FIGURE 4–2: Composite photograph of last Ranger IX P frames showing primary impact crater 46 m (=150 ft) in diameter and some 52 rocks scattered nearby.

The Ranger IX data have now confirmed this expectation. Figure 4–2 shows a photographic composite of the last P frames on which black arrows indicate rocks. About 52 have been marked. They are about 1 m² in size, project 10–30 cm out of the lunar surface, as derived from their shadows, and extend to 120 m = 400 ft from the primary crater, which is 46 m = 150 ft in diameter. The fact that Ranger IX impacted so close to a primary crater is, of course, fortuitous; only three or four craters of this general size are seen on the last Ranger IX A frame. Since the distance from each rock to the center of the crater may be measured, its approxi-mate impact velocity is known. If a bulk density of 2 is assumed (because the rock mass did not disintegrate) the limiting bearing strength of the upper 1–2 ft of the lunar surface is found to be 1–2 kg/cm² (or 1–2 tons per square foot). If the trajectory of the rock did not start off with a 45-degree elevation, the impact velocity was greater and the lunar surface more resistant.

The crater floor of Alphonsus is presumably somewhat softer than a true mare surface. At any rate, the measure obtained is not unsatisfactory in view of soft-landing requirements.

5.

The Moon's Surface

T. Gold

*Center for Radio Physics and Space Research,
Newman Laboratory of Nuclear Studies, Cornell University, Ithaca, New York*

As was everyone in this field, I was elated at the enormous success of the last three Ranger shots. I feel strongly that we who try to interpret the features of the moon cannot match the success of the technology involved in the Ranger shots. We will still leave the scene confused after all the discussions; there will still be as many theories at the end as there were at the beginning. Nevertheless, it is true that we now know a great deal more, but what we know still does not allow us to tighten up our viewpoints completely. Before we do this, devices such as Surveyor will have to investigate the structural nature of the ground.

The degree of toughness of the moon's surface is a matter of great concern to those in technology, and optical investigations just are not capable of giving us that answer. They can give us various hints and clues that I shall discuss, but they cannot tell us whether it is material we can walk on and will sink only a few inches in, or something we would wallow in and have great difficulty in even taking a step in. At the moment we cannot make that distinction. When I say that, I have in mind a lot of data in addition to the optical discussions, and even with that I feel we cannot yet make this important distinction.

Let me first refer to information that comes from sources other than the Ranger pictures in order to present the Ranger discussions against the background of other things we already know. I will start with radar information of the moon's surface, which has now progressed a great deal, owing largely to the work of Pettengill and his associates at Arecibo. It has become possible, at a wave length of 70 centimeters, to chart in fine detail individual regions with a resolution of a few kilometers.

The mean reflectivity of the moon to radio wavelengths is about 7 per cent, which is much less than the radio reflectivity of any solid rock would be, and which would require a layer of rock of some meters in thickness in which the density is not more than half that of solid rock, and probably less.

The general radar information therefore is that the moon as a whole has a layer of at least a few meters, and possibly much more, in which the density is not that of solid rock but a factor of two or more down.

The young craters, however, like Tycho, Copernicus, and many others, give much larger radar reflectivity which would be compatible with solid rock. This correspondence between young craters and radar highlights is now very well established (see Figures 5–1, 5–2, and 5–3).

In general the radar highlights are strongly related to the optical albedo. The bright regions seen at full moon are the regions that coincide with the enhanced radar reflectivity. This suggests that there is more solid rock in the vicinity of such places.

A radar map of the area of the Ranger VII impact (Figure 5–4) shows a featureless region. It is uniform, as is the rest of the mare ground.

Tycho of course was the first crater to be known, from Pettengill's work, as an outstanding radar reflector. Its relative radar reflectivity goes up to 900, compared to 40 or so in the surrounding areas.

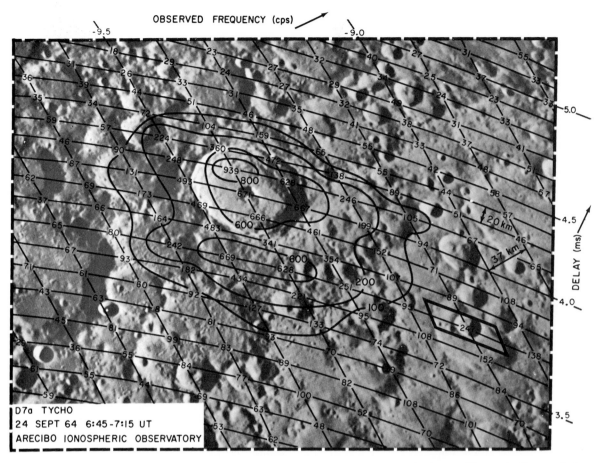

FIGURE 5–1: Tycho—radar scattering intensity contours superposed on an optical picture. Figures indicate relative scattering intensity.

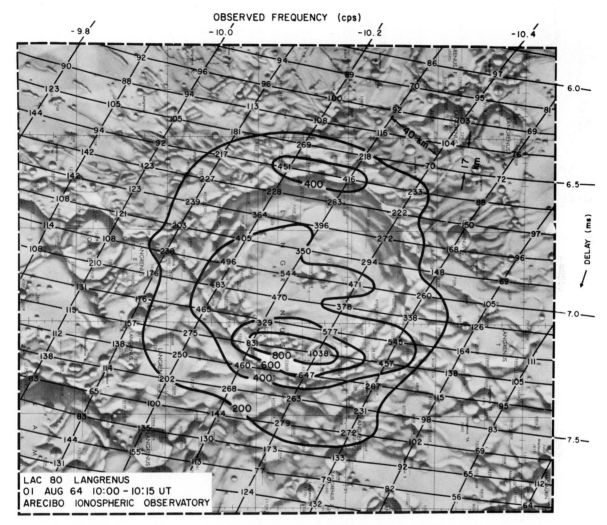

FIGURE 5–2: Langrenus—radar scattering intensity contours. Intensity reaches 800 in the crater, compared with 60, 70, or 80 in surrounding terrain.

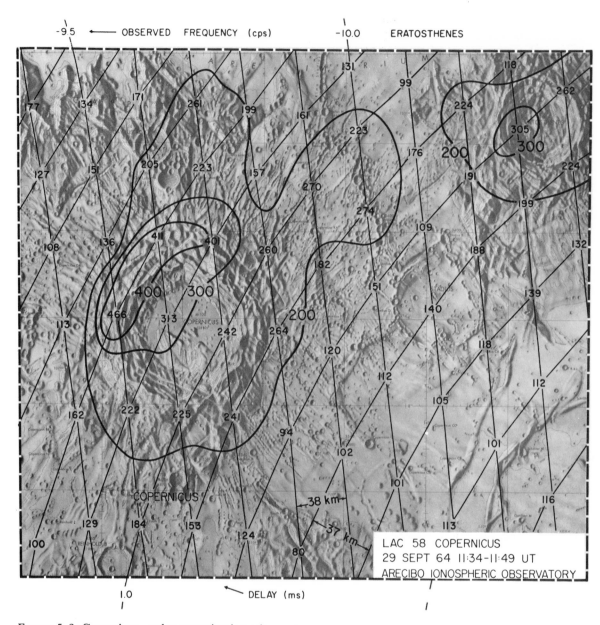

FIGURE 5-3: Copernicus—radar scattering intensity contours.

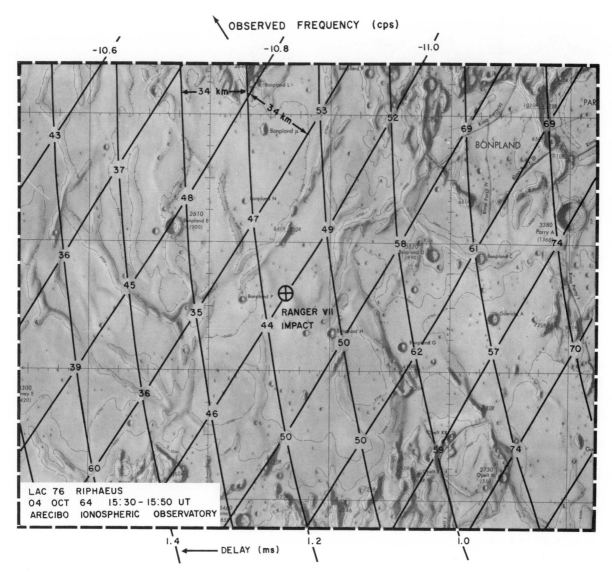

FIGURE 5–4: Ranger VII impact area. Radar scattering map. Grid and figures indicate resolution and scattering intensity.

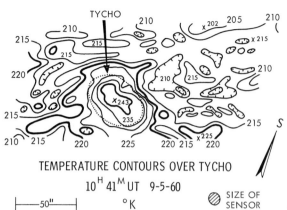

FIGURE 5–5: Temperature contours over Tycho.

Tycho, like other apparently young craters, is also outstanding in being a thermal anomaly.

The Shorthill and Saari contours taken during an eclipse of the crater Tycho (Figure 5–5) show that there is a clear correspondence with the young craters in all the thermal data. If we try to interpret this, the radar tells us that most of the lunar ground is very porous, down to a depth of several wavelengths, in which the radar waves get absorbed. On the other hand, the same kind of information on porosity of the ground is contained in the thermal data, namely that the moon cools much too fast to be solid rock.

The exception to the above are the young craters which behave more like solid rock in every respect. After all, if a crater is made by a large impact, the pressures that are exerted on the ground during the excavation are in the region of megabars. At these pressures the rock is not only completely compacted, it actually suffers pressure-induced phase changes as well. Whatever the original composition of the area was when the crater Tycho was blasted, its walls were made into some absolutely solid compacted material. The same must be true of all the other craters if they were made by the same process. The anomaly of the moon is not that the young craters behave more like solid rock. The anomaly is in the rest of the moon, which has suffered some degradation down to a depth at which radar waves get absorbed—quite some meters—and in which the material has become much more loosely compacted. All the old craters which must once have been hard are now radar absorbent. The highland regions are full of craters

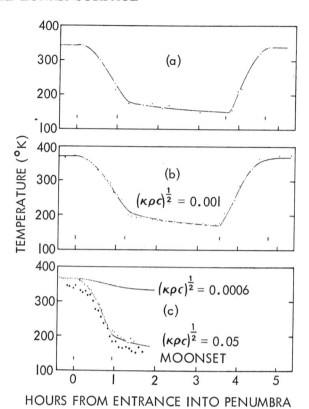

FIGURE 5–6: Temperature measurements of lunar eclipses. (a) June 14, 1927, observed by Pettit and Nicholson; the curve is one drawn through the points. (b) October 27, 1939, observed by Pettit; the curve is a theoretical one computed by Wesselink. (c) July 26, 1953, observed by Strong and Sinton; the curves are theoretical ones taken from Jaeger. The large dots are for a point 2 minutes from the east limb and the small dots are for a point 9 minutes from the east limb. The vertical marks on the diagrams indicate the times of entering and leaving the penumbra and umbra. $(k\rho c)^{1/2} = 0.05$ would correspond to some rocks.

the size of Tycho but none of the old-looking craters are good radar reflectors and none have these pronounced thermal anomalies.

We have to understand, therefore, how over a period of time a degradation process took place on the moon, thereby making the craters radar absorbent and thermally softer—a process that must have occurred down to a depth of at least a few meters and probably much more.

The Pettit and Nicholson data (Figure 5–6) show the enormity of the difference of the moon's surface from that of rock.

The Murray and Wildey cooling curves (Figure 5–7) show a very interesting effect. One curve shows the moon's surface as a function of length of

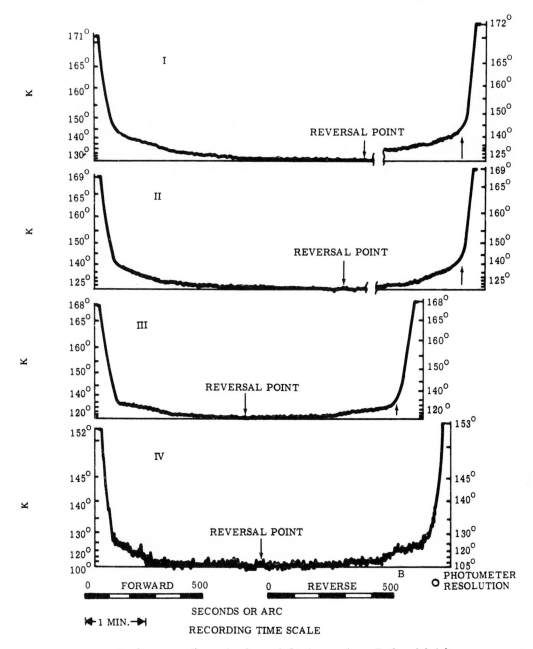

FIGURE 5–7: Tracings of strip-chart recordings of voltage deflection vs time. Reduced brightness-temperatures in degrees Kelvin are also shown in the vertical scale. The time base of the recordings is shown along with a scale illustrating the angular displacement in seconds of arc. The terminator, with a brightness-temperature of 200°K or greater, is off scale on the tracings (Murray and Wildey).

time for which it has been in the dark. The curves generally have a sharp kink at about 140 or 120 degrees absolute. Such a kink can be understood in terms of a mixture of solid or near-solid rock occupying a small fraction of the surface area, and some much less dense material occupying the rest of the area. A similar thing is true for the radar data; they also can be understood in terms of a mixture of two types of ground.

If the return signal is observed as a function of range or angle of incidence (Figure 5–8), then there is a big peak at the nearest range, which means that the moon is largely a specular scatterer to this 70-cm wavelength; but at lower power a little bit of a

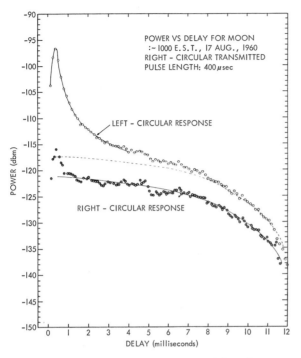

FIGURE 5–8: Lunar echo power vs delay for two orthogonal-received polarizations (Pettengill and Henry).

tail can be seen, which goes out to the limb. A diffuse scattering is going on most of the way out to the limb. Most of the area is smooth but a small fraction is very rough. On the opposite polarization to the one that would be expected back from a specular scatterer, the specular peak is much decreased. In other words, the nose of the moon gives very little depolarization proportionately. Most of the power it gives back comes in the right polarization and not the wrong one. But the tail gives almost as much in the depolarized component as in the correct component.

Thus the fraction of the ground that behaves as a rough scatterer has mostly a roughness on a scale comparable with the wavelength used—70 cm. The tail is not due to a few large areas that are steeply inclined to the local horizontal, but to many small areas which depolarize at the same time as they scatter.

This means, then, in an over-all view, that young craters possess hard rock at or near the surface; that the roughness which the radar reflection indicates is mostly concentrated in small areas a few meters across; that the hard ground is highly correlated with the young craters, and, therefore, that there would be a lot of hard ground connected with craters that are too small to be individually

seen from the earth. With craters like Tycho we could find this out; with smaller ones we could not. The sum total of the smaller craters no doubt will add to this radar tail and would add to the effect, shown in the Murray and Wildey curve, of a certain fraction of the lunar ground being hard.

Figure 5–9 is Evans' plot of the radio reflectivity, again as a function of delay or angle of mean slope to the line of sight. It shows that as one goes to shorter radar wavelengths the ground appears to become rapidly rougher. It is a very specular reflector, as a whole, at 3 cm, but at 1 cm or 8 mm it already is almost uniformly rough. This indicates that as one looks at the lunar ground in fine detail, i.e., at a scale of a centimeter, it will look very rough; but on a scale of tens of centimeters it will still appear very smooth.

Next I shall refer to the optical measurements (Figure 5–10), which will be reported in greater detail by Dr. Hapke. As a result of the investigations with rock powders, we feel very confident that the entire range of optical properties of the moon, the scattering function, color, albedo, and polarization, can all be reproduced by dust. But it can only be reproduced by rock dust which on the surface must be loosely packed and which has to be very dark and opaque, as when subjected to the proton bombardment that the moon suffers in a period of a few tens of thousands of years. Most rock powders become dark and opaque when irradiated by 1 kev

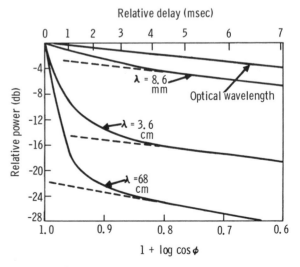

FIGURE 5–9: The average echo power plotted as a function of $1 + \log \cos \phi$ ($\phi = i =$ angle of incidence and reflection) at wavelengths of 8.6 mm, 3.6 cm, and 68 cm. A clear wavelength dependence in the scattering behavior of the moon is apparent.

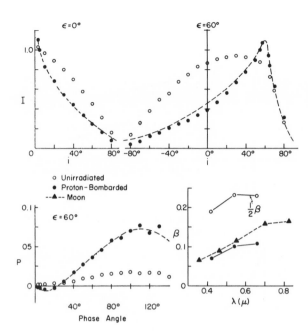

FIGURE 5–10: *Top*: Optical scattering law. Scattered power as a function of angle between ingoing and outgoing ray, shown for 90-degree incidence and for 60-degree incidence. The dashed curve is for the moon, showing in each case the sharp backscatter. The open circles are dunite powder (as example of a rock powder) before proton irradiation; full circles, after. *Lower left*: optical polarization against angle of ingoing to outgoing ray (phase angle for the moon). The dashed curve is for the moon. Open circles are for dunite powder before irradiation; full circles, after. *Lower right*: Reflectivity against wavelength. The dashed curve is the average of the moon. Open circles are dunite powder before irradiation; full circles, after.

protons. Both properties are needed, in addition to their physical structure, in order to achieve the scattering and polarization law.

The coloration and the albedo situation will be explained later. They are very satisfactory in that most proton-irradiated rock powders achieve the coloration and albedo of the moon. It is unfortunate that it doesn't work the other way—one might have hoped that very few rocks would have these properties, in which case we might have been able to specify which rocks the moon is composed of. This is not the case. The variety that comes close to the actual behavior in all these respects is quite large, and I, therefore, hesitate to say we can specify the detailed nature of the chemistry of the moon.

There is in all this an important argument to be constructed, which must be made absolutely clear.

Assume that the old craters that look quite similar to, except not so sharp and not so high as, Tycho and Langrenus, for example, were mostly made in the same way, except that they are older. Then it follows that since optically the entire moon's surface, with hardly any exception, has the optical scattering property that these curves give, it is therefore necessary to suppose that all craters, even the youngest, are already overlaid with some layer of such dust. Then if the older craters have lost the radar and thermal hardness which the younger ones exhibit, we must suppose that the older craters have suffered a destruction not only in having been leveled down, for they are not as high as the young craters of a given size, but they have suffered a destruction of their present surface material to a depth of at least some meters, for they are now much more radar absorbent than they must once have been. This is true not only for their interiors but also for their rims.

That process has not only leveled the craters down, but has changed the surface material from the solid rock which it must once have been to a material that is only half as dense, or less. This is important, for it rules out completely the suggestion that perhaps the low density of the lunar surface material is essentially due to a frothy lava or to a deposition of ash flows in a very loose way. For either of these possibilities one would need to suppose that this process had happened everywhere, including the rims of old craters. The low-density material cannot be thought of as the original material or as anything of internal origin.

Lava is absolutely out of the question because one would have to suppose that the lava climbed up on the slopes of the craters after their formation. When the explosion happened that made the crater, it certainly made it solid. If it is not now solid to a depth of several meters, one would have to suppose that whatever froth it was climbed up and covered the hillsides, not just the interior, as indicated by the radar information.

Frothiness in lava might explain the lowlands' being underdense, but it cannot explain how the hilltops of old craters are underdense, even if one supposed that those mountains were originally lava allowing that the craters in them were made by big explosions. With ash flows, of course, the same would apply because we can't imagine them to flow uphill on a big crater wall and cover every hill and every hilltop. On the other hand we might say ashes were widely distributed and covered everything. This argument is perfectly compatible with supposing that everything simply acquired a thick overlay, but the thick overlay concept, of course, has to face up to other arguments, chiefly that of coloration. If everything obtained a thick overlay, why are the highlands lighter in color when they are chemically homogenous with the lowlands?

I have tried to discover which erosion process could be responsible for transporting material over the moon's surface, leveling down old craters, filling flat regions with deposits, and at the same time could allow the possibility of a differentiation of color between highlands and lowlands, as well as between mare ground of different levels.

The first possibility is just the random shooting around from impacts, which tends merely to homogenize the surface material, and if too much of it occurs there is a uniform blanketing all over the moon. If all the old craters were leveled down by material shooting around at random, then there would be a very homogenous material all over the moon.

It is important to exercise extreme caution when we discuss these transportation processes, for the kind of erosion rates we are discussing are of the order of one micron per year, enormously slow compared with any terrestrial erosion. We must decide which of these very slight processes that might occur at such a slow rate is the dominant one.

If we consider what the electric field in the vicinity of the moon's surface is, then we discover that in the sunlight and over a height of some centimeters above the surface there must be a distribution of the electric potential of the nature of that shown in Figure 5–11. I have normalized it at the moon's surface, but since I don't know what the charge on the moon as a whole is, I don't know how to show the potential at infinity. Nevertheless, in the first few centimeters we have a steep gradient of potential because electrons are being boiled off the surface by the sunlight. That field is likely to be of the order of a few volts per centimeter close to the surface (Figure 5–12).

The uncertainty of the potential of the moon as a whole does not matter very much for the estimate of this field close in. Consider what a particle of micron size might do in such an electric field if it were torn loose from the surface. It would have an electric charge on it just from frictional electricity. In general any dust grain picked up from the

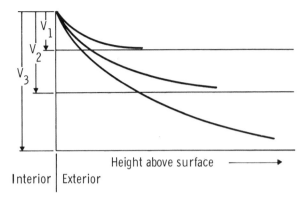

FIGURE 5–11: Distribution of electric potential above lunar surface. (Schematic only.)

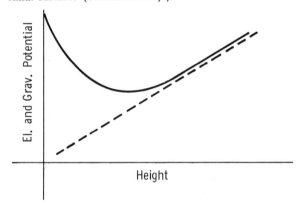

FIGURE 5–12: Potential of charged grain in combined electric and gravitational field.

surface will have quite a substantial charge, say, of a few thousand electrons, from the fact that at the tearing apart the frictional electricity charged it.

Also, if the dust grain were removed by a little bit of hot gas, such as from a micrometeorite explosion, a charge would be deposited on the grain, and it would take some seconds for this charge to be annihilated by the conductivity of the electron layer above the moon's surface. For a few seconds, or at any rate some substantial part of a second, a dust particle would be charged and would find itself in this electric field close to the surface.

We can then draw for a particular value of e/m of the particle the combined potential in the gravitational field of the moon and the electric field that is hugging the moon's surface, and that combined potential has a minimum. In other words, the particle would float stably at a particular height, sitting in this potential minimum.

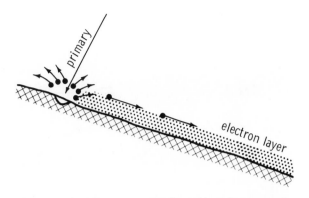

FIGURE 5–13: Floating grains released by micrometeorite.

What I envision taking place is that each time a micrometeorite hits the surface it throws out a number of secondaries. Some high-speed secondaries, but many more very low-speed secondaries, are knocked out of the surface. The moment that they are mechanically removed from the surface, some of them will have the right sign and an adequate value of charge, so as to float in this electric field.

Accordingly, if we stood on the moon's surface and were fortunate enough to observe a micrometeorite hit (Figure 5–13), we would see a little puff, but instead of that puff distributing itself evenly back to the ground, both up and downhill, a little bit more tended to go downhill every time, because about half the particles would have the right sign of charge not to settle down to the ground until they had been discharged. Consequently there would be a tendency for every little impact to slide a bit of material downhill.

This is a rather sophisticated type of mechanism, and it might well be said that this stage of knowledge does not warrant our going so far. However, when one has tried over the years to think of downhill transportation processes and finds that one of them really does stand out as superior, while at the same time requiring processes that are all fully expected to be there, one can begin to take it seriously.

With a process of this kind, which takes material downhill, and with the knowledge that we now have, that the albedo and coloration are dictated very largely by the bombardment that the moon has suffered from outside, it becomes possible to account for the albedo differentiation between highlands and lowlands. In this case the highland ground composed mainly of hillsides has the property that the material is constantly being denuded as that material is being slid downhill. This is the place where the ground is seen with the last optically absorbing layer recently removed. Therefore it is ground that has been bombarded by protons only since the last millimeter or so of surface was removed. That period of time may then be one which has not yet fully darkened the high ground. On the low ground, on the other hand, we see material which has been dancing around on the flat ground for long periods of time, not having anywhere to go, and therefore all the material that has accumulated there has already been fully darkened by radiation.

Also, this kind of a process is one that is capable of going down to a very flat deposition, and I would not feel sure at the moment that I could exclude the possibility that slight differences of level in the mare ground might not be responsible for channeling such fluidized dust and restraining it to flow to one side of some visible step, and not climbing over it. It would appear that this is perfectly possible.

The kind of scale with which we are concerned here is the scale of this charge above the surface, and that is centimeters. Hence, we would expect that whatever deposition process results from this would not be smooth on a scale much finer than that. I therefore find this kind of a process quite in accord with the startlingly strong effect shown in the radar data (Figure 5–9)—that between 3-cm and 1-cm wavelength the ground changes from quite specular, quite smooth, to almost completely rough.

Now I shall discuss a few of the Ranger pictures, mentioning particular points that have not been raised here. I agree that it is very likely that many features which we have known before, and which we now see in more detail, are the result of some stretching of the underlying ground. Let us consider what processes might be responsible for this stretching. I would like to emphasize that most of these features are in the mare type of ground and cease when they come to the highlands; there are few features of that nature in the highland ground. We therefore must worry about what stretches the ground in the mare much more than that in the highlands.

There are many other features, like the ridges and the collapse features in the mare ground (Figure 5–13), which all suggest that something is going on there. On the basis of an erosion process

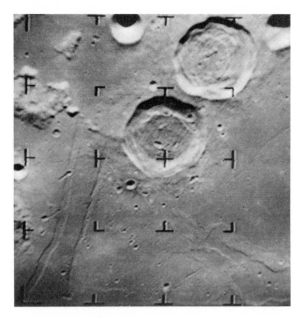

FIGURE 5–14: Ranger VIII photograph. Note irregular depression lower right.

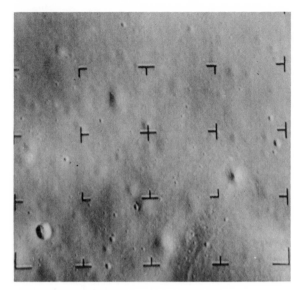

FIGURE 5–15: Ranger VIII photograph, showing "dimple crater."

and of a steady state of small craters on the mare ground, when the rate of formation is balanced by the erosion, one is required to suppose that all these not very high features on the mare ground must also be replenished in the life of the mare, and that they also have to be in a steady state. The mare ground has to be in some unstable condition, which evidently is not shared by the highland ground, because on the highlands there is no accumulated distortion to be seen which would clearly have been shown by the old craters. That is not the case, which is in marked contrast to the earth.

I would also like to refer to the odd-shaped features that seem to have lowered themselves, such as the parallel walled rilles (Figure 5–14). It is possible that a phenomenon occurs whereby sediment eventually slumps under its own weight as it is deposited to greater and greater depths, and that when it slumps it does so in a rather unstable fashion and in randomly shaped patches.

Some structures of manufactured powder, and indeed even snow, are capable of suddenly slumping disastrously, not only gradually. It may be that some of these features are triggered by an underlying structure, or are not triggered and thus are more random-shaped, being merely the consequence of the compression when too much has been piled on. Furthermore the Ranger pictures appear to show drain holes, as if there is a hollow under-

neath (Figure 5–15). Of course, considering the previous discussion of the erosion transportation process, there is no reason why this kind of a shape should not result in the presence of a subterranean cavity. I see no reason why a hole that is bottomless should not produce just exactly that shape. Also, since certain regions on the moon seem to have a much higher density of such features than others, I would suggest that they are connected with an underlying structure rather than having been externally impressed.

The dark patches that were seen require a detailed discussion. Many years ago I discussed the question of whether dark areas could be due to the exhalation of certain chemicals from the interior, coming either diffusely through the material or puffing out in particular places. And of course these dark patches seen in Alphonsus are the ones to which that discussion might best apply. It is also possible that the coloration differences which cover large regions in the mare ground themselves, like the sharp dividing line in Mare Tranquillitatis, are connected with a much larger scale of the same kind of phenomenon, which may be the diffusion of a chemical throughout the ground; and a quite small admixture of another chemical can very profoundly change the surface color. The dark patches in Alphonsus are clearly not a ray pattern thrown out by the explosion. Yet, as has been pointed out, some craters that have a dark halo

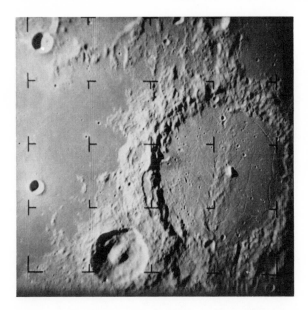

FIGURE 5–16: Ranger IX photograph. Note dark areas in the crater Alphonsus, and central dome in the other crater.

FIGURE 5–17: Terrestrial pingo.

look quite clearly like explosion craters. It may well be that some darkening material comes out of the ground as a consequence of an impact creating a hole which thereby gives access to a source of gas or vapor.

I would also like to refer to a class of features where there is a convex dome, not like the central peak but a much larger convex object inside the crater. There are many craters that have such a feature (Figure 5–16).

I suggested at the Leningrad Conference in 1960 that these raised features correspond to terrestrial pingos. MacRae published recently in *Sky and Telescope* a series of pingo pictures which showed their correspondence to the lunar features that are convex bumps with a dimple on top (Figure 5–17). As I believe in the likelihood of ice under the surface, I am inclined to attribute much of the deformation to ice and its various tricks; we will come to that point later.

The peculiar fuzziness of the appearance of the highlands is remarkable, compared with the clear and definite features of the low surfaces. The same fuzzy appearance is possessed also by the central peak of Alphonsus. This may be a consequence of an erosion process on a hillside where the erosion transportation is, of course, a great deal faster than on a flat surface. When we look at a flat surface we have a large accumulation of effects

showing because there has been a longer period of time in which the sample has been retained. On the highland ground the material quickly slides downhill and features are therefore more quickly eradicated.

The radar data indicate that the high ground, though apparently rather featureless, is in fact a better radar reflector than the low ground. If the high ground is, on an average, made of harder material, that of course would explain it. Erosion would make it likely that the high ground has a greater proportion of hard ground close to, or nearly showing through the surface, which is milled away gradually by the erosion destruction. In this way the high ground is firstly less sensitive to cratering because it is in general made of harder material and a given meteorite makes a much smaller hole, and secondly the erosion tends to fill in depressions somewhat more quickly.

Now I shall say a few words about the question of ice as suggested, for example, by Figure 5–18. All these many features in Alphonsus that seem to have a pattern like a fishbone structure might, of course, be attributed to an underlying motion. As in all the rille systems, they are sharply concentrated to the mare or to the flat ground inside craters. One is inclined to say that whatever deformation is taking place in these regions, it has to be a material that is deforming which is much less strong than the material making the surrounding highlands. If we suppose that the entire flat floor is filled by solidified lava, then we would have the problem that, when all this distorts many times over, these distortions must not be associated with a force that will be enough to distort a lot of the other features around it. This is not just true here but it is true also of the mare ground in general. A process must be found to account for the distortion, uplifting and collapsing, which is specific to the mare ground.

I personally believe that the most likely reason for this is that on the moon ice underlies low regions of the ground because water has slowly come up from the interior. Owing to the very low temperature, more than 30°C below freezing, that the moon's subsurface has, the water cannot penetrate freely to the surface, but will always accumulate underneath and make subterranean ice, possibly in great quantity. For the temperature conditions and overlying debris conditions that would obtain there, this ice will not readily evaporate. It will be able to maintain itself with an overlay of 100 m, for example, for long periods of time comparable with geologic time. The distortions that are characteristic of mare ground may be due to the underlying crevasses and the underlying movement of plastic ice, which is in contrast to the solid of the surrounds. If we were to place a large overlay, 100-m thick, over Greenland, what would it look like? We would certainly observe a lot of collapse features, fissures and so on, and see patterns of crevasses showing through to the top. The other point is that these lunar glaciers would be mostly contained in basins and would therefore not tend to be flowing nearly as much as terrestrial glaciers, which are generally deposited at a high point and flow down to the low points. The lunar glaciers are fed from underneath.

Figure 5–18 shows the expected ice configuration. If water had come up in substantial amounts from the interior of the moon it would have tended to flood the low regions, the mare, and the big craters, and in so doing would have become bottled up under a sedimentary layer. The sedimentary layer is cold and water freezes and cannot get through. It therefore makes an internal ice sheet which may or may not have water underneath it, and such an ice sheet will be a much less stable object than a sheet of debris or rock. It will be capable of suffering deformations in time, especially if the amount of water that is available there itself changes in the course of time as more water comes up or as water disappears by evaporation.

Figure 5–19 shows the temperature structure that one might expect. This is with a terrestrial-like

FIGURE 5–18: Schematic drawing of a proposed lunar crater.

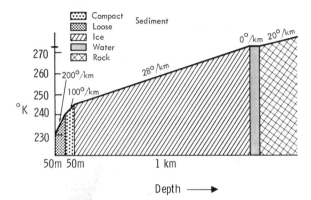

FIGURE 5–19: Proposed surface structure of the moon.

heat flow coming out of the lunar ground, which would give it gradients as shown. Going down, we would start off with a surface temperature of 230 or so, then we would have gradients that in the outer sediments might be like 200 degrees per kilometer, or in a more compacted sediment, 100 degrees per kilometer.

In stationary ice (this does not apply to terrestrial glaciers because they all move too fast, but to a glacier that is in a hollow and therefore does not flow away) the temperature gradient for terrestrial heat flow is 28 degrees per kilometer, and one can therefore tolerate something like a kilometer of ice underneath the debris layer, which, in turn, will be enough to prevent the rapid evaporation of ice. There may be water underneath the ice. There is nothing to tell us that underground glaciation to a depth of the order of a kilometer would run into conflicts with any of the thermal or evaporation knowledge that we have.

Some have said that on the earth the water has come up in volcanoes, and if there aren't volcanoes, how can the water come up? I think that this is not a problem, but it is just a question of the impedance of the flow. If the impedance to the flow of water coming from the deep interior of the earth is very high, there will be a certain rate of outflow, irrespective of the additional slight impedance of the outer layers. If the volcanism of the crust lets it through most easily then of course it will all come out in that manner. If there is no volcanism, exactly the same amount of water will still come out, so long as the impedance is mostly in the deep interior.

For the moon, even if volcanism on the surface is not common, I would think that the chance of water coming up is hardly affected; it would then come out wherever it could and would then underlie the mare and big craters mainly as ice, making that mobility of the surface that we see—a mobility which, and this must be re-emphasized, seems to be subsequent to the formation of the main mare ground. One cannot suppose that a solid was laid down there with wrinkles in it; it must have been something which subsequently wrinkles many times over.

Last, I shall point out a remarkable effect that has not yet been referred to. If one looks through the P frames on the Ranger IX and goes through the series of them, a remarkable change of sharpness of features becomes apparent. The pictures that were taken from farthest away look, if one doesn't know the scale, quite similar to the pictures that were taken nearest in, while the ones that are in between look very different (Figure 5–20).

I have looked through many of the pictures and it is, in general, true that there is a particular range of scales which is less populated with features than others. I feel that has a lot to say about the rates of erosion and the rates of crater formation of different sizes. The scale on which features are absent is of the order of a few hundred meters across. They are not absent in reality, but are much less numerous. Of course, the very large craters take a long time to be eroded away. When we come to an intermediate size the rate of erosion is, comparatively speaking, faster, whereas the rate of generating them is not that much faster. Then we come to a still smaller scale, for which, apparently, the rate of new craters being made is high enough to overcome the effect of erosion, so that when we observe them from close in, we see the heavy population of features again.

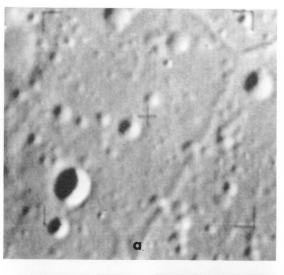

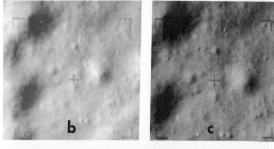

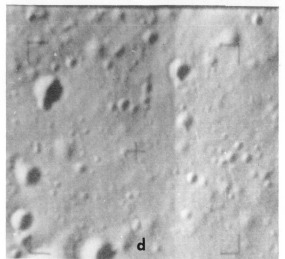

FIGURE 5–20: Ranger IX photographs: P_1 frames (a) 1161, (b) 1798, (c) 1802, and (d) 1826.

Part II.

Crater Formation and Surface Structure

6.

Interpreting Ranger Photographs from Impact Cratering Studies

Donald E. Gault, William L. Quaide, and Verne R. Oberbeck

*National Aeronautics and Space Administration, Space Sciences Division,
Ames Research Center, Moffett Field, California*

ABSTRACT

Impact cratering experiments have been performed at the Ames Research Center and field studies of missile impact craters have been carried out at White Sands Missile Range, New Mexico. The studies show that projectiles with velocities typical of secondary bodies on the moon produce craters with geometries which are sensitive to target strength and angle of impact. The principal conclusion obtained from the application of these findings to the interpretation of Ranger photographs is that the lunar surface consists of materials of low cohesive strength. The thickness is not certainly known, but it is probably measured in meters or tens of meters.

I. INTRODUCTION

Cratering experiments at the Ames Research Center and field studies of missile impact craters at White Sands Missile Range, New Mexico, have provided information of great value for interpreting Ranger photographs. Before considering the evidence and interpretations, it is necessary to point out that projectiles with velocities in the range of 1 km/sec were used. It is known from earlier work that the geometry of craters produced by hypervelocity impact is insensitive to material strength of the target and independent of the incidence angle of the projectile. Recent studies have shown, however, that target strength is an important

parameter in determining the geometry of a crater produced by a low-velocity impact. The purpose of this paper is twofold. First, experimental evidence is presented which shows the relationship between such parameters as target strength, projectile velocity, and angle of impact, and the resulting crater geometry. Secondly, the results of the studies are used to interpret the properties of the lunar surface by considering the geometry of craters of probable secondary (low-velocity) origin seen on the Ranger photographs.

Secondary is used here in a genetic sense to indicate an origin from low-velocity projectiles created by an impact of a primary, extra-lunar body. It is not used in a morphological sense as is done by Shoemaker (Chapter 2). The results of our work indicate that secondary craters cannot be distinguished from primary craters on the basis of geometry alone.

II. EXPERIMENTAL EVIDENCE

The experimental facility, the vertical gun, used to obtain most of the evidence presented here, is shown in Figure 6–1. It is made of an A-frame straddling a large vacuum tank. The gun rests on a boom, which can be raised to allow the gun to fire into the target medium inside the vacuum chamber at various angles from the horizontal to the vertical in 15-degree increments. The pressure inside the chamber at the time of impact is of the order of 100 microns of mercury.

FIGURE 6–1: The vertical gun.

FIGURE 6–2: Craters in weakly bonded sand (*top*) and loose sand (*bottom*) produced by 1 km/sec projectiles impacting at an angle of 30 degrees from the horizontal.

Target materials used to date are loose sand, bonded sand, clastic pumice debris, pumice blocks, and two-layer models of pumice blocks overlaid by clastic pumice debris. Impacts into loose sand and pumice debris produce circular craters regardless of the angle of impact. The craters are surrounded by a raised rim, characteristic rays, and granular ejecta. Impacts into sand which has small amounts of binder produce craters which differ strongly in geometry, morphology, and size. Figure 6–2 illustrates this point graphically. The two craters shown were formed under identical conditions, except that the top one was produced in bonded sand whereas the bottom one was noncohesive. The binder is Portland cement, one part in 40 parts sand. The unconfined compressive strength of the bonded material is about 10^6 dynes/cm^2 and crumbles between the fingers very easily. The angle of impact (from the horizontal) was 30 degrees in both cases. The projectiles were aluminum with velocities of approximately 1 km/sec. The craters differ in both morphology and size. The crater in

EJECTA PATTERNS
PUMICE

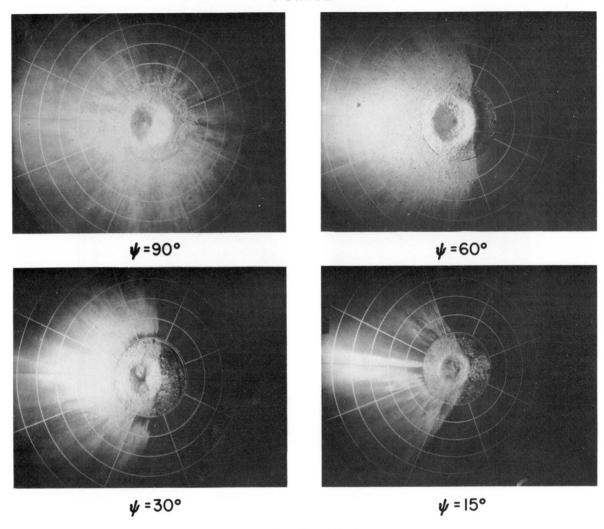

$\psi = 90°$

$\psi = 60°$

$\psi = 30°$

$\psi = 15°$

FIGURE 6–3: Ejecta patterns produced by impacts of varying obliquity.

unbonded sand is circular, whereas the crater in the bonded sand is elongate. The character of the ejecta and the morphology of the crater rims are strikingly different. The rim of the crater in bonded sand is poorly developed and contains numerous blocks of ejecta. The rim of the crater in loose sand is strongly developed and smooth. Ray systems are developed in both cases. It was found that the ray system produced under a given set of experimental conditions provides information regarding the angle of impact of the projectile, as shown in Figure 6–3. The target material in this experiment was poorly sorted fragmental pumice with a bulk density of

1.1 g/cm³, the impact velocity was 1 km/sec, and the projectiles were composed of lexan, a plastic. Impacts with incidence angles of 90, 60, 30, and 15 degrees from the horizontal are shown. The projectiles entered from the left in all oblique cases. The ray system produced by the normal impact is strikingly symmetrical. Oblique impacts focus the ray patterns, the focusing being very pronounced in the case of shallow-impact angles. It must be emphasized, however, that projectile velocity, projectile density, and target strength also have an effect on the degree of focusing of the ray ejecta, the details of which remain to be worked out.

CRATER SHADOW PATTERNS
UNBONDED QUARTZ SAND, $\psi = 15°$

 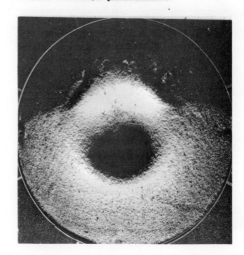

FIGURE 6–4: Shadow patterns in an asymmetrical crater produced by light at right angle (*left*) and parallel (*right*) to the projectile trajectory.

We have observed another geometrical feature of the craters that is related to the degree of obliquity of the impacting projectile. This feature is the shape of the crater in profile in the direction of the projectile trajectory. At low entry angles, 30 degrees and below, the crater wall is steeper on the side from which the projectile enters the target. The asymmetry in profile is revealed by shadow shapes when incident light is at a large angle to the line of the projectile trajectory. This feature is illustrated in Figure 6–4. The crater shown in the two photographs was produced by a projectile with a velocity of 1 km/sec impacting an unbonded sand target at an angle of 15 degrees from the horizontal. The crater is circular in plan but asymmetric in cross section. The two photographs were taken with the incident light parallel and at right angles to the projectile trajectory. The shadow pattern is asymmetrical when the incident light is at right angles to the line of flight, revealing the asymmetry of crater profile. Many such shadow patterns can be seen on the Ranger photographs. These, combined with focused ray patterns, provide a powerful tool for identifying the direction from which crater-producing secondary bodies came. Figure 6–5 summarizes the change in crater profile with angle of incidence for craters produced in unbonded sand by lexan projectiles with velocities of 1 km/sec. Displaced mass and crater diameter are also

shown. At 90 degrees the crater is roughly conical with a raised rim; at 60 degrees the shape is nearly the same; at 30 degrees the crater wall on the projectile entry side is noticeably steeper; and at 15 degrees the profile asymmetry is even more pronounced. Displaced mass is a factor of 40 greater than that produced previously by impacts of the same energy in cohesive or massive rock targets; it decreases very quickly with decreasing angles of impact. At angles of 30 degrees ricochet sometimes takes place, whereas at 15 degrees it always occurs.

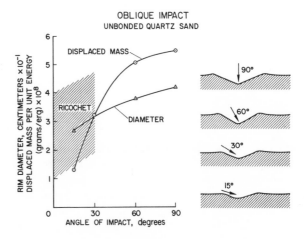

FIGURE 6–5: Crater profile; displaced mass and crater diameter produced by impacts at various angles.

IMPACT IN PUMICE

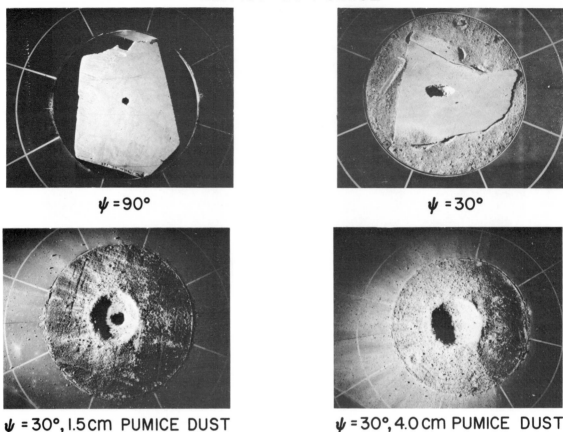

FIGURE 6–6: Impact craters in pumice and in pumice overlaid by granular pumice debris.

Experiments have been performed with pumice blocks and two-layer models of pumice blocks overlaid by granular pumice debris. Results for impacts of lexan projectiles with velocities of 1 km/sec are shown in Figure 6–6. A normal impact into a pumice block with a density of 0.6–0.7 g/cm³ is shown at the upper left. Such an impact produces a cylindrical hole with a diameter only minutely larger than that of the projectile and 10–12 cm deep. An oblique impact into the same material is shown at the upper right. A simple elliptical groove is produced. The two lower photographs are of layered targets. Both are composed of machined pumice blocks overlaid by varying thicknesses of granular pumice fragments. The crater at the lower left was produced by a projectile traveling at an angle of 30 degrees from the horizontal and impacting a target with a surface layer 1.5 cm thick, a thickness less than the diameter of the 20-mm projectile. The crater is circular in plan but

has a flat floor where the fragmental materials were cleared from the rock surface. The center of the crater is occupied by a deep cylindrical hole. The target on the lower right has a surface layer 4 cm thick. The crater depth in this model is exactly the same as though the pumice rock were not present. There is only a vague indication of crushing of the block at the bottom of the crater. Much experimentation remains to be done with layered models, but the preliminary information presented here certainly has a bearing on the current interpretation of Ranger photographs.

One other observation which may be of value for interpreting Ranger photographs is the occurrence of slumps in some of the craters with surface textures similar to the tree-bark texture observable on all high-resolution Ranger photographs. An example is shown in Figure 6–7. This crater was produced in loose sand by a 30-degree impact of an aluminum projectile with a velocity of 1 km/sec.

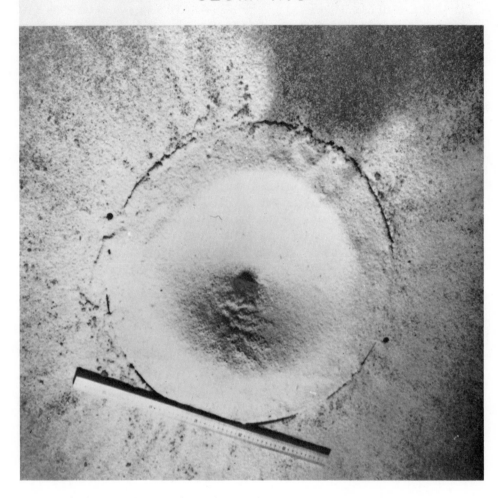

FIGURE 6–7: Crater in loose sand containing a slump with tree-bark texture.

The texture is the result of post-cratering slumping of finely crushed quartz. The same kind of texture has been observed in impact craters produced in 15-micron-diameter particulate material. It is possible that the tree-bark texture visible on the high-resolution Ranger picture is not a lava surface feature but instead is a result of slumping.

There is always a question of how well the small laboratory craters can be compared with larger lunar counterparts. Relevant information is being gathered in a program of field studies of craters produced by the impact of unarmed missiles at White Sands Missile Range, New Mexico. The craters studied are of the same size as those visible on the higher-resolution Ranger photographs. One such crater is shown in Figure 6–8. The angle, velocity, and energy of the impacting missile are known. The velocity is typical of a lunar secondary impact. The target material is a strongly cemented gypsiferous sand. The crater is elongate with a length to width ratio of 2. The projectile entered from the right. The crater shape and the ejecta distribution are both asymmetric; one large block

is visible on the crater rim. This view also shows very clearly the focusing of the ejecta.

Missiles with the same entry angles, velocity, and energy produce significantly different craters when they impact weakly cohesive materials. Figure 6–9 is a photograph of a crater in very weakly cohesive material. The crater, produced by a missile entering from the left, is 10 m in diameter and is almost perfectly circular. Figure 6–10 shows another crater of similar size produced in weakly to non-cohesive material. It is circular in outline and again block ejecta is rare. The one large block visible was produced by shock compression and was not a part of the original target material. It is extremely weakly bonded, so weak that it can not be picked up. Impact craters produced in very strong rock targets are significantly different; they are of irregular shape with hardly any rim and have abundant, large blocks of ejecta.

Field studies thus confirm the results obtained in the laboratory. Circular craters are produced by low velocity impacts in weakly to noncohesive materials regardless of the angle of impact. Impacts

FIGURE 6–8: A crater in strongly cemented gypsiferous sand, White Sands Missile Range, New Mexico.

FIGURE 6–9: A crater in weakly cohesive material, White Sands Missile Range, New Mexico.

FIGURE 6–10: A crater in weakly cohesive to noncohesive material, White Sands Missile Range, New Mexico.

WSMR IMPACT CRATERS

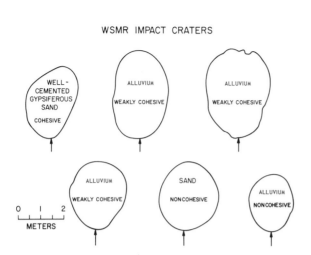

FIGURE 6–11: Plan views of craters at White Sands Missile Range with descriptions of the degree of cohesion of the target materials.

WSMR IMPACT CRATER

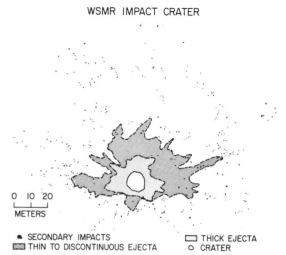

FIGURE 6–12: Map of the ejecta of a crater at White Sands Missile Range, New Mexico.

in cohesive materials produce asymmetric craters surrounded by blocky ejecta. Figure 6–11 shows the outlines of a number of craters at White Sands, New Mexico, along with a qualitative description of the degree of cohesiveness of the impacted materials. Impacts in very strongly cohesive rock targets produce irregularly shaped rimless craters with a profusion of large ejecta blocks in and around the crater.

One other result of the White Sands study presents evidence relevant to the problem of identifying secondary craters. Figure 6–12 is a map of the ejecta of a White Sands Missile Range impact crater prepared by Dr. Henry Moore of the U.S. Geological Survey. The significant features are the patterns exhibited by the secondary impact craters produced by low-strength ejecta clots. All imaginable patterns can be observed: random, clusters, straight radial lines, straight lines which are not radially directed, and looping lines. The secondary craters are all small but of various shapes. All were produced by very low-speed projectiles with velocities of meters per second, and some of the impacting projectiles remained intact in the crater. These impacts were nonviolent in that the clots of material did not break up and were not thrown free of the crater. Other craters must have been produced by projectiles of about the same ejection and impact velocities, but the clots were weaker and broke apart. The shapes of these craters are

functions of both the strength of materials and impact velocities. The details of their geometry cannot be compared directly with the geometry of secondary lunar craters produced by projectiles with velocities of kilometers per second. The cratering process in the latter case is significantly different; however, the distribution pattern is applicable. The pattern suggests that secondary craters on the moon may occur randomly, in which case they may not differ sufficiently in geometry to distinguish them from craters produced by primary bodies. They may also occur in clusters or linear arrangements. Such arrangements are not suspected for primary bodies and it is on this basis only that secondary craters may be identified. Even this criterion is subject to error because volcanic craters most frequently occur in clusters and linear arrangements.

III. INTERPRETATION OF RANGER PHOTOGRAPHS

The physical characteristics of the lunar surface can be interpreted from the Ranger photographs using the results of the low-velocity cratering studies only when low-speed (secondary) impact craters can be identified. It was stated earlier that the geometry of craters produced by primary and secondary projectiles may not be sufficiently different to be distinguished. Craters of secondary origin can be identified unequivocally only when they

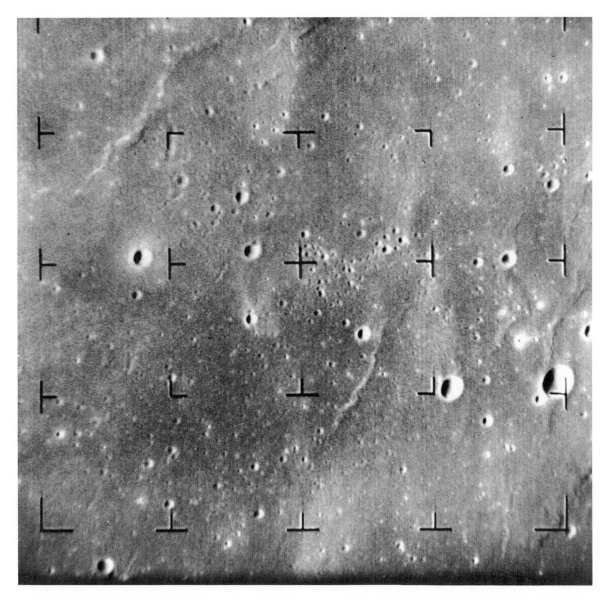

FIGURE 6–13: Ranger VII photograph showing a cluster of secondary craters with an ejecta plume.

occur in clusters or in linear arrays within ray patterns emanating from prominent craters or when focused ejecta patterns and asymmetric shadow patterns indicating cross-sectioned asymmetry can be seen.

On the basis of the above evidence, secondary craters can be identified on Ranger VII and VIII photographs. Ranger VII impacted in an area traversed by rays from the craters Copernicus and Tycho. Figure 6–13 is a Ranger VII, A-camera photograph showing a cluster of craters (near the top of the photograph) from which an ejecta plume extends generally northward. The direction of the plume lines up exactly with a line through the source craters to Tycho. The same cluster of craters appears in a closer view in the upper left corner of Figure 6–14. Since rays from Tycho do extend through the area and the ejecta plume does line up, it is most likely that the cluster was produced by blocks thrown from Tycho. The acute angle formed by the plume suggests that the projectiles struck the surface at low angles from the horizontal, probably in the neighborhood of 15 to 30 degrees. This angle, together with the distance from Tycho, suggests that the impact velocity was of the order of 1.3 km/sec.

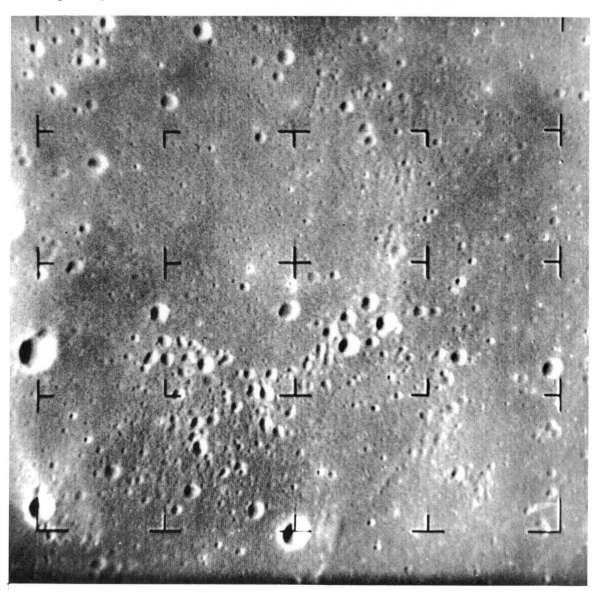

FIGURE 6–14: Ranger VII photograph of clusters of secondary craters.

Another cluster can be seen a few tens of kilometers to the southeast. This cluster has generally been considered to have been produced by impacts of secondary bodies from Tycho. There is doubt that ejecta plumes can be unequivocally linked with these craters, but there is no doubt that asymmetric shadow patterns are present in some of the craters and that the steep slope is on the side of the crater closest to Copernicus. In keeping with experimental evidence, it is proposed that this cluster of craters had as its origin low-angle impacts of secondary bodies derived from Copernicus. The fact that the craters are predominantly circular indicates that the surface is composed of materials at least weakly cohesive. The weakly cohesive layers must extend to a considerable depth. Many of the craters of the cluster are well-formed, bowl-shaped structures having no geometrical features

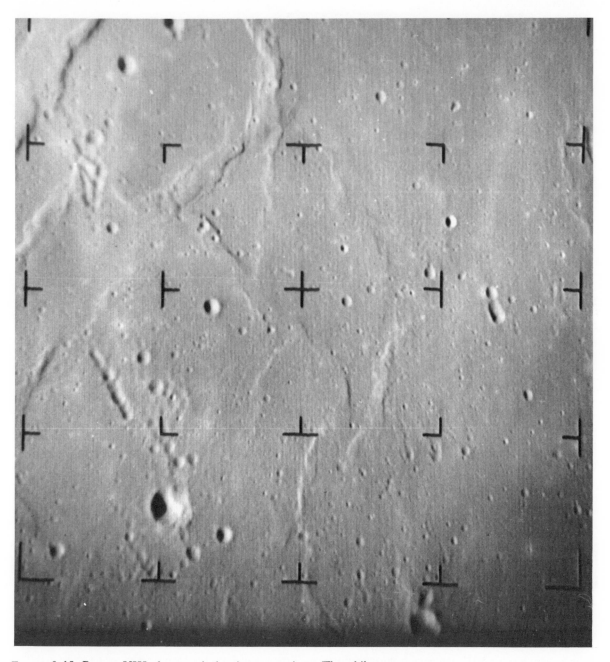

FIGURE 6–15: Ranger VIII photograph showing a prominent Theophilus ray.

indicative of a two-layer model where the upper layer is thin, relative to the crater depths observed (30–150 meters).

Ranger VIII photographs contain similar evidence. The Ranger VIII spacecraft traversed a large portion of Mare Tranquillitatis, a region whose surface is crossed by many prominent rays emanating from the crater Theophilus. Figure 6–15 shows one of the more prominent rays. The higher density of craters within the ray, the linear chains of some craters in the ray, and the asymmetric shadow patterns all support the concept that these craters were produced by low-angle impacts of secondary projectiles from Theophilus. Figure 6–16 is a view of the region slightly to the south of that shown in Figure 6–15. Elements of the same ray can be seen (A). Other rays of probable Theophilus origin can be observed and at least one of probable

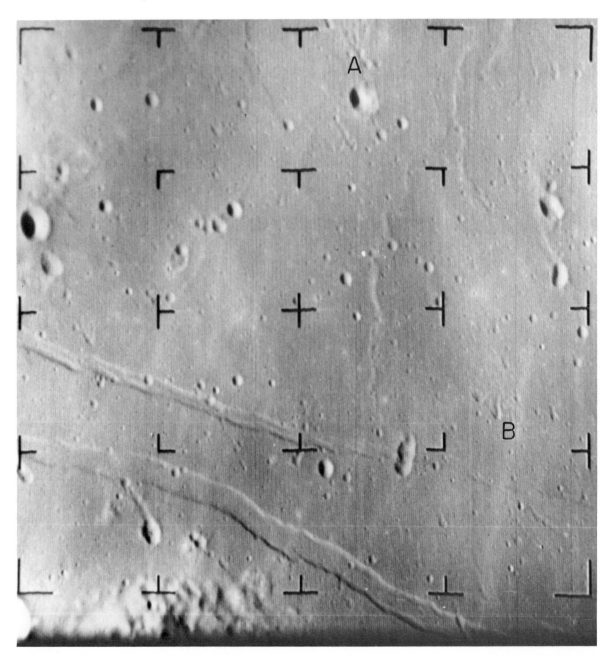

FIGURE 6–16: Ranger VIII photograph showing a portion of the ray of Figure 6–15 and other ray elements.

Tycho origin is present. In addition to the rays, a few gouges can be seen (B). Others can be seen on other photographs. Ricochets were observed in the laboratory in low-angle, low-velocity experiments where the projectiles have grooved the surface near the crater in an irregular manner. The grooves observed may well result from such projectile behavior.

If craters in the ray systems are produced by secondary projectiles as proposed, and if the number of randomly distributed secondary craters is as great as size-frequency plots of the craters suggest, the prevalence of circular craters and the rarity of positive relief features in Ranger VIII photographs suggest that surface materials in this area are weakly cohesive. The thickness must be measured in meters or tens of meters to account for the geometry of the larger secondary craters.

In view of the foregoing conclusions it is necessary to explain the variety of crater forms observed in the Ranger photographs. It is suggested that the continuum of crater forms observable from sharp, well-defined ones to faint, shallow, hardly recognizable ones was not the result of different mechanisms of formation but instead was produced by modification of the craters. It must be remembered that impact does not clean the surface; it litters the surface. Each high-speed impact excavates a thousand or ten thousand times the mass of the incoming projectile. Of this, only an amount equal to a few projectile masses leaves the moon. The rest is spread over the surface and must modify previously existing topographic features, preferentially eroding positive relief features and covering all with ejecta. The deposited material is thickest in depressions, but the whole of the surface is blanketed. The inevitable result of this process repeated again and again is the production of a continuum of crater shapes.

Photographs from the Ranger IX mission have provided new and different information to be considered in an interpretation of the lunar surface properties. These photographs indicate that the floor of the crater Alphonsus has not only been bombarded by projectiles but that a significant amount of the surface detail has been created by explosive outgassing from vent craters. Details of the surface suggest that the products of this probably volcanic activity are clastic blankets with physical properties similar in all respects to impact ejecta deposits. Collapsed craters appear to be

present as well. It is not known if such explosive outgassing activity is restricted to certain large craters or if it has been widespread on the maria surfaces. If it has occurred on the maria the thicknesses of fragmental material suggested by this investigation may be more readily understood than if the ejecta had been formed by impact alone.

IV. CONCLUSIONS

Experiments indicate that the geometry of craters produced by hypervelocity impact is generally insensitive to the strength of the target material and angle of impact. The geometry of those produced by the impact of projectiles with velocities similar to those of secondary bodies on the moon is quite sensitive to target strength and angle of impact. Projectiles with these velocities produce circular craters in noncohesive materials regardless of the angle of impact, and elongate or irregular craters in cohesive materials when the angle of impact is low. In addition, ejecta from noncohesive targets has a maximum block size equal to the maximum diameter of the components of the target, whereas ejecta from cohesive targets has a large range of sizes. Oblique impacts produce focused ejecta patterns and crater asymmetry in the plane containing the projectile trajectory, enabling identification of the direction of origin of the impacting body.

Secondary craters can be recognized in Ranger photographs when they occur in clusters or in linear arrays within well-defined rays, or when they have focused ejecta patterns and cross-sectional asymmetry. Analyses of such craters indicate that they are predominantly circular in outline, that there is an absence of blocks with large dimensions, and that their geometry is so regular that the lunar surface must be composed of weakly to noncohesive materials with thicknesses measured in meters or tens of meters. The surface materials may be composed entirely of impact ejecta or they may be of mixed origin, part impact and part explosive outgassing.

DISCUSSION

VOICE: I would like to make one suggestion for consideration of the experimenters. Even in the maria the distribution and clusters of large craters versus small craters elsewhere may be indicative of variations such as already buried regions under the build-up of rubble that you have described. It may be that strength and depth contribute as much or more to the variety of groups of craters clustered in patches than do differences from randomness of the impact of the debris that produces them. This may be a clue for investigating what is below the general level of the debris.

GAULT: I would certainly agree. I don't think anyone would expect to find a uniform layer material sitting on top of a nice plain surface below the mare. This is certainly not the typical model one would assume for the mare.

UREY: You talk about erosion and deposition. Isn't that what we all mean? That is what I mean.

GAULT: Erosion is the removal of mass.

UREY: That's right. It is a combination of erosion and deposition, isn't it?

GAULT: Yes.

UREY: That is what I thought. You seemed to present it as though you were disagreeing with somebody. Are you?

GAULT: No.

7.

Optical Properties
of the Moon's Surface*

B. W. Hapke

Center for Radiophysics and Space Research, Cornell University, Ithaca, New York

I would like to discuss the optical properties of the moon's surface and what can be deduced concerning the outermost millimeter or so of the lunar surface when these properties are combined with appropriate laboratory studies. I believe that the optical evidence gives very strong indications that the lunar surface is covered with a layer of fine dust of unknown thickness.

The moon's surface is characterized by a number of rather unusual optical properties, which are summarized in Figure 7–1. The brightness peaks at full moon, when the source is directly behind the observer, that is, when the sun is directly behind the earth. The brightness decreases sharply as the phase angle increases from 0 degrees. This is true no matter what part of the lunar disk one is observing. The upper part of Figure 7–1 gives two typical curves illustrating how the brightness of two areas vary as the angle of incidence i changes. The upper left curve is for an area on the 0-degree meridian of longitude, ϵ being the angle of observation. The upper right curve corresponds to an area on the 60-degree meridian of longitude as the angle of incidence changes. The shape of the curves is apparently independent of latitude; one gets a similar sort of photometric function for any lunar latitude as long as the longitude remains the same. There is some scatter about these mean curves for various areas on the lunar surface, but the departures from the mean curves are not nearly as significant as the range of values which the reflec-

tion law for a variety of different kinds of surfaces can take.

In the lower left of Figure 7–1 are shown curves of polarization as a function of phase angle ϕ. The shape of the curve is very nearly independent of position on the lunar surface. The polarization is negative for phase angles less than about 23 degrees, and goes through a negative maximum of about 1.0 per cent. Then at 23 degrees or so, the polarization becomes zero and the plane of polarization rotates 90 degrees. The polarization then goes through a positive maximum, when the phase angle is around 90 to 110 degrees, depending on the area observed. The brighter areas, such as the highlands, generally have lower positive polarization and the darker areas, such as the maria, have higher polarization. The position of the maximum may be shifted a little bit toward larger phase angles for the darker areas.

The color of the moon is rather significant: it is redder than sunlight. The lower right part of Figure 7–1 shows color differences on a magnitude scale versus wavelength for various areas on the lunar surface. These data have been corrected to a color difference of zero at a wavelength of 5,600 Å, which is the wavelength of the green filter we use.

It is convenient to characterize these photometric curves by a number of parameters. The first parameter is the normal albedo A_n, which is the brightness of the surface relative to the brightness of a perfectly reflecting, perfectly diffusing surface, both areas viewed and illuminated normally. The

* This research has been sponsored by a grant from the National Aeronautics and Space Administration.

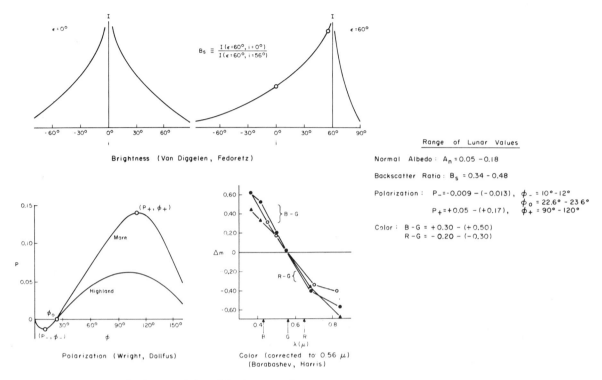

FIGURE 7–1: Photometric characteristics of the lunar surface.

range of normal albedo on the moon is from 5 per cent to about 18 per cent.

The shape of the backscatter curve can be characterized by the backscatter ratio, B_s, which is the ratio of the brightness at $\epsilon = 60$ degrees and $i = 0$ degrees to the ratio of the brightness at $\epsilon = 60$ degrees and $i = 56$ degrees. These two points are shown in the upper right curve of Figure 7–1. The backscatter ratio for the moon is about 0.34 to 0.48.

The polarization can be characterized by three parameters: the point of negative maximum, the inversion angle, and the point of the positive maximum. On the moon the negative maximum has the range of about 0.9 to 1.3 per cent, the inversion angle from 22.5 to 23.5 degrees; the value of polarization at positive maximum ranges from about 5 to 17 per cent, and the position of the positive maximum is 90 to 120 degrees.

We measured the color of our laboratory surfaces at 4,250 Å, 5,600 Å, and 6,450 Å. *B-G* refers to the brightness of the surface at the position of the blue filter on a magnitude scale relative to the brightness at 5,600 Å, and similarly for *R-G*, which refers to the red filter. Lunar values of *B-G* are from

about $+0.30$ to $+0.50$, and of *R-G*, from -0.20 to -0.30.

The fact that every area on the lunar surface possesses these unusual optical characteristics shows that they must be exogenous and are not due to some peculiar property of lunar lavas or to some other internal cause. I suggested a few years ago that these rather remarkable lunar photometric properties could be explained as the result of micrometeorites impacting the lunar surface and pulverizing it to a very high degree. The resulting dust would be acted upon by the solar wind, darkening and otherwise altering the optical characteristics of the dust. If this suggestion is correct, then if we take a rock of the proper composition, grind it up, and irradiate it with protons of a few kilovolts energy to simulate the solar wind hitting the moon, the resulting material should possess the proper photometric properties.

Figure 7–2 shows the photometric properties of hydrogen-ion-irradiated dunite powder. If the solar wind is impacting the moon at the same flux as measured by Mariner II, the radiation dose which this powder has received would be equivalent to something like 100,000 years on the moon.

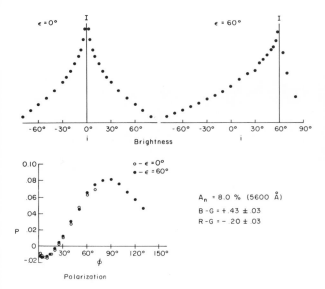

FIGURE 7–2: Photometric properties of dunite powder (size $< 7\mu$) after 65 coul/cm² of H-ion irradiation.

The photometric properties of the dunite powder reproduce those of the moon quite well.

The remainder of this paper is devoted to a discussion of some of the properties of surfaces that affect their photometric characteristics; whether one can deduce that other types of surfaces could not have these photometric properties; and the composition of the lunar surface. In the laboratory studies about to be described, I have been helped by Hsiu Yung Chow and Eddie Wells, graduate students at Cornell University.

Figure 7–3 is a schematic diagram of the process which I believe is responsible for the darkening of the lunar surface by the solar wind. Ions from the solar wind strike the particles that make up the lunar surface and sputter atoms off these particles. Assuming that the moon is composed of a silicate rock material, the sputtered atoms will consist of oxygen, silicon, and various kinds of metals. Some of these sputtered atoms will leave the surface completely. However, when one has a rather complex surface, some of these sputtered atoms may fly over and stick to the undersides of adjacent rock particles. Because oxygen is a more volatile element it will have a lower sticking coefficient than the other types of atoms and fewer oxygen atoms than the silicon or metal atoms will stay on the undersides of these particles. This process results in a coating of a dark material on the underside of a rock particle; the coating is probably a non-

stoichiometric silicate compound (or glass) which is deficient in oxygen.

The sputtering action of the solar wind, of course, will also make etch pits in the surfaces of the particles and it will generally clean their upper surfaces. But the primary mechanism responsible for the darkening is the coating of the underside of the particles with a thin, absorbing, nonstoichiometric compound. Lattice vacancies in such compounds would be highly efficient in producing absorbing effects.

Figure 7–4 shows an experiment we did in the laboratory. We put an aluminum oxide ball inside an aluminum oxide crucible and bombarded it from above with 2 kev He ions. The middle photo shows the bombarded ball and the unirradiated ball. The unirradiated ball is shiny and the upper surface of the irradiated ball has been cleaned and roughened by sputtering but the undersurface is darkened. The right hand photo is a photomicrograph of the interface between the dark and light areas of the irradiated ball. Dark streaks were formed under the little asperities sticking out from the ball. These streaks are the geometric shadows of the asperities to the ion beam.

Figure 7–5 shows how the ion bombardment affects the photometric properties of some large materials, rocks and chunks of rocks. As in Figure 7–3, the darkening is much more efficient on a rough surface than on a smooth surface.

In many of the following figures a white disk appears on a black square as an albedo reference. The dark square is black velvet, which has an

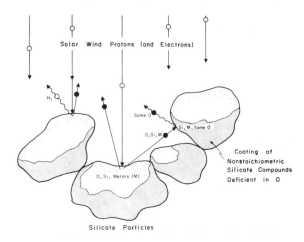

FIGURE 7–3: Schematic diagram of process responsible for darkening of lunar surface by solar wind.

Al₂O₃ Ball Irradiated in Al₂O₃ Crucible by 2 Kev He - Ions

Dose = 90 coul/cm²

FIGURE 7–4: Effect of ion bombardment on aluminum oxide ball (laboratory experiment).

albedo of about 1 per cent; the disk is magnesium oxide powder, which has an albedo close to unity.

Figure 7–6 is a photo of some coarse olivine basalt powders, showing the effects of particle size and different types of irradiation on the appearance of a rock powder. The top row is the unirradiated material. The middle row is after 10^6 roentgens of gamma ray irradiation from Co^{60}. The bottom row is after hydrogen ion irradiation. The gamma radiation had no effect at all, nor were the color, albedo, or other photometric properties appreciably affected. In contrast, the hydrogen ion radiation is very efficient for changing the photometric properties. Observe that the coarser materials darken less than the finer materials.

Finer particles are much more efficiently darkened than coarser particles by the mechanism shown in Figure 7–3 because there are so many more free surfaces. All naturally occurring rocks and minerals are partially absorbing. Large particles, whether or not they are bombarded, have their optical properties dominated by the absorbing and reflecting properties of the rock itself. But, when finely ground, the particles become translucent, and if an absorbing coating is put on the undersides the optical properties are controlled by the coating rather than by the optical properties of the rock itself.

Figure 7–7 is a photo of fine olivine basalt powders, which were handled somewhat before the picture was taken. When first taken out of the vacuum system they were quite uniformly darkened. Again the gamma irradiation had no effect at all; however, the hydrogen ion bombardment had a remarkable effect. Also shown is a sample bombarded with helium ions. The effect of helium ion irradiation is about the same as hydrogen ion irradiation, except the efficiency is better. The same dose of helium ions will produce the same effects in a much shorter time than an equivalent dose of hydrogen ions.

It is important to show that the darkening effects of ion irradiation are not due to cracked pump oil or to some other spurious effect. We have several independent indications that the effects are real, but the most dramatic proof is shown in Figure 7–8.

The materials in each row of Figure 7–8 were irradiated simultaneously side by side in the vacuum system. The top row is untreated material; the center row was irradiated by hydrogen ions; the lower row by helium ions. The powders are pure magnesium oxide, aluminum oxide, and silicon dioxide, plus mixtures of these three powders. The mixtures are physical only and not chemically combined. Note that the two mixtures of SiO_2 with Al_2O_3 and with MgO both darkened appreciably under hydrogen ion irradiation, whereas the pure materials darkened only very slightly. This figure illustrates the nonlinear effect, so to speak, of the ion irradiation, in that one cannot deduce from

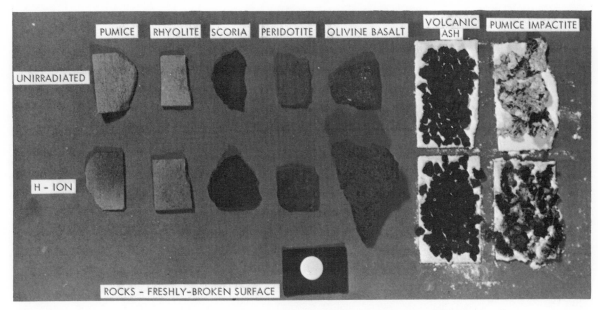

FIGURE 7–5: Effect of ion bombardment on the photometric properties of rocks and chunks of rocks.

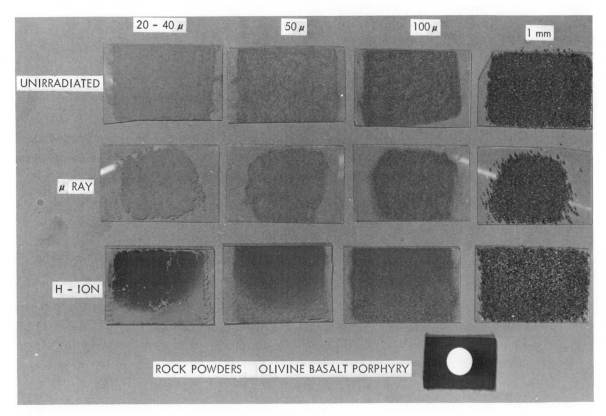

FIGURE 7–6: Effects of particle size and different types of irradiation on the appearance of a rock powder.

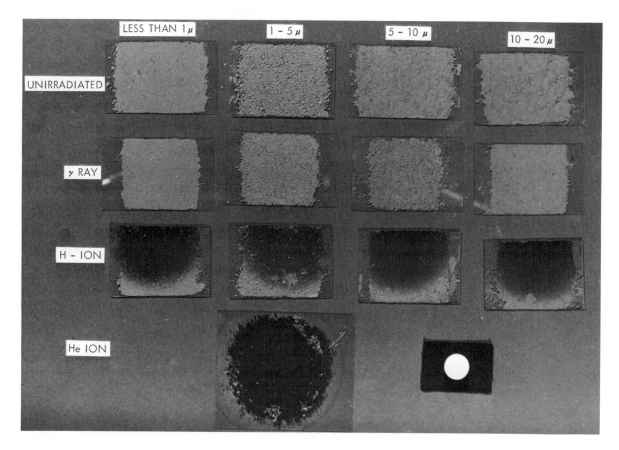

FIGURE 7–7: Effect of hydrogen-ion bombardment on fine olivine-basalt powders.

FIGURE 7–8: Effect of irradiation on mineral oxide powders.

FIGURE 7–9: Effect of irradiation on mineral oxide powders.

FIGURE 7–10: Effect of irradiation on mineral powders.

bombarding pure materials what the optical properties of a mixture of materials would be. Bombardment by helium ions darkened all the powders quite a bit, with the exception of the magnesium oxide which darkened very little. Even so, the mixtures have a lower reflectivity than the pure materials.

This figure also illustrates that the SiO_2 lattice has a strong role to play in this phenomenon since the mixture of the aluminum oxide and magnesium oxide did not darken nearly as much as did the mixtures which contained the silicon dioxide.

Figure 7–9 shows some pure metal oxides which were bombarded. The ferric oxide was darkened in both cases by both hydrogen and helium ion irradiation to about the same extent. Hydrogen ion irradiation reduced cupric oxide to pure metal but helium ion irradiation did not greatly affect it. This illustrates that, particularly in the case of copper oxide, there evidently are some chemical effects

occurring which are important for certain pure materials, in addition to the mechanical effects of sputtering. The other oxides were affected only slightly by irradiation.

Figure 7–10 shows some rock-forming mineral powders before and after hydrogen irradiation. Quartz is not changed much. There is a strong correlation between composition and the amount of darkening. Basic materials generally turn darker and bluer than acidic materials.

Figure 7–11, also illustrating the effect of composition, shows enstatite, which is mainly magnesium silicate, and hypersthene, in which some of the magnesium atoms are replaced by iron. The hypersthene is darker and bluer than the enstatite. In other words, the iron content has an effect on the optical properties of an irradiated mineral. This probably has something to do with the fact that iron is a transition metal and itself forms non-stoichiometric compounds.

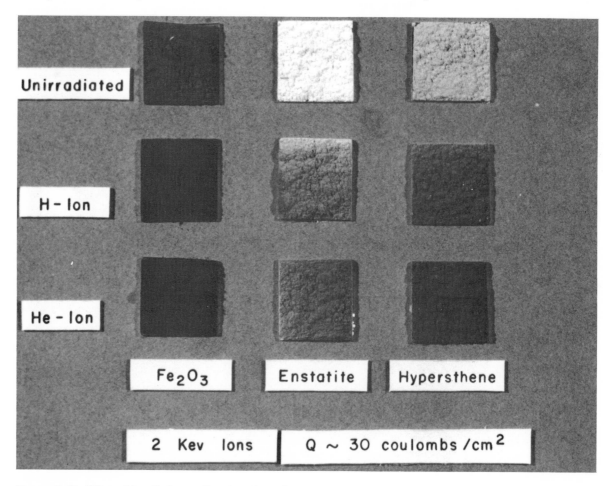

FIGURE 7–11: Effect of irradiation on Fe mineral powders.

Figure 7–12 shows some irradiated igneous rock powders. Again the effect of composition is striking —the basic materials are generally darkened and less red than the acidic materials. Note that the irradiated chondrite is much darker than any of the igneous rocks.

Figure 7–13 gives the effect of particle size on the quantitative photometric properties of olivine basalt powder; these are the photometric characteristics which were defined in connection with Figure 7–1. The gray bands on all the curves are the range of lunar values.

The normal albedo of the unirradiated material increases as the particle size decreases, but for the irradiated material the albedo is roughly constant. The backscatter ratio decreases drastically with particle size. The amount of positive polarization is a strong function of particle size and decreases in a very striking manner as particle size decreases. For large materials the polarization is far too great for the moon. For large particles the phase angle of the positive polarization peak is shifted to much higher values than is true for the moon. In our apparatus we can measure only up to a maximum phase angle of 130 degrees and the ϕ_+ curve is still

rising at 130 degrees for the particles which are labeled with an arrow. Thus the polarization provides another indication that large chunks of material are not exposed at the lunar surface. This has been emphasized by Dollfus. The material labeled as being one centimeter in size on the figure actually refers to the freshly broken surface of solid rock and is not pulverized material.

It is clear from this figure that only particles which are of the order of 1 to 10 μ in size can simultaneously reproduce all the lunar photometric characteristics. It is possible, however, to reproduce one or two of the lunar photometric properties in other ways. For instance, a high backscatter ratio can be obtained by using chunks of vesicular rock formed into a jumbled surface that is riddled with tunnels pointing in all directions. However, in general, such large chunks will have a high polarization, far too high for the moon, and the phase angle of the maximum polarization has too large a value. Also, Lyot and Dollfus found in their investigations that certain varieties of volcanic ash would have the correct polarization curves, but these ashes do not have the correct brightness functions. It may be inferred that the size distribu-

FIGURE 7–12: Effect of irradiation on rock powders.

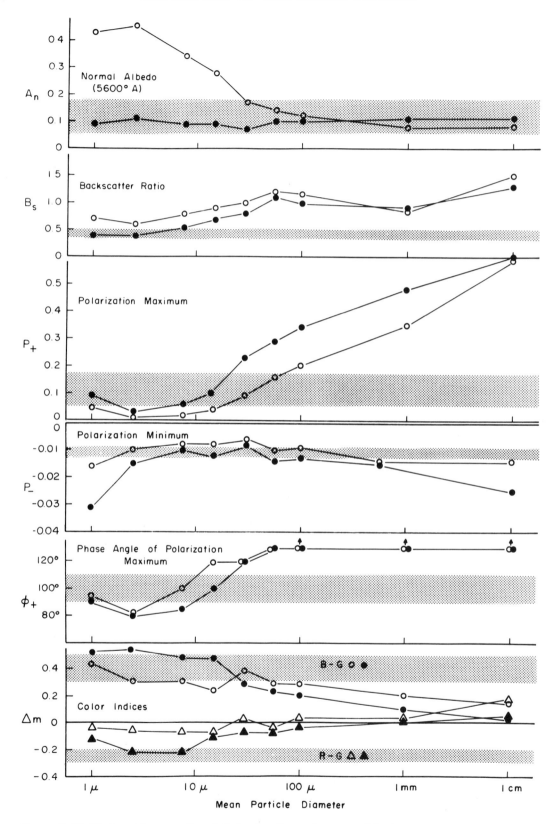

FIGURE 7–13: Photometric properties of olivine-basalt powders vs particle size.

tion of particles composing the lunar soil peaks somewhere between 1 and 10 μ.

Figure 7–14 shows the effect of radiation dose on the photometric properties of olivine basalt powder. All the curves saturate in a time on the order of a hundred thousand years or so on the moon. The effect of gamma ray irradiation is also shown in this figure; the optical properties of material treated with gamma rays is virtually the same as for the unirradiated material. The effect of helium ion irradiation to a dose of about 88 coulombs/cm^2 is essentially the same as a three to five times larger dose of hydrogen ion irradiation.

On the basis of these curves, it is reasonable to

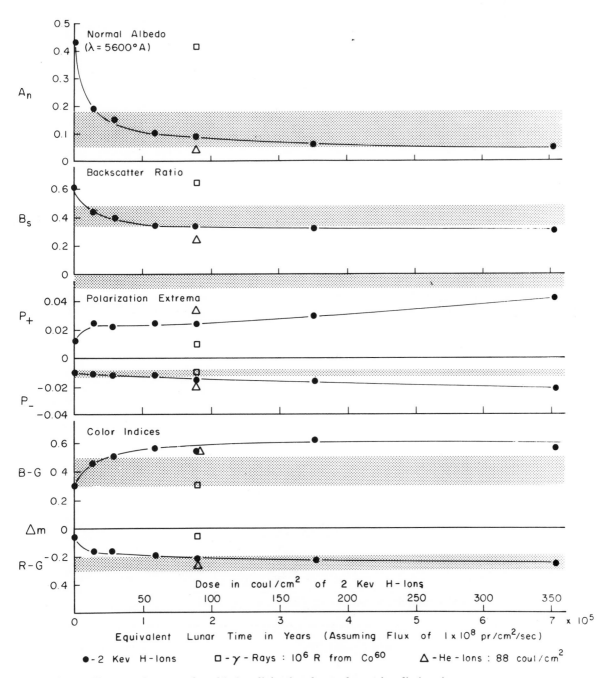

FIGURE 7–14: Photometric properties of 1–5 μ olivine-basalt powders vs irradiation time.

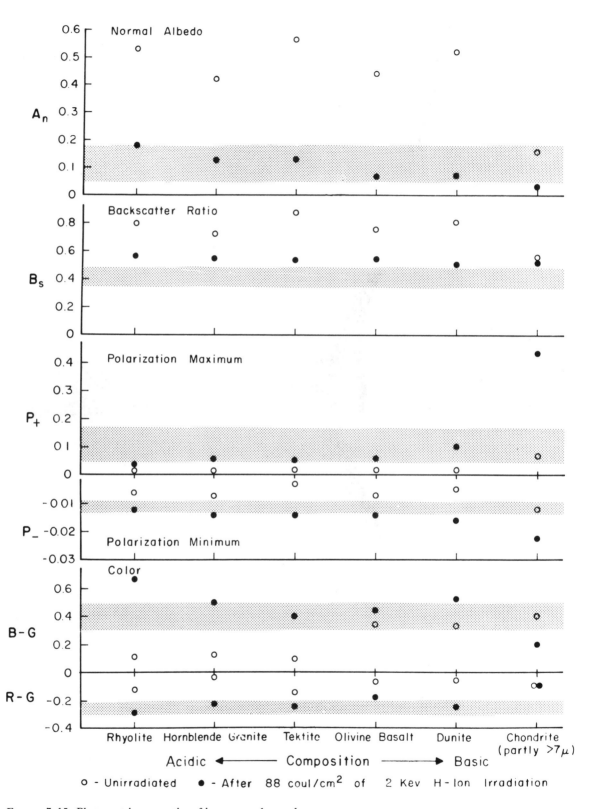

FIGURE 7–15: Photometric properties of igneous rock powders.

state that the average particle on the lunar surface has been exposed for a time on the order of a hundred thousand to a million years. This is much less than any estimates of the ages of most features on the lunar surface.

After prolonged irradiation the albedo of most rock powders actually gets lower than the lunar surface albedo. Hence, it appears that some agent is counteracting this darkening process. Some irradiated rock powder was placed in a vacuum furnace and heated to 450°C and held there for over a day. There was no appreciable change in the photometric properties; these coatings appeared to be quite stable.

A reasonable explanation for the higher lunar albedo is that micrometeorites are continually stirring up the surface and exposing undarkened materials. At all times one sees a mixture of darkened material and undarkened material such that the average particle must have been exposed on the lunar surface for something like a hundred thousand years.

We also investigated the effect of porosity of the surface. We took the same powder and formed surfaces, by pressing, pouring, and sieving the powder, in order to see the effect of compaction on the photometric properties. In the case of both the pressed and the sieved material, the photometric peak is wider than that of the moon, but for different reasons. Only the poured material gave the correctly shaped backscatter peak. The pressed powder does not have a sufficiently complex and open structure to backscatter well. The sieved powder has the requisite complex structure, but even after irradiation the particles are somewhat translucent. In a loose structure, light can shine through the particles and the powder will appear too bright at large phase angles. When a powder is poured, the particles form clumps sufficiently complex to backscatter well, but which will also block some of the transmitted light. Surfaces with the correct optical properties can be made by pouring the powder in a vacuum as well as in air.

Apparently the lunar surface is not porous to the extreme extent that Hugh Van Horn and I suggested previously. The surface does not have an extremely underdense, fairy castle structure, but rather consists of loose clumps of fine particles which are quite complex and capable of backscattering strongly. The porosity is about 80 per cent, instead of being something like 90 per cent,

which would be the case for the fairy castle structure. This is still quite underdense. The powder is very compressive; it has the consistency of baking flour. If there were even a few feet of this material, an astronaut would sink into it up to his knees and would have great difficulty churning his way through. The dust would stick to him and would likely be quite a nuisance in a number of ways, and I imagine that on the moon anything which is a nuisance is dangerous.

Figure 7–15 shows the effect of chemical composition on the photometric characteristics of rock powders. In all cases, the effect of irradiation is to bring the photometric properties close to those of the moon. The backscatter ratio of the irradiated powder shown here is a little bit high. However, in this experiment we were more interested in investigating the effect of composition on the photometric properties than in duplicating those of the moon, so uniformity of the samples was our primary concern. As shown in Figure 7–2, the backscatter ratio, at least for basic rock powders, can be reduced by a proper preparation of the surface.

Most rocks, when ground to a fine particle size, are relatively colorless, so their color indices are quite low. Only after irradiation does the color move toward the lunar values. The general tendency of the effect of irradiation is to redden rock particles. Evidently the dark compounds that coat these particles absorb more heavily in the blue than in the red.

There are a few remarks that can be made about the effect of composition. As one goes from an acidic to a basic material the albedo of the irradiated powder decreases. There is a slight tendency for the backscatter ratio to decrease. The polarization maximum tends to increase, and this is directly connected with the decrease in albedo. The polarization is dependent on two things: (1) the light which is reflected from the surface of the particles, which is positively polarized; and (2) the light which is refracted through the particles, which is negatively polarized and which tends to cancel out some of the positive polarization. As the material is darkened and made more absorbing, some of the refracted, negatively polarized light is cut out and the positive polarization is enhanced. Roughly speaking, irradiated acidic materials are redder than the basic materials, especially their blue-green index. However, there is not much variation in the red-green index. These are rough trends, but there are departures from these trends.

Now consider the powdered chondrite shown at the extreme right of Figure 7–15. This is a sample of Plainview, a bronzite chondrite. After receiving the same amount of irradiation as the rest of the rocks, its albedo dropped far below that of the lunar surface. The polarization positive maximum was way up to 43 per cent. The negative minimum dropped down to 2.5 per cent. The colors remained much bluer than the moon. There may be some indication here that at least chondritic meteorites do not come from the moon. One might say that perhaps the flux of micrometeorites at the lunar surface which is stirring up the material is bigger than we think it is, so that the average particle is irradiated just very slightly. This would tend to keep most of the optical properties of the chondrite within the range of lunar values except for the red-minus-green index, which would stay too low.

The primary reason that the photometric properties of the chondrite are different from those of igneous rocks is the high metallic iron content of the meteorites. Adding 15 per cent by weight of metallic iron to any rock has the effect of increasing the positive and negative polarizations and of decreasing the color, making the material much bluer.

Finally I would like to see if these studies can give any indication concerning the composition of the highlands and the maria. As you know, there are two theories. One says that the highlands are of different composition than the maria. Figure 7–15 is consistent with the theory of the highlands' being more acidic than the maria.

The other theory, due to Gold, is that the material on the highlands contains a larger admixture of unirradiated material than that in the maria. That is to say, the material exposed on the surface of the maria is older on the average than the material covering the highlands. However, Figure 7–14 shows that all the photometric properties are monotonically changing functions of radiation dose. That is, as one increases the dose, the positive polarization rises, the negative polarization decreases, the color indices all change monotonically.

Thus, if the *only* effect were one of exposure age, then we would expect that the differences in photometric properties of the highlands and the maria would always be in the same direction everywhere on the moon. It is known that this is not always the case. As far as I know, there is no correlation between, for instance, the amount of negative polarization and the albedo, although there is a correlation between positive polarization and albedo. In general, the maria tend to be somewhat bluer than the highlands; this is just the opposite from what would be expected of most rocks, which tend to get redder with increasing dose, rather than the other way around. However, this last argument is not a very strong one because igneous rocks which are rich in ferric oxides are initially red and become bluer under irradiation. Nevertheless, I feel that there are some tentative indications that the differences in the photometric properties of the light and dark areas of the moon are at least partly due to real differences in composition and not just to differences in exposure age.

8.

The Application of Polarized Light for the Study of the Surface of the Moon

A. Dollfus

Observatoire de Paris, Section d'Astrophysique, Meudon (Seine et Oise), France

The curve of polarization of the moon is the plot of the percentage of polarization as a function of the phases of the moon. The polarization is called positive if the largest component is perpendicular to the plane of the sun, the earth, and the moon, and is called negative if this component is parallel to this plane. Figure 8–1 shows the fringe polari-

meter that we used for those measurements. The accuracy is 10^{-3} or 0.1 per cent in the amount of polarization.

Figure 8–2 is the original curve that Lyot produced more than half a century ago. It shows the maximum of polarization for the evening moon and the morning moon. We immediately see that we

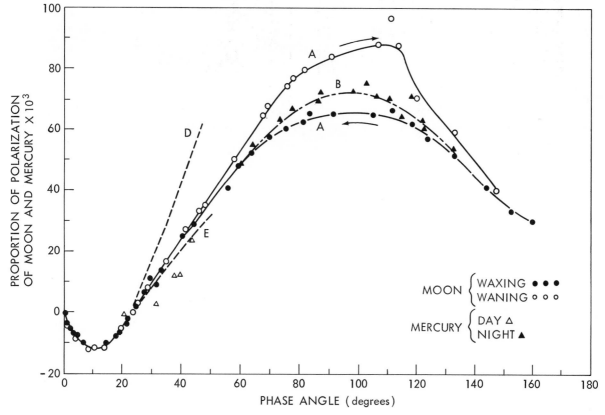

FIGURE 8–1: Polarization of the moon (after Lyot, 1929).

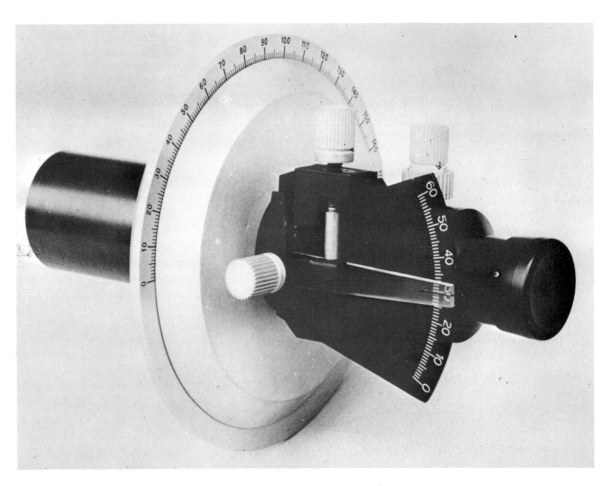

FIGURE 8–2: The fringe polarimeter used to measure the polarization of the moon.

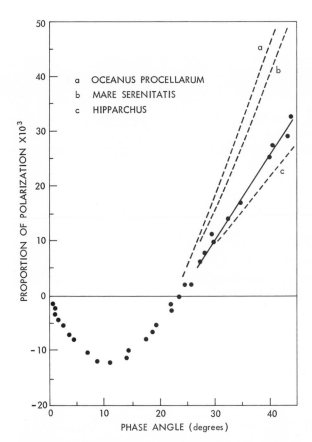

FIGURE 8–3: Measurements on the negative branch of the polarization curve.

on powders of dark grains of all sizes stuck together, the grains completely absorbing the light in a few wavelengths of thickness.

Figure 8–4 shows laboratory results about the negative branch of the polarization on different kinds of material, in cases of loose deposits and compact deposits of powders. The loose deposits of dark grain materials reproduced exactly this property of the surface of the moon. The compact deposit did not. We measured many other kinds of materials. I published some results in the past in Professor Kopal's book and in Professor Kuiper's book; therefore, I prefer not to discuss it extensively here.

The case is that all the surface of the moon gives the same curve. So we must conclude that the surface of the moon is covered in all the parts by a layer of dust at least a millimeter thick, which is made of small grains and is completely absorbing. I would like to be very positive on this point. It is certain, because fortunately the curve is very specific for powder of dark grains for all the areas of the surface of the moon.

I have another proof of this property: it is depolarization of the light. If a sample is lighted by a completely polarized light, the scattered light is partly depolarized; the residual amount of polarization is a function of the reflecting power of the surface. We measured the depolarization factor of the surface of the moon by measuring the polarization of the ashen light, because the ashen light is the scattering by the moon of the light coming from the earth. The light of the earth at 90 degrees is very highly polarized. Therefore, if we measure the polarization of the ashen light, and if we know the initial polarization of the light of the earth, we can deduce the depolarization factor of the surface of the moon.

Figure 8–5 is a picture that we took at Pic-du-Midi with the coronagraph, which was used for this study of the moon. The occulting disk of the moon has a window with absorbing glasses, showing the edge of the moon, namely Mare Crisium. The ashen light is passing at the edge of the disk. The last mountains at the terminator are seen at the edge of the disk. This plate was taken before the full moon, with a phase angle of about 20 degrees. We are able to detect and measure the ashen light very close to the full moon.

Polarimetric measurements on the ashen light selected with the coronagraph enabled us to give

can split the problem in two: first, the explanation of the negative branch, and second, the discussion of the maximum of polarization.

Figure 8–3 shows measurements of the negative part of the curve. It is close to the full moon, just before an eclipse, when the phase angle is very small. The dispersion of the measurement is less than 0.1 per cent. Curves are given for the darkest maria, like Mare Procellarum, and for the bright areas. Departures for the near curves occurred for phase angles larger than 20 degrees, but all of these measurements are the same in the negative branch. For measurements made in all spectral ranges from visible light to infrared at $1.05\ \mu$, the negative branch remains exactly the same. This branch is not sensitive to the wavelength nor to the albedo of the surface.

At the laboratory we are able to reproduce such a curve and it appears that it is very specific of a special kind of surface; it can be reproduced only

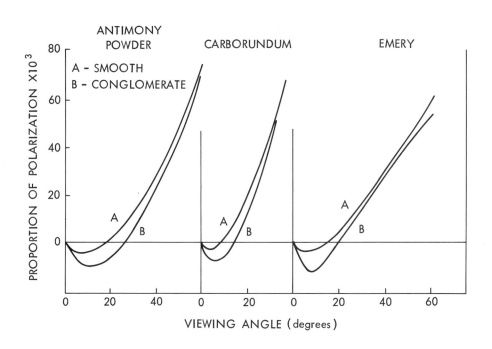

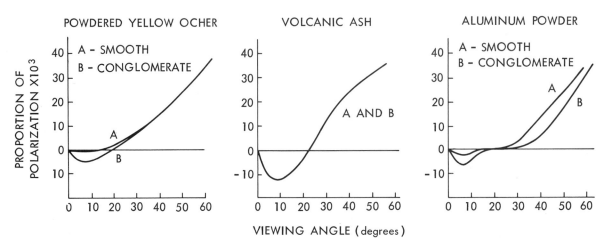

FIGURE 8–4: Laboratory-derived negative-branch polarization curves for various materials.

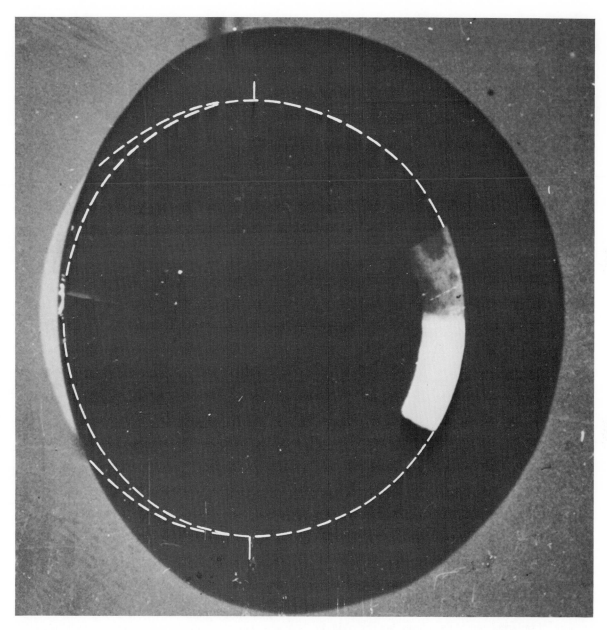

FIGURE 8–5: Coronagraph picture of the moon, creating an artificial lunar eclipse with the ashen light shown at the edge of the occulting disk.

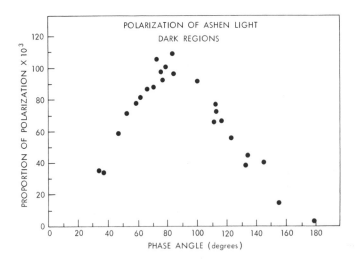

FIGURE 8–6: Polarization of ashen light on the dark regions of the lunar surface.

the complete polarization curve of this ashen light, as seen in Figure 8–6. The plots are the phase angle vs the amount of polarization. If we compute the polarization of the earth we conclude that most of the polarization is due to scattering by the atmosphere of the earth. The maximum polarization is expected to be about 80 degrees. It is exactly what we found. But we have to know the exact amount of polarization given by the earth at an 80-degree phase angle. We did that by measurements from balloons.

Figure 8–7 shows one of the flights we made in France. This particular flight we launched from the Meudon Observatory. On the occasion of such flights using a visual polarimeter we measured the polarization at several heights and were able to extrapolate the polarization as if the earth were seen from the outside.

Figure 8–8 shows some of the results. One of the two curves at the top is a measurement of the surface from the ground. The other is taken at a height of 1 km. With similar measurements we are able to extrapolate to infinity. If we know the amount of polarization of the moon, we are able to deduce the depolarizing factor of the earth.

In Figure 8–9 the open circles are the measurements from the ashen light, of the depolarization factor of the surface of the moon as a function of the albedo of the surface. The dots and crosses are measurements obtained on dark powders made of small grains of completely absorbing material. It fits exactly the measurements; other kinds of materials like bare rocks or powders of transparent grains did not. All other kinds of surfaces are completely different. Again, the depolarization factor is a very crucial result, characteristic of a powder of very absorbing grains. This powder covers all the surface of the moon.

Figure 8–10 is a microscopic picture of a typical powder that is probably similar to the surface of the moon; the size is 3 mm square. This powder gives exactly the negative range of polarization and the depolarization coefficient of the surface of the moon.

The next problem will be to try to improve these results by studying the maximum of polarization. The variation of maximum polarization may enable us to give additional data about the composition and the nature of these powders. Unfortunately we have not yet completed this work; we are doing the measurements. I would like to report here the current state of the measurements. We tried to extend the polarization to a larger range of wavelengths because the maximum of polarization is wavelength dependent.

We divided the moon in selected areas in which we are taking polarimetric measurements (Figure 8–11). We have two kinds of areas: 14 large ones for the infrared to ultraviolet measurements in all the ranges accessible from earth, and smaller areas for careful studies of variations of polarization on special features. We have several polarimeters for several wavelength ranges. Figure 8–12 shows the telescope itself, with photoelectric polarimeter. It is a Cassegrainian coudé, and the light is reflected by a mirror, introducing a spurious polarization that we have to take into account in the final study of the measurements.

FIGURE 8–7: Flight launched from Meudon Observatory.

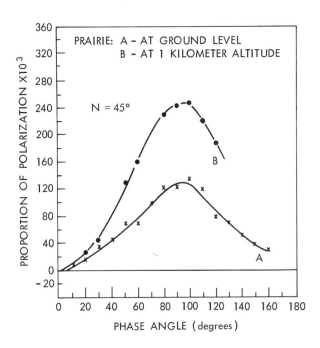

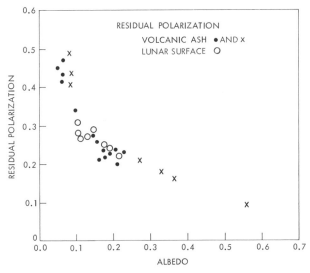

FIGURE 8-9: Measurements of the depolarizing factor of ashen light and dark absorbent powders.

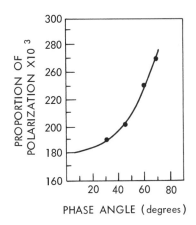

FIGURE 8-8: Polarization measurements of the ground from near the surface of the earth and from a height of 1 km.

Figure 8-13 shows the infrared measurements for special areas, at 0.85 μ wavelength. One of the two curves is for one of the darkest surfaces of the moon, Mare Tranquillitatis. An idea of the accuracy of the measurements is given by the scattering of the dots. The other curve is for one of the lightest areas on the surface of the moon, northeast of Mare Crisium; the negative branch is

completely independent of the albedo, but the maximum of polarization is strongly dependent on the albedo.

Now going more to the infrared, Figure 8-14 shows the curves for 0.95 μ. The polarization is reduced in the two cases. Again the negative branch is still the same. Figure 8-15 is for 1.05 μ, and again the scattering of the measurements is good, but the polarization is still reduced. The maximum of polarization is a function of the albedo and of the wavelength.

Now we have enough measurements to disentangle these two factors; Figure 8-16 shows the variation of the maximum of polarization as a function of the brightness or the albedo. For visible light we have a very sharp variation of the maximum of polarization with brightness, and for infrared measurements we have again this variation, but with a lower amount.

Figure 8-17 shows the variation of the maximum of polarization as a function of wavelength. This is for four different areas: Mare Tranquillitatis, one of the darkest areas on the moon; Mare Crisium; and two light continents. The curves show the strong variation of maximum of polarization as a function of wavelength, from 1.1 to 0.6 μ.

So we have now about all the quantitative information about the properties of the variations of the maximum of polarization with albedo and wavelength. We can make comparative measure-

FIGURE 8–10: Picture (3 mm square) of a typical powder viewed through a microscope.

Figure 8–11: Selected areas under polarimetric measurement study.

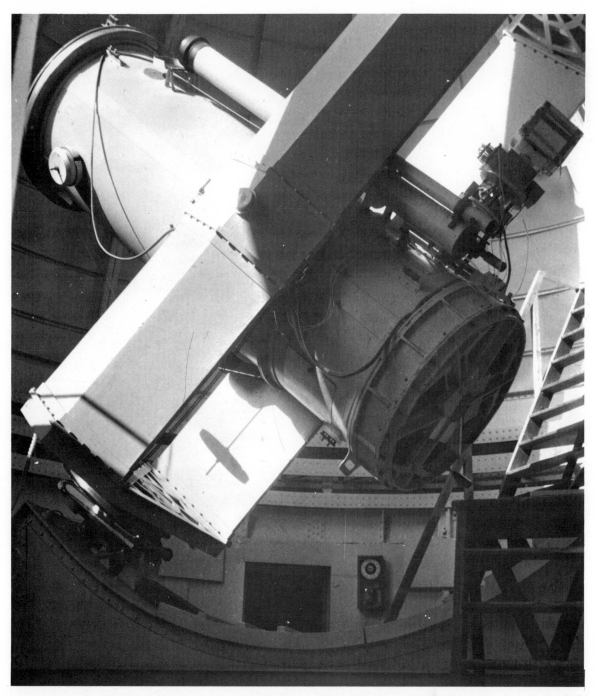

FIGURE 8–12: 40-inch Telescope at Meudon Observatory used for photoelectric measurements of polarization on planets and the moon.

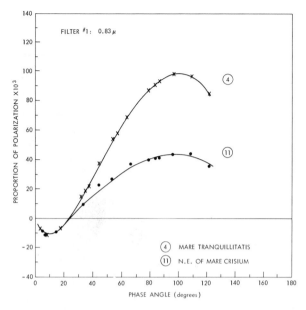

Figure 8–13: Infrared measurement, 0.83 μ of wavelength; *4* is Mare Tranquillitatis, *11* is northeast of Mare Crisium.

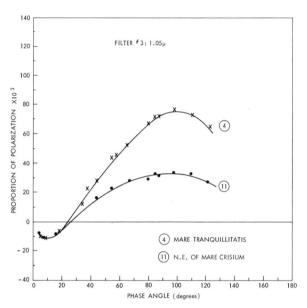

FIGURE 8–15: Infrared measurement, 1.05 μ of wavelength; *4* is Mare Tranquillitatis, *11* is northeast of Mare Crisium.

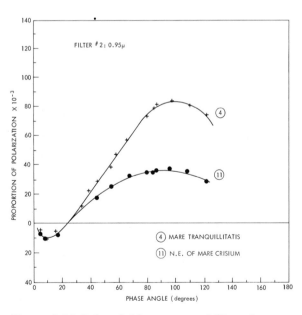

FIGURE 8–14: Infrared Measurement, 0.95 μ of wavelength; Mare *4* is Tranquillitatis, *11* is northeast of Mare Crisium.

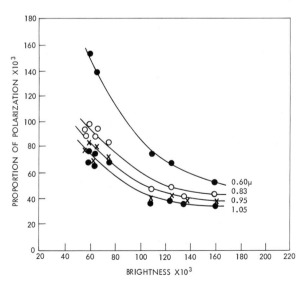

FIGURE 8–16: Variation of the amount of the maximum polarization as a function of brightness.

ments at the laboratory to see if some information can be deduced from these data. We measured at the laboratory with the photoelectric polarimeter and with the visual polarimeter.

Figure 8–18 shows one of the instruments we used for laboratory measurements. The sample is put in the holder, and we can change the angle and azimuth of the light. We can also tilt the direction of observation of the photoelectric polarimeter.

Figure 8–19 shows some characteristic curves with materials selected for obvious theoretical reasons: two kinds of obsidian, the brecciated coating of the Meteor Crater in Arizona, tektites, etc. We measured also several other powders of this kind. It can be seen immediately that all these curves are too flat; they are not able to reproduce the negative branch. This is because these surfaces are not dark enough. I pointed out that we must use powder of grains in loose deposits of very absorbing material; all these surfaces are not dark enough.

Figure 8–20 shows different kinds of achondrite meteorites with enstatite. The negative branch cannot be reproduced because the powders are far too light. The maximum of polarization is completely out of scale.

Figure 8–21 shows the results for the different kinds of chondritic meteorites with enstatite, bronzite, hypersthene. Most of these have not enough negative branch. But one of those is closer to the moon; it is a meteorite very rich in olivine, and the blackness is 0.045. This curve, number 1, is not very far from the moon, but cannot reproduce exactly the surface of the moon. Olivine meteorites give a first indication about the interpretation.

The best fits are given by ash flows of broken lava. If we pulverize lava samples, or if we measure ashes from volcanic areas, we are very close to the surface of the moon, as can be seen in Figure 8–22. Different varieties of Vesuvius and other volcanic ashes are represented. If you accept the last one as being far too light with an albedo of 0.56, and the other having brightness of the same order as in the case of the moon, it can be seen that the negative branch fits the curve of the moon well and the variation of P maximum with albedo is of the same order as indicated on the moon. We are proceeding at the laboratory to make more complete investigations about the ash and lava powders.

However the situation is far more complicated, as Dr. Hapke pointed out, because we must take

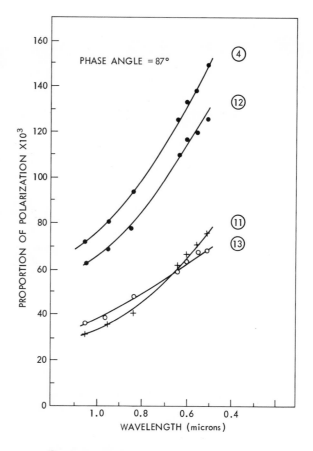

(4) MARE TRANQUILLITATIS

(11) N.E. OF MARE CRISIUM

(12) MARE CRISIUM

(13) ZAGUT

FIGURE 8–17: Variation of the amount of the maximum polarization as a function of wavelength.

into account the effect of the proton-darkening. This proton-darkening confuses the situation because some of the light materials of which we gave curves may fit the curve of the moon after appropriate bombardment by radiation.

In Figure 8–23 is a measurement we made with Dr. J. Geake at Manchester. The sample is an enstatite achondrite; the darkest area was darkened by 16-kv bombardment. Figure 8–24 shows that the properties of the darkened part of this material give a very good reproduction of the negative branch of the polarization. The curve of polariza-

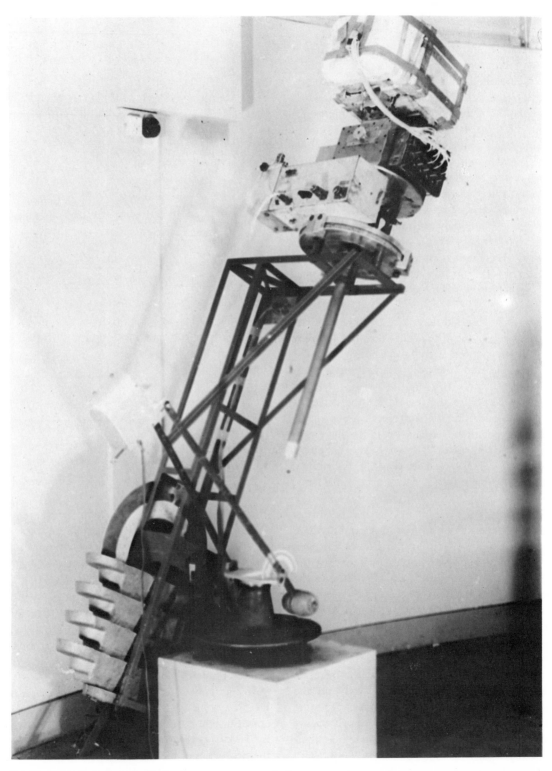

FIGURE 8–18: Laboratory polarization measurement device.

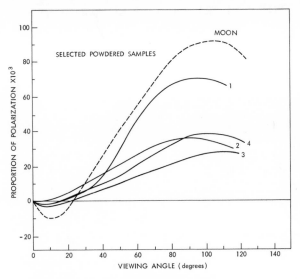

① OBSIDIAN, BLACK (ASCENSION ISLAND)

② OBSIDIAN, RHYOLITIC (CANTON LAKE, OREGON)

③ METEOR CRATER, ARIZONA - INTERIOR BRECCIA

④ TEKTITES (DALAT, INDOCHINA)

FIGURE 8–19: Characteristic polarization curves of two kinds of obsidian, the breccia of the Meteor Crater (Arizona), and tektites.

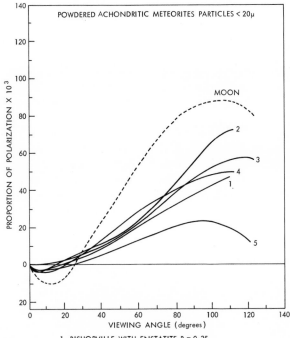

1 BISHOPVILLE WITH ENSTATITE B = 0.35

2 BUSTEE - AUBRITE WITH ENSTATITE B = 0.30

3 KHOR TEMIKI WITH ENSTATITE (NO IRON) B = 0.30

4 JUVINAS - EUCRITE RICH IN CADMIUM

5 TATAOUINE - DIOGENITE WITH HYPERSTHENE

FIGURE 8–20: Comparison polarization curves of the moon and achondrite meteorites.

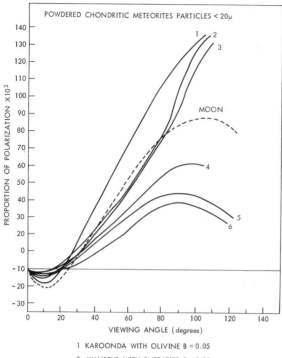

1 KAROONDA WITH OLIVINE B = 0.05

2 KHAIRPUR WITH ENSTATITE B = 0.15

3 DANIELS KUILS WITH ENSTATITE B = 0.18

4 OCHANSK BRONZITE

5 OUBARI WITH HYPERSTHENE

6 PUTULSK BRONZITE

FIGURE 8–21: Characteristic polarization curves of different kinds of enstatite, bronzite, and hypersthene chondritic meteorites.

tion of the enstatite achondrite is shown before darkening; it has no negative branch at all. The curve measured after the darkening is identical to the curve of the moon. The reproduced negative branch is unexpectedly good. This seems to give a definite fit of optical properties.

A great number of powders may fit this negative branch, provided they are darkened enough by proton bombardment. Furthermore the variation

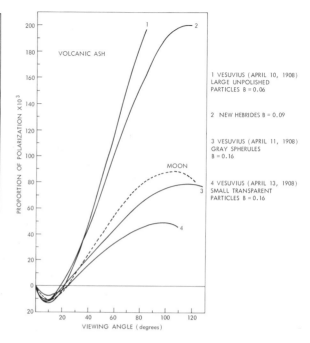

FIGURE 8–22: Characteristic polarization curves of ashes from various volcanic areas.

of maximum of polarization with wavelength is very good. The only point in the sample shown is that the variation of brightness with wavelength is too high in this special kind of sample.

As a conclusion I would like to point out that polarization gave a very certain result, that the surface of the moon is covered in all parts, in all areas, by a coating, a layer of dust of small particles, not compacted, and very absorbing. For the explanation of the nature of the powder, polarization techniques are not so specific because several possibilities occur. The proton bombardment may have something to do with the result. But we follow in two ways now: a careful study of the igneous volcanic rocks, broken into powder, and the effect of darkening by protons.

FIGURE 8–23: Sample of an enstatite achondrite, part of which has been darkened by proton bombardment.

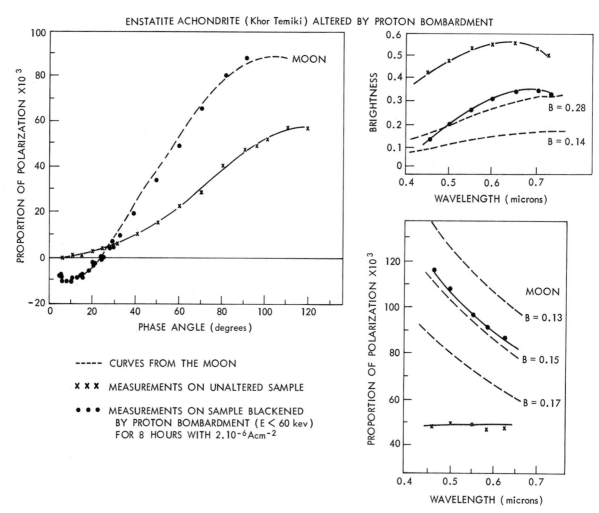

FIGURE 8–24: Characteristic polarization curves of enstatite achondrite before and after being darkened by proton bombardment.

9.

Luminescence of the Moon and Solar Activity

Zdeněk Kopal

Department of Astronomy, University of Manchester, Manchester, England

ABSTRACT

Observations of transient luminous phenomena on the lunar surface, which can scarcely be accounted for by anything other than ground luminescence, are reviewed and their quantitative aspects discussed. It is pointed out that several (though not all) conspicuous instances of such phenomena followed (though did *not* coincide with) transient disturbances on the sun, by intervals which ranged from several hours to a few days—a fact which would indicate that the sources of excitation are likely to be corpuscular, rather than electromagnetic, radiation.

Anomalous phenomena indicative of luminescence have been observed to occur on the bright as well as the dark side of the moon. In the latter case, the interplanetary magnetic field permits access to the dark hemisphere only to protons with energies in excess of 10 Mev. The flux of such particles in the energy range of 10–100 Mev during major flares on the sun is indeed more than adequate to increase appreciably the surface brightness of the earthlit moon for reasonable quantum efficiency of the luminophor; but to do so in daytime by direct action of solar protons would call for greater particle densities than are actually observed.

It is pointed out that luminescent phenomena on the moon may be allied with the anomalous brightening (and reddening) of the zodiacal dust cloud observed to follow the solar flares.

As has been known (or at least conjectured) since the days of Anaxagoras in the fourth century B.C., the visible light of our moon is essentially sunlight, scattered from the lunar surface by a process which leaves its spectral composition largely unchanged; but measurements of more recent data have revealed that only about 7 per cent of incident light is scattered in this way. Moreover, most incident sunlight which is not scattered must be absorbed by the lunar surface, converted into heat, and re-emitted in the infrared to which our atmosphere is only partially transparent. This kind of light is emitted at wavelengths generally too long to be visible to the human eye and to contaminate the white color of the "silvery moon" of our songs and romances, due to the scattered component, even though it represents some nine-tenths of transformed sunlight; but Anaxagoras knew nothing about it.

The question can, however, be asked: does scattered sunlight, plus that absorbed and re-emitted at generally lower frequencies, represent the sum total of the observable moonlight? The answer to this question is now known to be in the negative, and the aim of the present address is to elaborate on it in more detail. It is, of course, well known that—apart from visible sunlight—the lunar surface (unprotected by any atmosphere, to speak of) is continuously exposed to high-energy UV- and X-ray quanta, as well as to all corpuscular radiation from the sun. This surface is bound to give rise to X-ray emission (*bremsstrahlung*) at wavelengths mostly between 10–100 Å*—radiation which cannot be observed from ground.

---

* In other words, the sun can be likened to a "hot cathode" and the surface of the moon to the anticathode of an "ion tube" of cosmic dimensions, whose glass walls (i.e., our atmosphere) enclose the observer rather than the apparatus.

Suppose, however, that the recombination of atoms ionized by energetic particles or quanta does not occur (as in the *bremsstrahlung*) by a single transition, but by a cascade process giving rise to luminescent emission at longer wavelengths, which can penetrate our atmosphere and become observable on the ground. Indications that such an emission may indeed make an appreciable contribution to the total light of the moon have been forthcoming from different directions for many years. To begin with the most elementary manifestations: it has been noted by experienced observers in the past[1] that the apparent brightness of the moon at a given phase (when due regard is paid in reductions to its instantaneous distance from the earth, libration, etc.) is not quite the same from month to month or year to year; and these fluctuations appeared, moreover, to be correlated with the cycle of solar activity (though their amplitude seemed larger than that manifested by the solar constant). According to the most recent and reliable photometric studies by Gehrels and his associates,[2] between 1956 and 1959 (i.e., near the maximum of the last cycle of solar activity) the lunar surface was between 10 and 20 per cent brighter in visible light than between November, 1963, and January, 1964, when solar activity was near its minimum.

Another indication of solar influence on global brightness of lunar eclipses was brought to light first by Danjon.[3] By a discussion of the data extending over three and a half centuries, Danjon proved that the brightness of the eclipsed moon is strongly correlated with the cycle of solar activity: the residual brightness of totality is found to increase with advancing cycle and to drop abruptly at the time of the minimum of solar activity in such a way that eclipses just preceding a minimum are more than two and a half times as bright as those following it (Figure 9–1). One other feature comes to mind which also changes abruptly at the minimum of solar activity: namely, the location of sunspots and associated disturbed areas on the apparent solar disk, which are known to expire near the

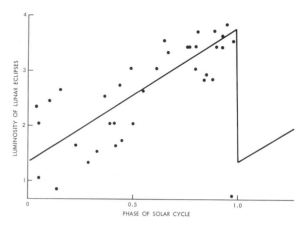

FIGURE 9–1: The residual brightness of the moon when it is eclipsed by the earth varies with the 11-year cycle of solar activity. The luminosity of the eclipse is greatest not in mid-cycle, at the peak of solar activity, but just before the minimum point at the end of the cycle.

equator at the end of a cycle, to reappear in high latitudes at the commencement of the next one.

Inasmuch as the corpuscular streams emanating from the disturbed regions of the sun do not follow the same optical path as visible light, they could reach the moon during eclipse and evoke luminescence; and as the spots move to lower latitudes (coinciding more nearly with the ecliptic) their luminescent effect may increase. However, when the sunspot latitudes change suddenly (as they do around the time of the minimum of the cycle) the excitation caused by them may be expected to drop in intensity, and the residual brightness of a lunar eclipse around that time should be markedly smaller. This is just what Danjon found; and the case for the corpuscular excitation of the luminescence of the lunar surface thereby receives indirect support.

Evidence for it was strengthened further by the results of the photometric studies by Link[4] and others[5] of partial phases of lunar eclipses, during

[1] G. Rougier, "Photométrie Photoélectrique Globale de la Lune," *Ann. de l'Obs. de Strasbourg*, 2 (fasc. 3), 1933, pp. 205–399.

[2] T. Gehrels, T. Coffeen, and D. Owings, "Wavelength Dependence of Polarization: III. The Lunar Surface," *Astron. J.*, Vol. 69 (1964), pp. 826–852.

[3] A. Danjon, "Sur une Relation Entre l'Eclairement de la Lune Eclipsée et l'Activité Solaire," *Comptes Rendus Hebdomadaires des Séances de l'Academie des Sciences*, Vol. 171 (1920), pp. 1127–1129.

[4] F. Link, "Sur la Luminescence de la Lune," *Acad. des Sci., Paris. Compt. Rend.* Vol. 223 (1946), pp. 976–977.

[5] M. Cimino and T. Fortini, "Fotometria Fotografica dell'Eclisse Totale di Luna del 29 Gennaio 1953 e la Luminescenza del Suolo Lunare per la Radiazione Ultravioletta Solare," *Atti della Accademia Nazionale dei Lincei, serie 8va, Rendiconti, Classe di Scienze Fisiche, Matematiche e Naturali*, Vol. 14 (1953), pp. 619–626; Cimino and A. Fresa, "Fotometria Fotoelettrica dell'Eclisse Totale di Luna del 13–14 Maggio 1957," *Atti della Accademia Nazionale de Lincei, serie 8va, Rendiconti, Classe di Scienze Fisiche, Matematiche e Naturali*, Vol. 25 (1958), pp. 58–64; N. Sanduleak and J. Stock, "Indication of Luminescence Found in the December 1964 Lunar Eclipse," *Publ. Astr. Soc. Pacific*, Vol. 77 (1965), pp. 237–240.

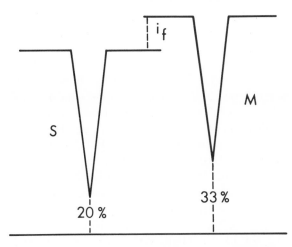

FIGURE 9–2: The line-depth method of detecting luminescence calls for comparing profiles of absorption lines
in the spectra A of the sun (*left*) and moon (*right*).
An increase in the residual intensity in the case of the
moon is a measure of the light (*i_F*) attributable to lunar
luminescence—in this example 16.67 per cent of the
total moonlight.

which the loss of light was found to be less than
that expected from the geometry of the sun-earth
disks, indicating that significant effects are being
produced by illumination of the moon by un-
eclipsed parts of the solar corona. The latter emits
only a negligible amount of visible light, but, on
account of its high temperature of 1–2 million
degrees, a much larger proportion of X-rays. Link
conjectured that this hard radiation produces
luminescence on the lunar surface which accounts
for the optical anomaly.

Photometric studies of the global light of the
moon clearly foreshadowed the existence of a
relationship between these optical phenomena and
solar activity, but revealed little or nothing about
possible spectral composition of the hypothetical
luminescence. In order to learn more about it, we
must obviously resort to some spectrometric
method, and this can be done in the following
manner. Consider the profile of an absorption line
of the solar spectrum—preferably a deep one (i.e.,
of small residual intensity)—as shown schematically
in Figure 9–2. The scattering of sunlight on the
lunar surface depends but very little on the frequency. Therefore, if moonlight consisted only of
scattered sunlight, the line profiles in its spectrum
should be exact replicas of those in the parent
sunlight. If, on the other hand, the surface of the
moon intercepted by the slit of the spectrograph)

luminesces, the lunar line profile should become
correspondingly shallower than those in the solar
spectrum. In more specific terms, if $D_{M,S}$ denotes
the ratio of the intensity of any particular point of
a line profile to that of the adjacent continuum in
the spectrum of the moon and the sun, respectively,
then the ratio

$$d = \frac{D_M - D_S}{1 - D_M} \qquad (1)$$

denotes the fractional intensity of the luminescent
radiation superimposed on scattered moonlight,
and expressed again in terms of the intensity of the
adjacent continuum taken as the unit.

There exists, furthermore, another way in which
the luminescence of the lunar surface could be
established from the observations: namely, by
measures of the polarization of moonlight and its
dependence on the wavelength. The basic idea goes
back to the fact that, whereas the fraction of the
moonlight which represents scattered sunlight
becomes distinctly polarized by this process (and
the degree as well as the position of the plane of
polarization varies with the phase), the luminescent
emission is, of course, nonpolarized. Therefore, any
temporary anomalies on the phase-polarization
curves which exceed the limits of observational
errors, manifested so clearly on Lyot's excellent
measurements,[6] may (as was recently pointed out
by Gehrels) have been caused by variable admixture of nonpolarized light of luminescent origin.
Moreover, the polarization of scattered moonlight
is known to vary but slowly with the wavelength.
Therefore, any irregularities on the curve of
polarization versus wavelength should indicate—
no less distinctly than the filling up of the absorption lines of the lunar spectrum—the presence of
luminescent bands at the corresponding frequency.[7]

The "method of line depths" for the determination of the fractional luminescence d of the lunar
surface from the measured ratios $D_{M,S}$ by means of
equation (1) was proposed by Link[8] and employed

[6] B. Lyot, "Recherches sur la Polarisation de la Lumiére
des Planétes et de Quelques Substances Terrestres," *Ann. de
l'Obs. de Paris, Section de Meudon*, 8 (fasc. 1), 1929, pp. 1–161.
[7] V. G. Teifel, "Spectropolarimetry of Some Regions of
the Lunar Surface," *Astron. Ẓh.*, Vol. 37 (1960), pp. 703–708
[in Russian].
[8] Link, "Variations Lumineuses de la Lune," *Central
Astron. Inst. of Czechoslovakia. Bullet.*, Vol. 2 (1951), pp.
131–133.

in practice first by Kozyrev[9] and Dubois.[10] Both found indications of luminescence to be present in different parts of the lunar surface and in light of different frequencies. According to Kozyrev, the maximum value of d observed for the crater Aristarchus in the profile of the H-line of ionized calcium (λ 3970 Å) amounted to 0.13 on October 4, 1955; but less than a month later (October 28 and November 4, 1955) it diminished to 0.03. Since, however, both Kozyrev and Dubois employed photographic techniques which were of marginal accuracy for the purpose, their results were regarded by contemporary astronomers as suggestive rather than conclusive; and it became obvious that in order to obtain results of indubitable significance, the methods of photographic spectroscopy would have to give way to photoelectric spectrometry.

This was first done at Manchester between 1960 and 1962 (as a part of a systematic program of lunar luminescence studies sponsored by the U.S. Air Force), when Grainger and Ring built a scanning-type photoelectric spectrograph which was used in connection with the 50-inch reflector of Padua University Observatory at Asiago. The actual observations of the moon began in 1961 and were largely confined to the scans of the profile of Ca II H line (λ 3970 Å), for reasons which previously led Kozyrev to the same choice: namely, the large half-width of this line (9 Å) and its low residual intensity—both factors facilitating the detection of luminescence. Spectroscopic work carried out in 1961–62 confirmed, on the whole, the previous results by Kozyrev and Dubois and established the reality of lunar luminescence beyond reasonable doubt.[11]

The strongest luminescence—amounting to 10 ± 1 per cent of intensity of the adjacent continuum—was at that time found to be emitted by the bright ray traversing Mare Serenitatis through the

crater Bessel (the mare itself appeared to luminesce no more than to 2 ± 0.7 per cent); and other regions (in particular, a region in the neighborhood of Plato at $\lambda = 3\,^0W$, $\beta = 56\,^0N$) proved to luminesce to almost the same extent. However, the actual value of d changed abruptly with any shift in position of the slit (indicative of pronounced localization of the luminophor on the lunar surface), and also with the time; but the data were too few to indicate any real correspondence between the lunar and solar phenomena. Moreover, Spinrad, working at the Dominion Astrophysical Observatory at Victoria, observed on September 16, 1962, a luminescence of intensity corresponding to $d = 0.13$ in the light of the calcium H line; and, as his slit trailed over the moon during exposure, the luminescence was probably widespread at that time.[12]

Since that time, a photoelectric scanning spectrometer of optical power comparable to that built by Grainger and Ring was employed in quest of lunar luminescence by the line-depth method by Scarfe at Cambridge (England), who extended this search to other spectral lines than the H line of Ca II (such as Hα, the sodium D lines, or a group of Fe I lines near λ 5450 Å) in the yellow and red part of the spectrum.[13] Near λ5450, a time-variable luminescence attaining as much as 30 per cent of the intensity of the adjacent continuum was observed on October 5, 1963, while at the wavelengths of Hα or of the sodium D lines no luminescence exceeding 2 per cent could be detected at the same time. The crater Aristarchus showed the strongest effect, Copernicus almost as strong, and Kepler was only somewhat less. These were the only places where the luminescence was marked. Moreover, normal integrated sunlight (diffused from a plate coated with magnesium oxide) exhibited line profiles virtually identical with those of diffuse moonlight—a fact supporting the idea that the time-independent radiation from the sun does not induce observable luminescence—and Scarfe concluded that the latter must be related to solar activity.

However, the last two months of 1963 (when a solar cycle expiring about that time was in its last

[9] N. A. Kozyrev, "Luminescence of the Moon and the Intensity of Corpuscular Radiation from the Sun," *Akad. Nauk SSSR Krymskaia Astrofiz. Obser. Izves*, Vol. 16 (1956), pp. 148–158 [in Russian].

[10] M. J. Dubois, "Sur l'Existence de la Luminescence Lunaire. Résultats Obtenus," *Bull. de la Société Française de Physique*, in: *J. de Physique et le Radium*, Vol. 18 (1957), pp. 13 S–15; "Contribution à l'Etude de la Luminescence Lunaire," *Rozpravy Československé Akademie Věd, Řada Matemattchých a Přírodních Věd*, Vol. 69 (1959), pp. 1–44.

[11] J. F. Grainger and J. Ring, "Lunar Luminescence and Solar Radiation," *Space Research—Proceedings of the Third International Space Sciences Symposium, Washington, D. C., May 2–8, 1962*, ed. by W. Priester, New York: Interscience Publishers (1963), pp. 986–996.

[12] H. Spinrad, "Lunar Luminescence in the Near Ultraviolet," *Icarus*, Vol. 3 (1964), pp. 500–501.

[13] C. D. Scarfe, "Observations of Lunar Luminescence at Visual Wavelengths," *Mon. Not. Roy. Astr. Soc.*, Vol. 130 (1965), pp. 19–29.

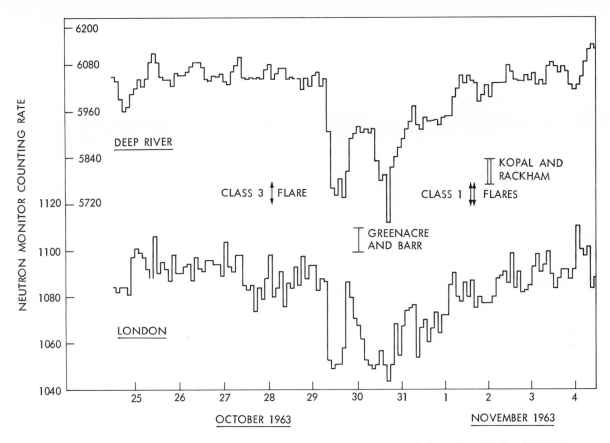

FIGURE 9–3: Solar activity affected the neutron flux at Deep River, Canada (*top*), and at London (*bottom*) late in 1963. Notable solar events were the great flare of October 28 and the small flares of November 1. Associated lunar events were the Greenacre-Barr sightings on October 30 and the author's observations of November 1–2. (Data are from T. Thambyahpillai of the Imperial College of Science and Technology in London.)

throes) had more surprises in store for the students of the moon. On October 28 at 1:58 UT the largest flare of the year (of class 3) made its appearance on the sun and was followed by a week of disturbed sun, with several minor flares, which profoundly affected the particle density in space at the distance of the earth (Figure 9–3). And—behold—not less than two transient luminous phenomena were observed on the moon during this time, which we shall briefly describe.

The first instance occurred on October 30 (between 1:30 and 1:50 UT, i.e., 48 hours after the flare of October 28), when Greenacre and Barr at the Lowell Observatory noted that three distinct spots in the neighborhood of the crater Aristarchus (in locations marked on Figure 9–4) flared up temporarily in reddish-orange light, intense enough to

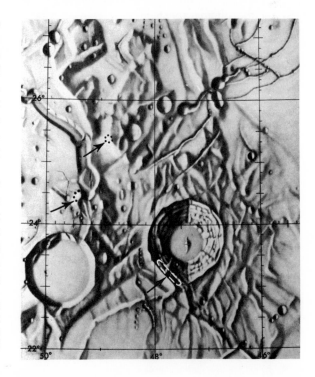

FIGURE 9–4: The arrows show the locations of transient color phenomena observed by Greenacre and Barr at Lowell Observatory on October 29, 1963, in the Aristarchus-Herodotus region. (The base map is ACIC Lunar Chart LAC 39 in draft form after Greenacre.)

be seen visually without the aid of a filter, and lasted approximately 25 minutes.[14] No photographs or spectra were obtained because of the short duration of the phenomenon; but its intensity was at least comparable with that of scattered sunlight and, moreover, seemed to change from minute to minute.

Less than three days later at the Observatoire du Pic-du-Midi, Rackham and I succeeded in securing photographs of another—and much larger—"lunar flare" on the night of November 1–2, which are reproduced in Figure 9–5. The first pair of photographs reproduced (left side) were exposed at 22:45 UT on November 1, within the same minute, in the focal plane of the observatory's 24-inch refractor through two different interference filters: one (bottom) centered at λ 6725 Å in the red, of 45 Å half-width; the other (top) centered at λ 5450 Å off the green, with 95 Å half-width. Photographic emulsions in both cases were the same (Kodak 1-F spectroscopic), as was their subsequent processing in the darkroom. A glance at Figure 9–5 reveals that the "red" photograph (bottom) of a portion of Oceanus Procellarum between the craters Aristarchus, Copernicus, and Kepler showed a striking enhancement of surface brightness of a large area around (and to the north of) the crater Kepler, which is normally of lesser albedo and bluer than the surroundings of Copernicus (to the left on the photographs), and which was totally absent in the green. Shortly after 23:00 UT the red enhancement had virtually disappeared, but it became once more very distinct on four pairs of plates exposed after midnight (between 00:20 and 00:35 UT on November 2), one of which is reproduced in Figure 9–5. The red enhancement of the Kepler region occurred therefore at least twice that night, each time lasting probably no longer than 15–20 minutes; no subsequent night for the rest of the month revealed anything unusual. It was, in particular, cloudy on the Pic on November 27, when Greenacre and Barr reported a recurrence of anomalous reddening of other spots in the proximity of Aristarchus, lasting 1 ¼ hour.[15]

Fuller details of Rackham's and my observations of November 1–2 were adequately reported else-where.[16] Some of their plates were calibrated for photometric purposes; and their microdensitometric analysis revealed that, at the peak of the second brightening, the relative red enhancement of the Kepler area corresponded to a value of $d = 0.86 \pm 0.03$, i.e., several times as large as those observed previously by the line-depth method. The "red spots" of Greenacre and Barr may have been equally intense (and possibly more), but the area covered by them—a few square kilometers of the lunar surface—was relatively minute in comparison with the enhanced area as shown in Figure 9–5, which covered more than 60,000 square kilometers of lunar ground. The transient enhancement shown on this plate represents the largest and most conspicuous example known of lunar luminous phenomena of this kind, and, together with the visual observations at Lowell, represents so far the best evidence we possess of the rapid time variations of these phenomena—all secured in the course of one month!

However, a scrutiny of older literature revealed that phenomena of this type were observed before. The first astronomer who indubitably noticed them on the moon in the neighborhood of the crater Aristarchus (and, as far as we can say now, on the same spots as Greenacre and Barr) was William Herschel,[17] who on April 18–19, 1787, noted there the appearance of spots glowing like "slowly burning charcoal thinly covered with ashes" (he thought these were active volcanoes on the moon). So certain was he of their existence that he invited his royal patron (King George III) to look at them through his telescope! May, 1787, was just about the time of the maximum of the solar cycle (an unusually active one); although no flare observations are, of course, available for that time, the fact that the sun must have been greatly disturbed is attested by the reported visibility (on both April 18 and 19, 1787) of polar aurorae as far south as Padua, Italy—an event which occurs scarcely once in a decade.[18]

[14] J. A. Greenacre, "A Recent Observation of Lunar Color Phenomena," *Sky and Teles.*, Vol. 26 (1963), pp. 316–317.

[15] Greenacre, "Another Lunar Color Phenomenon," *Sky and Teles.*, Vol. 27 (1964), p. 3.

[16] Z. Kopal and T. W. Rackham, "Excitation of Lunar Luminescence by Solar Activity," *Icarus*, Vol. 2 (1963), pp. 481–500; "Excitation of Lunar Luminescence by Solar Flares," *Nature*, Vol. 201 (1964), pp. 239–241.

[17] W. Herschel, "An Account of Three Volcanos in the Moon," *Phil. Trans. Roy. Soc. of Lond.*, Vol. 77 (1787), pp. 229–231.

[18] H. Fritz, "Verzeichniss Beobachteter Polarlichten," *Kosten der Kaiserlichen Akademie der Wissenschaften*, 1873.

FIGURE 9–5: Green (*upper*) and red (*lower*) photographs of the Copernicus-Kepler region at the time of the red enhancement, taken on the night of November 1–2 at 22:45 UT (*left*) and 00:22 UT (*right*). (East is at the top.)

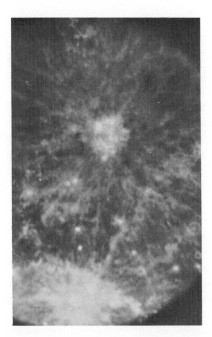

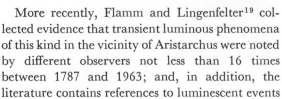

More recently, Flamm and Lingenfelter[19] collected evidence that transient luminous phenomena of this kind in the vicinity of Aristarchus were noted by different observers not less than 16 times between 1787 and 1963; and, in addition, the literature contains references to luminescent events

in the vicinity of several other lunar features. Thus there seems no room for doubt that temporary reddish enhancements represent characteristic recurrent features of the lunar surface demanding explanation.

How to account for them? First, let it be stressed that any explanation in terms of ordinary thermal phenomena are utterly out of the question. No

[19] E. J. Flamm and R. E. Lingenfelter, "On Lunar Luminescence," *Nature*, Vol. 205 (1965), p. 1301.

known matter could possibly heat up (by whatever process) and cool off on the observed scale in one hour or less. The light emission must obviously be nonthermal; but ordinary thermo-luminescence is likewise ruled out by the fact that (unlike lunar monthly temperature changes) the luminous phenomena are not recurrent each lunation, and occur also at times when local temperatures on the spot (i.e., the height of the sun above horizon) are very different. The fact that they seem to recur in certain (and often quite small) places indicates that they favor a certain type of ground; moreover, their spectra are obviously not continuous, but emission seems confined to broad bands concentrated in the yellow and red parts of the spectrum.

A possibility of photo-excitation of gas escaping from lunar interior at the spots near Aristarchus exhibiting the anomalous color phenomena was mentioned recently by Swings,[20] who conjectured that ammonia NH_3 is dissociated by sunlight into NH_2 and H_1 and that the amine radical fluoresces in the red, as is observed in comet tails. The quantitative consequences of such a hypothesis are, however, scarcely admissible any more, for too large a mass of gas would be required to account in this way for the luminous phenomena observed by Greenacre and Barr, and the discharge of such a mass of it into a vacuum would be bound to bring about structural changes of the surrounding landscape which have not been observed. The size of the Kepler enhancement as photographed by Rackham and me makes these difficulties overwhelming, and the fact that such phenomena (particularly around Aristarchus) occur also at night—when no sunlight is available to cause photo-dissociation or luminescence—rules out such a hypothesis altogether.

The manifestly transient nature of the luminescent enhancement is completely uncorrelated with the lunar thermal calendar, and its origin must, therefore, obviously be sought outside—and what, other than the sun, could control external events? The rhythm of the solar cycle has clearly made itself felt in the residual brightness of lunar eclipses (Danjon), and it is only natural to ask to what extent may the "lunar flares" be correlated with other specific aspects of solar activity.

Of these, the ones which immediately come to mind (because of their time scale) are solar flares.

[20] P. Swings, private communication, 1964.

These are known to be as short lived (from minutes to hours) as the observed lunar enhancements; and while they last they emit sufficient amounts of energy—both electromagnetic (X-rays) and corpuscular (mainly fast protons)—to disturb the inner precincts of the solar system for hours and days afterwards. It is, in principle, quite simple to differentiate between the effects produced by their electromagnetic and corpuscular emission: while the effects of X-ray emission, propagating with the speed of light, would reach the moon at virtually the same time as the flare is observed on Earth, particles emitted from the flare may reach us with a time lag of many hours (depending on their speed and directness of trajectories).

For many years now solar disturbances have been monitored by astronomers with close to 100 per cent efficiency. We know, in particular, almost exhaustively what the sun has been doing from day to day in October–November, 1963—from ground-based as well as satellite observations—when so many lunar phenomena have been noted by several observers. Thus the occurrence of the "red spots" near Aristarchus, as observed by Greenacre and Barr on October 30, followed by 48 hours the great flare of October 28; and the Kopal-Rackham photographs of the enhancement of the Kepler region followed the same flare by 118 hours (and a class 1 double flare of November 1 by 8½ hours). Herschel's observations of April 18–19, 1787, coincided also with a period of highly disturbed sun. On the other hand, the Greenacre-Barr observations of November 27 were made at a time when the sun was exceptionally quiet.

Whatever the general case may be, one feature emerges from the foregoing facts quite clearly: namely, that none of the lunar phenomena occurred simultaneously with any major manifestation of solar activity, and cannot, therefore, be due to electromagnetic (X-ray) excitation. When we add to it the fact that the Herschel "volcanoes" on the moon of 1787 (as well as, according to Flamm and Lingenfelter, the majority of the past lunar events) occurred on the dark (night) side of the moon—which could not be reached by direct sunlight—the case against direct electromagnetic excitation becomes overwhelming, and the corpuscular radiation seems to offer the only remaining avenue of approach to the solution of our problem.

This road is feasible in principle, for not only can corpuscular radiation reach the moon after a time lag of the observed order of magnitude (i.e., from several hours to a few days), but particles of sufficient energy (spiraling along the respective lines of force) can impinge on any part of the day—or night—hemisphere of the moon at any time, provided only that the gyro radius ρ of such spirals exceeds that of our satellite. As is well known, this gyro radius ρ is given by the formula

$$\rho = \frac{E}{300H} \, cm \qquad (2)$$

if E stands for the energy of the particles in question in electron volts, and H denotes the strength of the interplanetary magnetic field in gauss.

From recent extensive measurements performed aboard Mariner 2[21] the quantity H is known to fluctuate between 5 and 50γ, with a value of $10\gamma = 10^{-4}$ gauss representing a fair average. If so, however, the radius ρ of gyration for 1 Mev protons proves to be approximately equal to 3×10^7 cm, increasing to 3×10^8 cm (i.e., about twice the radius a of the moon) for $E = 10$ Mev, and 3×10^9 cm (i.e., about $17a$) for 100 Mev protons. Therefore, it is not until energies are close to 10 Mev that ρ attains the dimensions of the lunar globe and exceeds it for $E > 10$ Mev (i.e., for proton velocities in excess of 50,000 km/sec).

However, when we come to consider quantitative aspects of corpuscular stimulation of lunar luminescence, the situation becomes less clear. The 1963 observations by Rackham and me of the large enhancement of the Kepler region, which temporarily almost doubled its surface brightness in the red, indicated that the energy flux F of energy stimulating luminescent emission should have been of the order of 10^5 ergs/cm² sec;[22] and for typical observations of luminescence by the "line-depth method," it could have been down to 10^3–10^4 ergs/cm² sec. If protons alone had been responsible for its excitation, the energy balance would require that the ratio

$$\frac{F}{v} = \frac{1}{2} m_H v^2 N, \qquad (3)$$

where m_H denotes the mass of a proton; v, their velocity; and N, their number per cm³. For $m_H = 1.67 \times 10^{-24}$ g and $F = 10^5$ ergs/cm² sec, the

foregoing equation requires that the product $Nv^3 \sim 10^{29}$ sec⁻³ for the November 1–2, 1963, events, and about 10^{27} sec⁻³ for smaller events observed spectroscopically through narrower passbands.

These numbers seem rather large. For $v = 5000$ km/sec (about the maximum velocity of "slow" protons emitted by solar flares) the corresponding particle density N should be of the order of 10^3 cm⁻³ for the maximum flux of 10^5 ergs/cm² sec, and 10–100 per cm³ for more moderate events. Now the velocity of the quiet-sun solar wind (as measured during the historic flight of Mariner 2 in 1962, for instance) is only about 400–500 km/sec, and the particle density N only about 0.1 per cm³. During storm conditions both the values of v and N may increase, perhaps, 10 times; but even this leaves us with values of N which are 10–100 times too small—a fact which either discloses that the actual solar activity may still have surprises in store for us, or that solar particles may not be the primary source of energy for lunar luminescence emission, but merely trigger its release; we cannot as yet be sure.

If we turn our attention to the lunar night events—when luminescence has to compete in contrast only with earthshine which is (other things being equal) 10^4 times less intense than direct sunlight—luminescent glow concentrated in emission bands of 100–1000 Å width and evoked by incident energy flux as low as 1–10 ergs/cm² sec could become visually observable. Only protons with energies in excess of 10 Mev (i.e., velocities larger than 50,000 km/sec) can, to be sure, follow trajectories in interplanetary space which are curved enough to enable them to impinge on the dark side of the moon, and a class of fast particles emitted by solar flares are known indeed to move with speeds clustering around 150,000 km/sec.[23] A flux of 1–10 ergs/cm² sec of such particles would (for $v = 1.5 \times 10^5$ km/sec) correspond by (22-3) to a density N of the order of 10^{-7} to 10^{-6}. Now solar-flare proton events in the 10–100 Mev range do exhibit fluxes of 10^3–10^4 particles per cm² per sec—the one following the flare of November 12, 1960, attained almost 10^5 particles/cm² sec—[24]

[21] P. J. Coleman, Jr., et al., "Interplanetary Magnetic Fields," *Science*, Vol. 138 (1962), pp. 1099–1100.

[22] Kopal and Rackham, "Excitation of Lunar Luminescence."

[23] M. A. Ellison, "Solar Flares and Associated Phenomena," *Plane. and Space Sci.*, Vol. 11 (1963), pp. 597–619.

[24] P. S. Freier and W. R. Webber, "Exponential Rigidity Spectrums for Solar-Flare Cosmic Rays," *J. Geophys. Res.*, Vol. 68 (1963), pp. 1605–1630.

corresponding to $N \sim 10^{-7}–10^{-6}$ particles per cm³ or even more. This indeed seems to be of the right order of magnitude to account for the visibility of night luminescence on the moon, but the daytime luminescence still seems anomalously large.

This suspicion is strengthened by some events of the last solar cycle when, following the great flare (class 3+) of July 7, 1958, Blackwell and Ingham observed at Chacaltaya (in the high Bolivian Andes) a temporary reddening and general increase in brightness of the zodiacal light and obtained evidence which satisfied them that the extra emission came from the interplanetary dust cloud. This phenomenon could again be scarcely understood as other than luminescence produced by the corpuscular output of that flare; but for an assumed particle density of 300 protons per cm³ their velocity had to be in excess of 40,000 km/sec to make the luminescent process less than 100 per cent efficient.

Moreover, Blackwell and Ingham[25] pointed out that the surface brightness of the zodiacal cloud—another example of solid matter illuminated by the sun in vacuum, and from greater proximity than the moon—seems correlated with the planetary magnetic number K_p, and increased by 40 per cent when K_p was between 9 and 10. The relevance of the events represented by this number also to lunar luminescence has recently been emphasized by Cameron,[26] who pointed out that a daytime luminescence as intense as that photographed by Rackham and me on November 1–2, 1963, could have been caused, not by the direct impact of primary solar particles associated with a flare, but indirectly by their effect on the terrestrial magnetosphere. The interaction between solar wind and the earth's magnetosphere is very complex, but it seems that in solar direction the pressure of this wind compresses the magnetosphere, and in the antisolar direction causes the formation of a long cavity in which the terrestrial magnetic field can expand. Since the velocity of the solar wind relative to the earth is supersonic, a standing shock must be produced beyond the magnetopause, causing extensive particle acceleration, or it may be that lunar luminescence is produced by particles

trapped in the distant tail of the magnetosphere. Cameron pointed out that, in all cases of intense daytime luminescence on record, the moon was never very far from full—which may be an important factor in the explanation; but it may also be that principal concentrations of suspected luminophors (around Aristarchus or Kepler, for instance) are localized in regions which do not become sunlit till shortly before full moon; we cannot as yet be sure. But more recent studies by Dodson and Hedeman[27] indicate the existence of an unexpected correlation between the proton and neutron events of solar origin and the lunar cycle, and, if so, our magnetosphere may indeed have something to do with it; but more work remains to be done before this suggestion can be placed on a more secure basis.

In summary, we may say that the existence of a transient luminescence of the lunar surface—long suspected from several independent lines of evidence—should now be accepted as an established fact. It seems to recur preferentially in certain parts of the lunar surface, and its spectrum is not continuous, but confined to certain emission bands clustering toward the red. A transient emission of intensity amounting to a small percentage of that of the adjacent continuum in narrow bandwidths seems, moreover, to be of fairly frequent occurrence. Instances of enhancements amounting from 30 to 80 per cent (and possibly more) are on record, but their frequency can only be guessed at.

Secondly, it is quite clear now that all known aspects of this luminescence seem (directly or indirectly) related with some activity of the sun. A correlation between the two may not be simple and very probably there is no mere proportionality with the sunspot numbers. Flamm and Lingenfelter[28] rightly contend that the majority of past lunar events occurred in the years of low sunspot numbers; the Greenacre-Barr and Kopal-Rackham events of October–November, 1963, are eloquent instances of this fact. But (as was shown by Danjon) the abrupt change in residual brightness of the eclipsed moon occurs also, not at the maximum, but the minimum of solar activity.

The establishment of one-to-one correlation between the corresponding solar and lunar events

[25] D. E. Blackwell and M. F. Ingham, "Observations of the Zodiacal Light from a Very High Altitude Station, III: The Disturbed Zodiacal Light and Corpuscular Radiation," *Mon. Not. Roy. Astron. Soc.*, Vol. 122 (1961), pp. 143–155.

[26] A. G. W. Cameron, "Particle Acceleration in Cislunar Space," *Nature*, Vol. 202 (1964), p. 785.

[27] H. W. Dodson and E. R. Hedeman, "An Unexpected Effect in Solar Cosmic Ray Data Related to 29.5 Days," *J. Geophys. Res.*, Vol. 69 (1964), pp. 3965–3971.

[28] Flamm and Lingenfelter, "On Lunar Luminescence."

still represents a goal facing us at some distance in the future; but from an obvious lack of simultaneous occurrence it is clear that luminescence on the moon is not stimulated by solar electromagnetic (X-ray), but corpuscular, radiation which reaches the moon after a transit time ranging from several hours to a few days. At daytime—in the absence of any appreciable magnetic field—particles of all energies can impinge on the lunar surface, while at night only energetic particles ($E > 10$ Mev) can do so. Their numbers, as measured by independent experiments, are just about adequate to produce visible luminescence at night; but for daytime events the number of direct solar particles seems deficient by two or three orders of magnitude, so that some secondary accelerating (or storage) facilities may have to be considered. Besides, there are other problems to face: for instance, if the luminescence observed on October 30, 1963, by Greenacre and Barr, or on November 1–2 by Rackham and me, was indeed excited by corpuscular emission of the class-3 flare of October 28, is it possible that these particles could have remained so long compacted in space as to give rise to transient phenomena lasting no more than 20–30 minutes after a time lapse of 48 or 118 hours?

Third, the luminescence of the moon is observed to recur in certain regions often limited to a few kilometers in size, suggesting strong localization of the luminophor. So far we can only guess at its chemical structure; and eventual identification of its mineralogical constituents may, in due course, provide a valuable tool for the petrographic prospecting of the outer crust of the moon at a distance. The argument that any appreciable quantum efficiency of this process would have been quenched by radiation damage over long intervals of time is belied by the demonstration that meteorites luminesce in the laboratory when excited at first by artificial corpuscular bombardment,[29] in spite of the fact that, while in space, they must have suffered as much radiation damage (from cosmic rays, etc.) as anything that lies exposed on the lunar surface. Like the laboratory scientist, Nature probably resorts to the same restorative of the luminescent power: namely, periodic heating of the sample; and on the moon this is no doubt taken care of by the diurnal cycle.

[29] C. J. Derham and J. E. Geake, "Luminescence of Meteorites," *Nature*, Vol. 201 (1964), pp. 62–63.

10.

A Critical Analysis of Lunar Temperature Measurements in the Infrared*

Hector C. Ingrao, Andrew T. Young, and Jeffrey L. Linsky

Harvard College Observatory, Cambridge, Massachusetts

I. INTRODUCTION TO THE OBSERVATIONAL PROBLEM

The type of lunar temperature measurements made and the methods used will be determined solely by the information we want to obtain from the data; the choice of high or low spatial resolution, high or low temperature resolution, and relative or absolute measurements can be made only on the basis of the answers we are seeking from our program. For example, to correlate radio with infrared measurements, instruments with resolution elements of several minutes have been used. With these instruments it is not a problem to locate the area under measurement within a fraction of the resolution element. However, if the program calls

* This work was sponsored by the National Aeronautics and Space Administration under grant NsG 64–60. A detailed analysis of the material presented in this paper is described in H. C. Ingrao, A. T. Young, and J. L. Linsky, "A Critical Analysis of Lunar Temperature Measurements in the Infrared," *Scientific Report No. 6*, Cambridge, Mass.: Harvard College Observatory (1965).

The authors wish to thank Mr. Roger Moore, former Head, Planetary Sciences, NASA, for his continued encouragement throughout the development of the first part of this program; Dr. A. D. Thackeray, director of the Radcliffe Observatory (Pretoria, Republic of South Africa) for allowing us to use the 74-inch reflector to observe the lunar eclipse on June 24–25, 1964; Mr. S. Roland Drayson of the University of Michigan for his assistance with the CO_2 band absorption data; and the members of the Infrared Laboratory at Harvard College Observatory for their assistance during various phases of this project. Thanks are also due to Mr. Richard Munro for supplying the original version of the computer program to integrate heat conduction equations.

The authors are particularly indebted to Professor Donald H. Menzel for his continuous interest in this phase of the project as well as for many fruitful discussions.

for high spatial resolution and one expects to correlate the thermal features with the lunar features, the observational and instrumental problem is completely different. For example, if the resolution element is $9'' \times 9''$ in size, its accurate location is very difficult and, in fact, requires us to solve an astrometric problem.

At high spatial resolution the stability requirements of the telescope also are more demanding. If the moon is being scanned, we must be able to determine the position of the resolution element on the lunar surface as a function of time and within a fraction of the resolution element size.

A scan during full moon and through the subsolar point must handle a signal power ratio of the order of 270, if we assume two extreme temperatures, 150°K and 400°K. If the noise level of the pyrometer is of the order of 5×10^{-11} watts and if we want the measurements to be limited only by this noise level, we have to resolve 3 parts in 1,000 at 400°K.

The probable error in the absolute temperature measurements will depend mainly on the accuracy of the measurements of the instrumental parameters and of the atmospheric attenuation. For measurements of relative temperatures, knowledge of the instrumental parameters is not so important and less accuracy is required in measurements of the atmospheric attenuation.

In our program of lunar temperature measurements, we have designed and built at Harvard College Observatory a radiation pyrometer.[1] This

[1] H. C. Ingrao and D. H. Menzel, "Radiation Pyrometer for Lunar Observation," *Scientific Report No. 4*, Cambridge, Mass.: Harvard College Observatory (1964).

instrument was designed to provide high spatial resolution, high accuracy in locating the resolution element on the lunar surface, and the highest temperature resolution that can be achieved with thermal detectors. This instrument has three channels: infrared, visual, and photographic. The three channels use the same telescope optics; an optical switching mechanism allows one to observe the moon 50 per cent of the time in the infrared and 50 per cent of the time in the visual range and photographically, at an adjustable rate from 10 cps to 70 cps. The pyrometer has the following data outputs: (1) a 35-mm film having a field of view 7.5' × 5.0', with a crosshair centered on the frame, which coincides with the barycenter[2] of the infrared detector; (2) a paper chart on which are recorded the infrared signal, marks indicating the time, marks indicating when a picture has been secured, and a mark indicating an event that needs to be recorded; and (3) a magnetic tape recording on which the observer describes the area under measurement or any event important to the data analysis, and the operator of the electronic equipment records relevant data; the tape also records WWV time signals and a tone indicating that an event needs to be recorded.

II. IDENTIFICATION OF THE RESOLUTION ELEMENT ON THE LUNAR SURFACE

Schemes can be devised to identify the resolution element on the lunar surface and, without discussing the advantages and disadvantages of the several methods used, we will describe the one used in our pyrometer. The basic principle is indicated in Figure 10–1. The telescope beam is chopped at a rate which can be adjusted from 10 cps to 70 cps. The chopper is made of glass and has evaporated aluminum on the front face and evaporated gold on the back.

When the chopper blocks the optical path to the detector in position D, the focal plane d-d' is transferred to the position k'-k. We can call the point R the homologue of the point D that defines the geometrical center of the detector. Coplanar with the image plane k'-k is a reticle with ½-mm divisions, illuminated at the edge. A flat mirror, E,

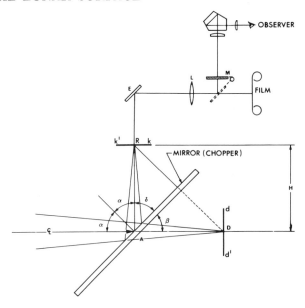

FIGURE 10–1: Basic principle of the mechanism for identifying the resolution element on the lunar disk in the radiation pyrometer developed at Harvard College Observatory.

folds back the beam and the photographic objective, L, images the plane k'-k onto the photographic film. A reflex system with mirror M and a pentaprism allows for visual observation of the same field of view.

Since we assume that the position of D is known for the alignment in the laboratory, a systematic error is introduced in the positioning of the reticle.

To determine the systematic error at the telescope, we scan the detector several times with a star image in a direction parallel to the diagonals (square detector) until we get the maximum response. If we correlate photographically the position of the star image on the reticle with maximum signal output from the detector, we can measure the systematic error very accurately.

With pictures of the moon taken with the photographic channel of the pyrometer, the task is to determine the orthographic co-ordinates of the reticle, which, in turn, are the co-ordinates of the barycenter of the detector. We obtain this information by projecting each frame on the proper plate of the *Orthographic Atlas of the Moon*.[3] A special projecting system has been constructed in our laboratory for this purpose.

The accuracy in determining the orthographic

---

[2] We define the barycenter as the point in the detector with the highest responsivity.

[3] *Orthographic Atlas of the Moon*, Tucson: University of Arizona Press (1961).

co-ordinates by this method depends upon the quality of the raw data. For pictures obtained under good conditions (seeing disk 2″ or smaller), having good contrast and a reasonable number of identifiable lunar features, the standard deviation in the determination of the orthographic co-ordinates is ±3″. For pictures obtained under poor seeing conditions (5″ or larger), having poor contrast and few identifiable features, the standard deviation will be between 6″ and 8″. Usually we take about four pictures per scan; thus, in the best case we know the position of the line of scan within 1.5″, and in the worst within 4.0″.

The previous discussion regarding the location of the resolution element applies only to the illuminated part of the moon. On the basis of these measurements, we can also determine the position of the line of scan on the shadowed areas, as shown in section III. The problem of location on the eclipsed moon is more complicated, especially during dark eclipses; yet in some cases it is possible to take the identifying pictures with high-speed film or an image converter, or to take the pictures at the limb of the moon against the star background.

III. SPATIAL RESOLUTION

In our radiation pyrometer the size of the resolution element will be limited by the telescope optics and the detector size. To obtain the instrumental profile experimentally we scanned the field of view of our radiation pyrometer with the image of a star, Alpha Scorpii, and the pyrometer mounted at the Newtonian focus of the 74-inch telescope in Pretoria (Republic of South Africa). In Figure 10–2, (a) shows the photographic record; each small division of the reticle is equivalent to 11″. Frames 1–4 show Alpha Scorpii crossing the field along the horizontal line of the reticle. The time when the picture was secured is indicated in the upper part of each frame. The output of the radiometer when Alpha Scorpii scanned the detector is shown in (b). The marks 1, 2, 3, and 4

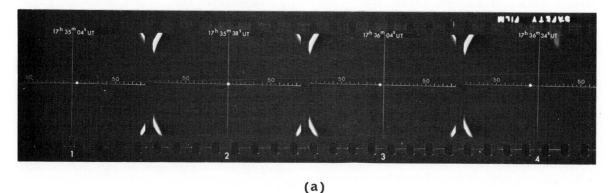

(a)

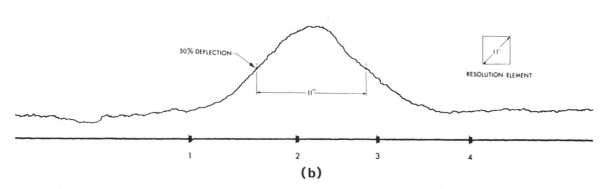

(b)

FIGURE 10–2: Measurement of the radiation pyrometer profile: (a) frames 1–4 show Alpha Scorpii crossing the field of view of the photographic channel of the radiation pyrometer along one line of the reticle; (b) output of the radiometric channel when Alpha Scorpii crossed the field of view.

indicate the times at which the identifying pictures were secured. The measurements were taken nearly along the diagonal of the square detector.

A series of similar scans was obtained parallel to the horizontal line of the reticle and spaced only a few seconds of arc apart. Using these measurements we determine the size of the resolution element, the responsivity diagram of the pyrometer-telescope combination, as well as any systematic error in the location of the barycenter of the detector with respect to the center of the reticle.

IV. ASTROMETRIC ANALYSIS OF THE OBSERVATIONAL DATA

From the different modes of scanning that can be used, we choose to fix the telescope relative to the earth and let the moon drift in the field of view of the detector. In this method, the spatial-time relationships can be easily determined if no perturbations are introduced in the mode of scanning. Under actual observing conditions, this is not always true; if we use a telescope with a plate scale of $25''/mm$ and a resolution element of $10'' \times 10''$, a displacement of 400μ at the end of the telescope will be equivalent to the displacement of one resolution element. With a telescope of 60 inches or larger, a displacement of 400μ can be produced by the wind, as well as by perturbations introduced by the observer and displacement of the telescope mirror objective.

For our modes of scanning, we analyzed the astrometric problem and worked out a computer program to give the orthographic co-ordinates of the resolution element as a function of time. It is clear from the physical situation that if the resolution element moves at a constant rate in each co-ordinate, the entire course of the scan relative to the moon, as a function of time, can be found from two timed photographs. If the telescope is held fixed relative to the earth, the rates are known exactly and only one photograph is needed.

The computer program logically consists of three blocks. The first block reads in the observer's co-ordinates, the Lunar Ephemeris, and other fixed data needed to solve the problem. The second block computes the hour angle and declination of the resolution element at the time of every photograph. The third block uses the photographic data to interpolate the hour angle and declination, and

hence the lunar co-ordinates of the resolution element, at any arbitrary time, for example, a time at which a temperature datum is measured.

In addition to the Lunar Ephemeris, the basic data consist of the observer's astronomical latitude, longitude, and height above sea level, and the effective wavelengths of the photographic and infrared detectors. The wavelengths are needed to compute the differential refraction due to atmospheric dispersion.

Ordinarily, the observer's co-ordinates are read first. The second set of fixed data consists of the effective wavelengths λ_P of the photographic and λ_D of the infrared detection systems; these are used to compute the differential refraction due to atmospheric dispersion between the two wavelengths.

The remaining fixed data consist of tables from the Astronomical Ephemeris. Three tables are required: the radial ephemeris (semidiameter and horizontal parallax, tabulated for every 0.5 day of ET); the geocentric angular ephemeris (apparent right ascension and declination, tabulated for every hour of ET); and the physical ephemeris (the earth's selenographic longitude and latitude, the sun's selenographic colongitude and latitude, and the position angle of the lunar axis, tabulated for 0^h UT).

Values required from the tables are interpolated to second differences by a subroutine which has special provisions for interpolating correctly across the discontinuity of 2π radians occurring in tabular values of angles that pass through zero without changing sign.

All data are checked for consistency and plausibility. No minutes or seconds greater than 60, no negative right ascension, no declinations greater than 90 degrees are acceptable. The tables must also agree as to month and year. As a final check, all fixed input data are printed out again so that a visual check can be made against the Ephemeris.

At this point, the first block of the program is completed. We now take up the second block, which computes the hour angle (h) and declination (δ) of the resolution element for each photograph.

The input data for each photograph are the Universal Time and the orthographic lunar co-ordinates ξ and η of the intersection of the cross-hairs, which is the optical conjugate of the detector. The Ephemeris Time, which is the argument of several tables, is computed from the Universal Time by adding the correction ΔT (currently 35″).

The first step in finding the topocentric h and δ of the point photographed is to find the topocentric co-ordinates of the moon's center—the topocentric librations. To do this, we first need the geocentric hour angle and declination of the center of the moon, referred to the local meridian.

The geocentric right ascension and declination of the moon's center, α_G and δ_G are interpolated from the tables, using the ET as argument. The geocentric hour angle is then computed.

At this point, the cosine of the geocentric zenith distance ($\mathcal{Z}_G$) is computed. If this value is negative, the moon is below the horizon at the specified time; the program gives an error message and terminates. We compute the sine of the geocentric zenith distance ($\mathcal{Z}_G$) to evaluate the differential corrections from geocentric to topocentric librations. The topocentric librations are computed by Atkinson's method.[4]

We first compute the "topocentric parallax" π_T, i.e., the angle subtended at the center of the moon between the observer and the center of the earth; this angle represents the difference between the geocentric and the topocentric lines of sight to the moon's center. For this calculation we can neglect the flattening of the earth; since the moon is about 60 Earth radii away, the error in π_T is less than $1/297 \times 1/60 \cong 1/18000$ radian. For our application, this error in the libration is reduced on the sky by an additional factor of about 240, since as seen from the earth the moon's radius is about $1/240$ of a radian. Thus the error resulting from this approximation is about 2×10^{-7} radian, or 0.05″, which can be neglected. All other approximations made by Atkinson in the formula for π_T produce smaller errors. The resulting formula for π_T is

$$\pi_T = \pi_G \sin \mathcal{Z}_G (1 + 0.0168 \cos \mathcal{Z}_G), \qquad (1)$$

where π_G is the geocentric horizontal lunar parallax interpolated from the table.

The sine and cosine of the parallactic angle Q are next found from Atkinson's formulae. The geocentric position angle of the moon's axis (C_G) is interpolated from the tables and the difference

[4] R. d'E. Atkinson, "The Computation of Topocentric Librations," *Mon. Not. Roy. Astron. Soc.* Vol. 111 (1951), pp. 448–454.

angle ($Q - C_G$) in Atkinson's formulae is computed.

The geocentric libration in latitude (the selenographic latitude of the earth) b_G is interpolated from the tables, so that it can be used in finding the topocentric libration in longitude (the selenographic longitude of the observer), ℓ_T.

Atkinson's formulae are intended to give an accuracy of 0.01 degree on the moon, which is about 0.15″ on the sky as seen from the earth. This is also the accuracy of the tables in the Ephemeris. Finally, the topocentric position angle of the lunar axis C_T is computed. The resulting error in C_T is less than 2×10^{-4}, or again about 0.15″ on the sky. Even if we have photographs from one limb only, the error at the opposite limb cannot exceed 0.3″, which is clearly acceptable for our work.

The topocentric hour angle (h_T) and declination (δ_T) of the lunar center are then calculated, by use of the auxiliary quantities given on p. 60 of the Explanatory Supplement and two other auxiliary quantities.

The next problem is to find the topocentric co-ordinates h_P and δ_P of the photographed point with orthographic co-ordinates (ξ, η).

This information suffices to determine the line of scan across the moon. However, we need some additional geometric information for the interpretation of the data, and the refraction correction is not yet included. We first compute the air mass, m, by the relation

$$m = \sec \mathcal{Z}_T [1 - 0.0012(\sec^2 \mathcal{Z}_T - 1)] \qquad (2)$$

which is accurate to 0.002 at $\mathcal{Z}_T = 75°$.

The differential refraction corrections are obtained by using the standard relations

$$h - h' = r \sec \delta \sin Q,$$

$$\delta - \delta' = -r \cos Q, \qquad (3)$$

where the primes denote the refracted co-ordinates and r is the refraction correction. In our case, we are correcting only for differential refraction. Using topocentric instead of geocentric co-ordinates, we obtain

$$h'_P = h_P - r' \sin h_P \cos \phi,$$

$$\delta'_P = \delta_P + r'(\sin \phi - \cos \mathcal{Z}_T \sin \delta_P), \qquad (4)$$

where

$$r' = \Delta(n - 1) \sec \mathcal{Z}/\cos \delta. \qquad (5)$$

Equation (4) gives the topocentric hour angle and declination of the resolution element at the time of the photograph, corrected for differential refraction. We do not correct for the whole refraction because only the *nonlinear* part of the change in refraction during a scan affects our results. On the assumption that the tangent term alone is a satisfactory representation, the second derivative of the refraction correction as a function of Z is $2(n-1)$ sec Z^2 tan Z. The error contributed by neglecting this term is $2(n-1)$ sec Z^2 tan $Z(\Delta Z)^2$; it amounts to about a second of arc for a scan one degree long at $Z = 75$ degrees, and decreases rapidly toward the zenith. In practice, the telescope moves only a fraction of a degree, or is stationary, during a scan, so that the curvature of the total refraction is not important. Ignoring the total refraction also simplifies the program by making accurate refraction corrections in both the second and third parts of the program unnecessary; only an approximate correction is needed for the second part, as explained above, and from this point on the refraction can be forgotten.

The only remaining task of the second part of the program is to compute the geometrical relations between the observer, the sun, and the point observed. The angles required are the *phase angle* (the angle at the point observed between the line of sight and the direction of illumination), the elevations of the sun and earth (observer) above the lunar horizon at the point observed, and the difference in azimuth between the sun and observer.

This completes the calculations for the second block of the program. The results are converted to practical units (degrees instead of radians) and printed out. The program remembers the values of h'_P and δ'_P, and looks for another record of photographic data.

When all the photographs referring to a single scan have been analyzed in this manner, the program goes on to the third block.

The first problem is to determine the course of h'_P and δ'_P with time. In some cases, we know that the telescope has been held fixed, so that h'_P and δ'_P must be constant. In other cases, we know that the telescope was moved and that time-dependent terms must be included. Occasionally we are not certain whether the telescope was moving or not, and we must ask the program to decide on the basis of the available data. In general, we cannot assume that motion takes place in only one co-ordinate, even if only one axis of the telescope is moving; for the effects of differential refraction, polar-axis error,[5] and perturbations on the telescope will in general produce a displacement in both co-ordinates.

We first compute the means of all the times, the

[5] We refer to the error in the alignment of the polar axis of the telescope.

LUNAR SCAN GEOMETRY

1964	DEC	U.T.	FRAME NO.	XI	ETA	HOUR ANGLE	DECLINATION	AIR MASS	ELEVATION OF EARTH	SUN	EARTH AZIMUTH FROM SUN	PHASE ANGLE	SCAN NO.
D	H M S												
19	1 21 50.18		181	0.828	0.178	-3 19 27.2	23 19 11.2	1.419	31.6	32.0	0.2	0.5	37
19	1 22 11.26		182	0.547	0.183	-3 19 27.4	23 19 18.7	1.419	54.3	54.6	0.4	0.4	37
19	1 22 34.21		183	0.242	0.177	-3 19 27.8	23 19 15.0	1.419	72.2	72.4	0.8	0.3	37
19	1 22 55.18		184	-0.032	0.177	-3 19 27.8	23 19 17.0	1.419	79.7	79.6	1.0	0.2	37

NOTE -- 'EARTH' MEANS OBSERVER (AGASSIZ STA)

COORDINATES AT MID-SCAN, 1 22 22.7 U.T.

SUBSOLAR POINT XI	ETA	CENTER OF DISC XI	ETA	HOUR ANGLE	DECLINATION
				H M S	O ' ''
-0.002	-0.002	-0.006	0.001	-3 19 27.5	23 19 15.4

DIFFERENTIAL REFRACTION CORRECTIONS   0.1   0.5

R.M.S. RESIDUAL IN POSITION 4.76 ARCSEC FROM DRIFT

RESIDUALS IN H.A. (H) AND DEC (D) FOLLOW

FIGURE 10-3a: Computer read-out of the astrometric analysis of the photographs of a scan during the eclipse of December 18-19, 1964.

h'_P's and the δ'_P's. For convenience, these means, together with the corresponding subsolar point and topocentric disk center (computed as described above), are printed out (Figure 10–3a.) The differential variables

$$\Delta h = h'_P - \bar{h},$$

$$\Delta \delta = \delta'_P - \bar{\delta},$$

$$\Delta t = t - \bar{t},\tag{6}$$

are then computed for each photograph. If the telescope was known to be fixed, a control card sets a switch in the program and $\bar{h}$ and $\bar{\delta}$ are adopted for all times. The values of Δh and $\Delta \delta$ are then regarded as residuals. On the other hand, if the telescope was either known or suspected to be moving, equations of the form

$$\Delta h = \frac{dh}{dt}\Delta t,$$

$$\Delta \delta = \frac{d\delta}{dt}\Delta t,\tag{7}$$

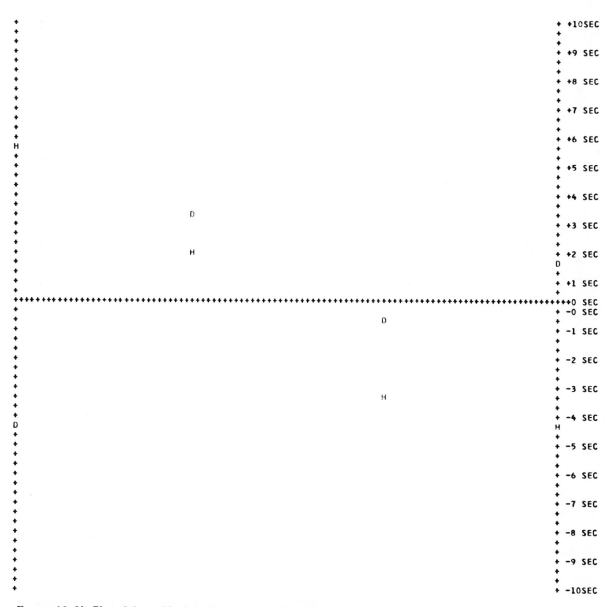

FIGURE 10–3b: Plot of the residuals in hour angle and declination as a function of time.

EPHEMERIS FOR SCAN 37, USING DRIFT METHOD BASED ON 4 POINTS.

1964	DEC	U.T.	XI	ETA	AIR MASS	ELEVATION OF EARTH	SUN	EARTH AZIMUTH FROM SUN	PHASE ANGLE	
D	H	M	S							
19	1	21	37.51	0.989	0.185	1.419	0.	0.	0.	0.	OFF LIMB
19	1	21	37.72	0.986	0.185	1.419	0.	0.	0.	0.	OFF LIMB
19	1	21	37.92	0.984	0.185	1.419	0.	0.	0.	0.	OFF LIMB
19	1	21	38.13	0.981	0.185	1.419	0.	0.	0.	0.	OFF LIMB
19	1	21	38.33	0.978	0.185	1.419	0.	0.	0.	0.	OFF LIMB
19	1	21	38.54	0.981	0.184	1.419	3.1	3.6	0.2	0.5	
19	1	21	38.74	0.978	0.184	1.419	5.2	5.7	0.2	0.5	
19	1	21	38.90	0.975	0.184	1.419	6.4	6.9	0.2	0.5	
19	1	21	39.10	0.972	0.184	1.419	7.7	8.1	0.2	0.5	
19	1	21	39.36	0.969	0.184	1.419	8.9	9.4	0.2	0.5	

D	H	M	S	XI	ETA	AIR MASS	EARTH	SUN	AZ	PHASE	
19	1	24	3.33	-0.913	0.171	1.419	21.8	21.8	0.1	0.1	
19	1	24	3.59	-0.917	0.171	1.419	21.3	21.3	0.1	0.1	
19	1	24	3.79	-0.919	0.171	1.419	20.9	20.9	0.1	0.1	
19	1	24	3.95	-0.921	0.171	1.419	20.5	20.6	0.1	0.1	
19	1	24	4.15	-0.924	0.171	1.419	20.1	20.2	0.1	0.1	
19	1	24	4.36	-0.926	0.171	1.419	19.7	19.7	0.1	0.1	
19	1	24	4.56	-0.929	0.171	1.419	19.2	19.3	0.1	0.1	
19	1	24	4.77	-0.932	0.171	1.419	18.7	18.8	0.1	0.1	
19	1	24	4.97	-0.935	0.171	1.419	18.3	18.3	0.1	0.1	
19	1	24	5.13	-0.937	0.171	1.419	17.9	17.9	0.1	0.1	
19	1	24	5.33	-0.939	0.171	1.419	17.4	17.5	0.1	0.1	
19	1	24	5.54	-0.942	0.171	1.419	16.9	17.0	0.1	0.1	
19	1	24	5.74	-0.944	0.171	1.419	16.4	16.4	0.1	0.1	
19	1	24	5.95	-0.947	0.171	1.419	15.8	15.9	0.1	0.1	
19	1	24	6.15	-0.950	0.171	1.419	15.3	15.3	0.1	0.1	
19	1	24	6.41	-0.953	0.171	1.419	14.5	14.6	0.1	0.1	
19	1	24	6.56	-0.955	0.171	1.419	14.1	14.1	0.1	0.1	
19	1	24	6.77	-0.958	0.171	1.419	13.4	13.5	0.1	0.1	
19	1	24	6.97	-0.961	0.171	1.419	12.7	12.8	0.1	0.1	
19	1	24	7.18	-0.963	0.171	1.419	12.0	12.1	0.1	0.1	
19	1	24	7.38	-0.966	0.171	1.419	11.3	11.4	0.1	0.1	
19	1	24	7.59	-0.969	0.171	1.419	10.5	10.6	0.1	0.1	
19	1	24	7.79	-0.971	0.171	1.419	9.6	9.7	0.1	0.1	
19	1	24	8.00	-0.974	0.171	1.419	8.7	8.7	0.1	0.1	
19	1	24	8.21	-0.977	0.171	1.419	7.6	7.7	0.1	0.1	
19	1	24	8.41	-0.979	0.171	1.419	6.3	6.4	0.1	0.1	
19	1	24	8.62	-0.982	0.171	1.419	4.8	4.8	0.1	0.1	
19	1	24	8.77	-0.984	0.171	1.419	3.2	3.3	0.1	0.1	
19	1	24	8.97	-0.987	0.171	1.419	0.	0.	0.	0.	OFF LIMB
19	1	24	9.18	-0.989	0.171	1.419	0.	0.	0.	0.	OFF LIMB
19	1	24	9.38	-0.992	0.171	1.419	0.	0.	0.	0.	OFF LIMB
19	1	24	9.59	-0.995	0.171	1.419	0.	0.	0.	0.	OFF LIMB
19	1	24	9.79	-0.998	0.171	1.419	0.	0.	0.	0.	OFF LIMB
19	1	24	10.00	-1.000	0.171	1.419	0.	0.	0.	0.	OFF LIMB
19	1	24	10.21	-1.003	0.171	1.419	0.	0.	0.	0.	OFF LIMB
19	1	24	10.41	-1.006	0.171	1.419	0.	0.	0.	0.	OFF LIMB
19	1	24	10.62	-1.008	0.171	1.419	0.	0.	0.	0.	OFF LIMB
19	1	24	10.82	-1.011	0.171	1.419	0.	0.	0.	0.	OFF LIMB

FIGURE 10–3c: Sample of the astrometric reduction of data points near each limb. All the computations pertain to the observational data shown in Figures 10–4a and 10–4b.

are fitted by least squares. Since all variables are measured from their means, no constant term is required and the least squares solution reduces to

$$\frac{dh}{dt} = \frac{\Sigma \Delta t \Delta h}{\Sigma (\Delta t)^2},$$

$$\frac{d\delta}{dt} = \frac{\Sigma \Delta t \Delta \delta}{\Sigma (\Delta t)^2}. \tag{8}$$

The residuals

$$\Delta h' = \Delta h - \frac{dh}{dt} \Delta t,$$

$$\Delta \delta' = \Delta \delta - \frac{d\delta}{dt} \Delta t, \tag{9}$$

are then computed.

In a large-volume automatic data-reduction program it is extremely important to reject faulty data, since enormous residuals frequently result from errors in transcription and keypunching, dropped minus signs, digit transpositions, and the like. Therefore the residuals are scanned to see whether any point falls more than $10''$ from its calculated position. If no point falls outside this tolerance limit, normal processing continues. If one or more points fall outside the rejection limit, the point with the *largest* residual in position $r^2 = [(\Delta h \cos \delta)^2 + \Delta \delta^2]$ is rejected, the residuals in both co-ordinates are printed with an error message, new means are taken of the remaining data, and the whole reduction process is repeated from that point on.

When a satisfactory set of residuals is obtained, the root-mean-square residual in position (both co-ordinates combined) is calculated and printed out, together with a graph of the residuals as a function of time (see Figure 10–3b).

If the program has been asked to decide whether or not the telescope was moving, it uses the following precepts. (1) The calculation is originally carried out on the assumption of telescope motion. (2) If, after bad data are rejected, only one or two points remain, the telescope is assumed to have been stationary, on the grounds that no test of the significance of dh/dt and $d\delta/dt$ is possible. The entire calculation from equations (6)–(9) is repeated on this basis. (3) If three or more points remain after bad data are rejected, the significance criterion

$$q = \left(\frac{dh}{dt} \right)^2 + \left(\frac{d\delta}{dt} \right)^2 - \frac{v}{(n-1) \Sigma (\Delta t)^2} \tag{10}$$

is evaluated, where

$$v = \Sigma r^2. \tag{11}$$

The sum of the first two terms in equation (10) is the square of the rate of telescope motion; the last term is the square of its standard deviation. If $q \leq 0$, the rate of motion is certainly not significant; the program concludes that the telescope was stationary, and the entire reduction beginning with equation (6) is repeated on this basis. If $q > 0$, the derived motion is accepted.

The last part of the program uses the equations

$$\Delta t = t - \bar{t},$$

$$h = \bar{h} + \frac{dh}{dt} \Delta t,$$

$$\delta = \bar{\delta} + \frac{d\delta}{dt} \Delta t, \tag{12}$$

to obtain the hour angle h, and declination δ, for each observation at time t. If the telescope was held fixed, $dh/dt = d\delta/dt = 0$. At this point we determine where the line of sight intersects the (spherical) moon.

First, we obtain the rectangular co-ordinates of the point at which the line of sight intersects the plane passing through the center of the moon and perpendicular to the line joining the observer to the center of the moon. We wish to have the point at which the line of sight intersects the unit sphere in this co-ordinate system, rather than the x,y-plane.

Clearly, the adopted arbitrary z-distance, R, is greater than the true distance from the observer to the point at which the line of sight intersects the lunar surface. If we express the true z-distance by

$$D = R(1 - \Delta), \tag{13}$$

then the correct values of x and y are

$$x_o = x(1 - \Delta),$$

$$y_o = y(1 - \Delta), \tag{14}$$

and

$$z_o = R\Delta. \tag{15}$$

The transformation to orthographic lunar co-ordinates uses the auxiliary quantity

$$c = z_o \cos b_T - y_o \sin b_T. \tag{16}$$

We then have for the co-ordinates of the detector at the time t, the values

$$\xi = x_o \cos \ell_T + c \sin \ell_T ,$$

$$\eta = y_o \cos b_T + z_o \sin b_T ,$$

$$\zeta = c \cos \ell_T - x_o \sin \ell_T . \qquad (17)$$

The auxiliary geometrical quantities referring to the relative positions of the earth and sun are calculated just as they were for each photographed point. The results are printed and also punched on cards for further analysis. Figures 10–3a, 10–3b, and 10–3c show the printed computer output for a typical scan during the eclipse of December 18–19, 1964. Radiometric data from this scan are discussed later in this paper.

V. TEMPERATURE MEASUREMENTS

The equations that we will use to reduce our radiant power measurements into actual temperatures are obtained for the radiation pyrometer developed at Harvard College Observatory, but our general conclusions will apply to the technique more than to the specific instrument.

If we use a reflector with two optical surfaces with a radiance reflectance $\bar{\rho}_A$, the radiance S produced by the moon at a temperature T_M will be expressed by

$$S(T_M) = \left(\frac{F_{eff}}{F_c}\right)^2 \left[\frac{(d_{M+S} - d_S)\bar{\epsilon}_c}{d_c \bar{\rho}_A \ \bar{\epsilon}_M \bar{\tau}_A(m, TM)}\right]$$

$$\left[S(T_c) - S(T_R)\right], \qquad (18)$$

given in watt ster^{-1} cm^{-2} and where

$(F_c/F_{eff})^2$ = ratio of the solid angles of moon and calibration blackbody seen by the detector; i.e., the F's are the effective f-numbers for the calibration blackbody and the moon,

d_{M+S} = moon plus sky amplitude of the power signal,

d_S = sky amplitude of the power signal,

d_c = calibration amplitude of the power signal,

$\bar{\tau}_A(m, T_M)$ = atmospheric radiant transmittance for radiation from a blackbody at

temperature T_M through m air masses,[6]

$\bar{\epsilon}_c$ = radiant emissivity of the calibration blackbody,

$\bar{\epsilon}_M$ = radiant emissivity of the moon,

$S(T_c)$ = radiance by the calibration blackbody at a temperature T_c,

$S(T_R)$ = radiance by the reference blackbody at a temperature T_R.

The radiances by the blackbodies are expressed by

$$S(T) = \int_o^\infty \mathcal{N}_\lambda(T) \ \tau_F(\lambda) \ \tau_D(\lambda) \ \epsilon(\lambda)d, \qquad (19)$$

where

$\mathcal{N}_\lambda(T)$ = spectral radiance of a blackbody at a temperature T,

$\tau_F(\lambda)$ = spectral radiant transmittance of the filter,

$\tau_D(\lambda)$ = spectral radiant transmittance of the detector's window,

$\epsilon(\lambda)$ = spectral radiant emissivity of the blackbody.

Error Analysis in Absolute Measurements

Let us examine the accuracy with which some of the instrumental parameters can be measured, and the maximum error we should expect in the absolute measurement of the lunar temperature T_M. From equation (18) we obtain the expression for $\Delta S(T_M)/S(T_M)$, from which can be derived the expression for the maximum error:

$$\left|\frac{\Delta T_M}{T_M}\right| = \left[\left(\frac{S(T_M)}{T_M}\right)\frac{dT_M}{dS(T_M)}\right]\left[2\left|\frac{\Delta F_{eff}}{F_{eff}}\right| + 2\left|\frac{\Delta F_c}{F_c}\right|\right.$$

$$+ \left|\frac{\Delta(d_{M+S} - d_S)}{(d_{M+S} - d_S)}\right| + \left|\frac{\Delta d_c}{d_c}\right| + 2\left|\frac{\Delta \bar{\rho}_A}{\bar{\rho}_A}\right| + \left|\frac{\Delta \bar{\epsilon}_M}{\epsilon_M}\right|$$

$$+ \left|\frac{\Delta \bar{\tau}_A(m, T_M)}{\tau_A(m, T_M)}\right| + \left|\frac{\Delta \bar{\epsilon}_c}{\epsilon_c}\right| + \left|\left(\frac{dS(T_c)}{dT_c}\right)\frac{\Delta T_c}{S(T_c) - S(T_R)}\right|$$

$$\left. + \left|\left(\frac{dS(T_R)}{dT_c}\right)\frac{\Delta T_R}{S(T_c) - S(T_R)}\right|\right]. \qquad (20)$$

[6] A discussion of the method used to obtain semi-empirically the atmospheric radiant transmittance may be found in H. C. Ingrao, A. T. Young, and J. L. Linsky, "A Critical Analysis of Lunar Temperature Measurements in the Infrared," *Scientific Report No. 6*, Cambridge, Mass.: Harvard College Observatory (1965). A more thorough discussion will be published elsewhere.

TABLE 10–1: Maximum and Probable Errors That Affect the Lunar Temperature Measurements

	$\dfrac{\Delta F_{eff}}{F_{eff}}$	$\dfrac{\Delta F_c}{F_c}$	$\dfrac{\Delta(d_{M+S} - dS)}{d_{M+S} - dS}$		$\dfrac{\Delta d_c}{d_c}$	$\dfrac{\Delta\rho_A}{\bar{\rho}_A}$	$\dfrac{\Delta\epsilon_M}{\bar{\epsilon}_M}$	$\dfrac{\Delta\bar{\tau}_A(m,400)}{\bar{\tau}_A(m,400)}$	$\dfrac{\Delta\bar{\tau}_A(m,175)}{\bar{\tau}_A(m,175)}$	$\dfrac{\Delta\bar{\epsilon}_c}{\bar{\epsilon}_c}$	ΔT_c	ΔT_R
			$(T = 400°\text{K}$	$(T = 175°\text{K}$								
Maximum Error	±4%	±2%	±1%	±20%	±1%	±5%	±5%	±10%	±12%	±3%	±0.3°K	±0.3°K
Probable Error	±2%	±2%	±1%	±10%	±1%	±2%	±3%	±3%	±5%	±1%	±0.1°K	±0.1

To compute the coefficients of propagation of errors we will assume $T_M = 400°\text{K}$ and the observing conditions to be $T_R = 260°\text{K}$ and $T_c = 270°\text{K}$. To bracket the error we refer to Table 10–1, which gives the estimated maximum errors in the measurements of the instrumental parameters, the observing conditions, and the data reduction process. The second row in Table 10–1 gives the probable errors in the measurements of the same quantities under good observational conditions.

Introducing the proper values into equation (20), we obtain

or
$$\frac{\Delta T_M}{T_M} = \pm 13\%$$
$$\Delta T_M = \pm 53°\text{K} , \qquad (21)$$

which is the *maximum* error to be expected near the subsolar point under good instrumental and observational conditions. This analysis applies only for signals with very high signal-to-noise ratio.

By taking the square root of the sum of the squares of the terms in equation (20), we can estimate the size of the systematic error to be *expected* under these conditions. We now take from Table 10–1 the estimated *probable* errors, and find that our determination of the subsolar point temperature is as likely as not to be systematically in error by ±2.1 per cent, or ±8.5°K.

For $T_M = 175°\text{K}$, typical of temperatures on the eclipsed moon, the factor $\left[\dfrac{S(T_M)}{T_M}\left(\dfrac{dT_M}{dS(T_M)}\right)\right]$ is reduced from 0.28 to 0.13. On the other hand, the atmospheric transmittance is somewhat more uncertain for such low-temperature radiation, and instrumental noise introduces an additional uncertainty in the measured signal $(d_{M+S} - dS)$. Allowing for these two effects, we estimate that the *maximum* error of one measurement at 175°K, including a 20 per cent instrumental noise contribution, is ±15°K, while the probable *systematic* error of the mean of 10 data points is ±2.1°K.

Error Analysis in Relative Measurements

Let us assume that we want to measure the ratio of the temperatures T_1/T_2 of two areas of the moon. In this case we obtain the general expression for maximum error:

$$\left|\frac{\Delta R}{R}\right| = \left|\text{D}(T_1) - \text{D}(T_2)\right|\left[2\left|\frac{\Delta F_{eff}}{F_{eff}}\right| + 2\left|\frac{\Delta F_c}{F_c}\right| + 2\left|\frac{\Delta\bar{\rho}_A}{\bar{\rho}_A}\right| + \left|\frac{\Delta\bar{\epsilon}_c}{\bar{\epsilon}_c}\right|\right] + \text{D}(T_1)\left[\left|\frac{\Delta(d_{M+S} - dS)_1}{(d_{M+S} - dS)_1}\right|\right.$$

$$+ \left|\frac{\Delta d_{c,1}}{d_{c,1}}\right| + \left|\frac{\Delta\bar{\epsilon}_{M,1}}{\bar{\epsilon}_{M,1}}\right| + \left|\frac{\Delta\bar{\tau}_A(m_1; T_{M,1})}{\bar{\tau}_A(m_1; T_{M,1})}\right| + \left|\left(\frac{dS(T_{c,1})}{dT_{c,1}}\right)\frac{\Delta T_{c,1}}{S(T_{c,1}) - S(T_{R,1})}\right|$$

$$+ \left|\left(\frac{dS(T_{R,1})}{dT_{R,1}}\right)\frac{\Delta T_{R,1}}{S(T_{c,1}) - S(T_{R,1})}\right|\right] + \text{D}(T_2)\left[\left|\frac{\Delta(d_{M+S} - dS)_2}{(d_{M+S} - dS)_2}\right| + \left|\frac{\Delta d_{c,2}}{d_{c,2}}\right| + \left|\frac{\Delta\bar{\epsilon}_{M,2}}{\bar{\epsilon}_{M,2}}\right|\right.$$

$$+ \left|\frac{\Delta\bar{\tau}_A(m_2; T_{M,2})}{\bar{\tau}_A(m_2; T_{M,2})}\right| + \left|\left(\frac{dS(T_{c,2})}{dT_{c,2}}\right)\frac{\Delta T_{c,2}}{S(T_{c,2}) - S(T_{R,2})}\right| + \left|\left(\frac{dS(T_{R,2})}{dT_{R,2}}\right)\frac{\Delta T_{R,2}}{S(T_{c,2}) - S(T_{R,2})}\right|\right]. \qquad (22)$$

In this equation, $D(T) = S(T)/T \, (dT/dS)$.

For a typical eclipse cooling curve with $T_1 = 175°K$ and $T = 400°K$ and $T_R = 260°K$ and $T_c = 270°K$, the maximum error in $\Delta R/R$ is ± 16 per cent, and the probable error is ± 2.2 per cent if we assume the values in Table 10–1. These figures are slightly too pessimistic because the systematic errors in atmospheric radiant transmittance and in calibration are likely to be in the same direction during one night, and should probably be put into the first term of equation (22) rather than be separated into the second and third terms.

In the case of observations made in a single scan to determine the brightness profile of the moon, the same calibration will apply to all parts of the scan, and the atmospheric transmittance errors are likely to be in the same direction at all temperatures. Putting these terms into the $\left| D(T_1) - D(T_2) \right|$ term, we have:

$$\left| \frac{\Delta R}{R} \right| = \left| D(T_1) - D(T_2) \right| \left[2 \left| \frac{\Delta F_{eff}}{F_{eff}} \right| + 2 \left| \frac{\Delta F_c}{F_c} \right| + 2 \left| \frac{\Delta \bar{\rho}_A}{\bar{\rho}_A} \right| + \left| \frac{\Delta \bar{\epsilon}_c}{\bar{\epsilon}_c} \right| + \left| \frac{\Delta d_c}{d_c} \right| + \left| \frac{\Delta \bar{\tau}_A(m; \, T_{M,1})}{\bar{\tau}_A(m; \, T_{M,1})} \right| \right.$$

$$\left. + \left| \left(\frac{dS(T_c)}{dT_c} \right) \frac{\Delta T_c}{S(T_c) - S(T_R)} \right| + \left| \left(\frac{dS(T_R)}{dT_R} \right) \frac{\Delta T_R}{S(T_c) - S(T_R)} \right| \right]$$

$$+ \left| D(T_1) + D(T_2) \right| \left[\left| \frac{\Delta(d_{M+S} - d_S)}{(d_{M+S} - d_S)} \right| + \left| \frac{\Delta \bar{\epsilon}_M}{\bar{\epsilon}_M} \right| \right]. \tag{23}$$

Adopting the usual values for high temperatures, we find that the maximum relative error is ± 6.7 per cent, and the probable error is ± 1.6 per cent. Finally, for observations made over both a short time interval and a very small temperature range (say 10°K), we can neglect $\left| D(T_1) - D(T_2) \right|$, and we have

$$\left| \frac{\Delta R}{R} \right| = D(T_M) \left[2 \left| \frac{\Delta(d_{M+S} - d_S)}{d_{M+S} - d_S} \right| + 2 \left| \frac{\Delta \bar{\epsilon}_M}{\bar{\epsilon}_M} \right| \right] \tag{24}$$

Setting $T_M = 400°K$, we obtain the maximum and probable relative errors of ± 3.4 per cent and ± 1.8 per cent, assuming the values in Table 10–1. We note that most of the uncertainty comes from possible variations in lunar emissivity.

Often we wish to deal with temperature differences. In such cases the error in $(T_1 - T_2)$ is then expressed by:

$$\Delta(T_1 - T_2) = T_2 D(T_2) \left[2 \left| \frac{\Delta(d_{M+S} - d_S)}{d_{M+S} - d_S} \right| + 2 \left| \frac{\Delta \bar{\epsilon}_M}{\bar{\epsilon}_M} \right| \right] + \left| (R - 1) \Delta T_2 \right|; \tag{25}$$

For $T_2 = 400°K$ and $(T_1 - T_2) = 10°K$, we have maximum and probable errors of $\pm 14.6°K$ and $\pm 7.1°K$, respectively.

If we assume that the lunar emissivity is constant and that the precision of measurement is limited only by the noise/signal ratio r of the pyrometer, we have for $R \approx 1$:

$$\Delta(T_1 - T_2) = 2T_2 D(T_2) \, r. \tag{26}$$

Equation (26) gives the minimum detectable temperature difference; for our equipment, operating with 4 seconds post-detection integration time and for $T_2 = 400°K$, this value is 0.67°K.

Finally, we may remark that the equations (24) and (25) may be written with $2\Delta(d_{M+S} - d_S)$ and

$2\Delta \bar{\epsilon}_M$ replaced by the equivalent expressions $\Delta(d_{M,1+S} - d_{M,2+S})$ and $\Delta(\bar{\epsilon}_{M,1}, \bar{\epsilon}_{M,2})$, respectively. This change makes explicit the dependence of the temperature differences on the differences in signals and emissivities.

Error Analysis for Relative Temperature Measurements during the Umbral Phase

Since the differences between lunar surface models are more evident in the cooling curve during the umbral phase of an eclipse, we will apply the previous analysis to bracket the maximum errors and to determine whether we will be able to distinguish between models solely on the basis of the cooling curve in the umbral phase.

For this analysis we will use equation (22) but since the lunar area is invariant, the emissivities can be grouped together. Taking $T_1 = 255°K$, $T_2 = 225°K$, which will apply for Tycho during total eclipse conditions, and the errors given in Table 10–1, we see that the errors of measurement are effectively somewhat smaller than the value given for 175°K in Table 10–1. The reasons are that in fact we are averaging together several successive data samples in determining a mean temperature, and that the signal-to-noise ratio is higher at the crater temperature than at 175°K. We find that the maximum error in the measurements of T_1/T_2 is ±0.061 and the probable error ±0.018, corresponding to maximum and probable errors in T/T_{Max} of ±0.037 and ±0.011 respectively. The main contribution to the error in this case comes from the uncertainties in the measurements of d_M and $\bar{\tau}_A(m, T_M)$.

We conclude that a model of the lunar surface must predict the observed decline in the normalized temperature T/T_{Max} during total eclipse within a very few per cent, if we are to accept the model as satisfactory.

Instrumental Conclusion

One instrumental conclusion from the previous error analysis is that the calibration blackbody should be placed in front of the telescope entrance stop. This arrangement will eliminate the contribution in the uncertainty of the measurement by the errors in the value of F_{eff}, F_c, and $\bar{\rho}_A$. In this case the value given in (21) is reduced to $\Delta T_M/T_M = ±7$ per cent. If we apply the above considerations to the relative measurements, the maximum relative error for temperature ratios, and for $T_2 = 400°K$, will be ±4.6 per cent. We can carry out the same analysis for other observing conditions with this instrumental modification to see what reduction in the maximum error can be achieved.

Observational Data

The radiation pyrometer and observational technique are fully described by Ingrao and Menzel.

In order to validate the data presented in this paper we will describe the observational conditions

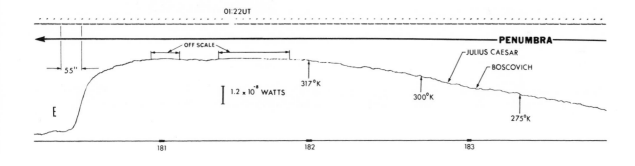

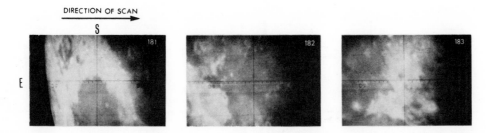

FIGURE 10–4a: First part of a complete lunar scan, time-identifying pictures, signal marks, and marks indicating when a picture has been secured. Data obtained during the lunar eclipse, December 18–19, 1964, with the 61-inch telescope at Agassiz Station. The total amount of precipitable water in the earth's atmosphere was 1.4 mm. This part of the scan went through an area in penumbra.

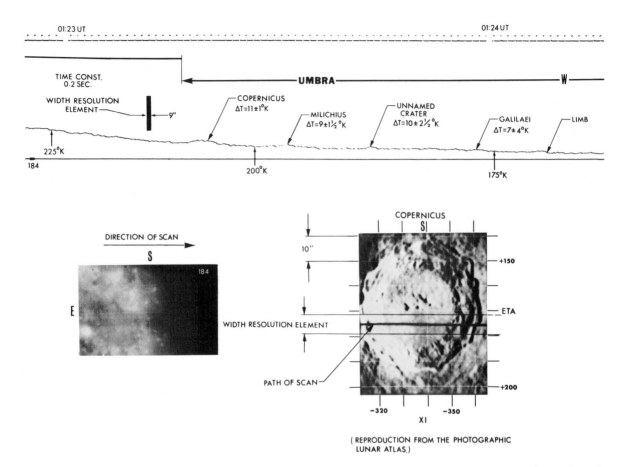

FIGURE 10–4b: Continuation of data and scan shown in Figure 10–4a. This part of the scan passed almost through the center of Copernicus during the umbral phase. The ΔT's are not corrected by instrumental profile.

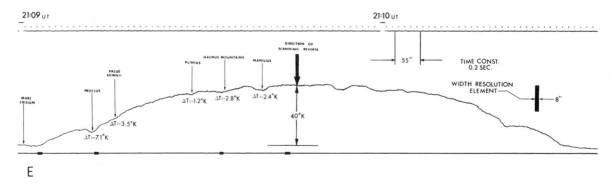

FIGURE 10–5: Scan and time marks a few hours before the lunar eclipse of June 24–25, 1964, observed from Pretoria (Republic of South Africa) with the 74-inch telescope. The scan shows several temperature anomalies. To show the validity of the measurements, the direction of scanning was reversed. The second scan shows an almost perfect mirror image of the first scan. The ΔT's not are corrected by instrumental profile.

and parameters of the equipment for the scans shown in Figures 10–4a, 10–4b, and 10–5.

The scan shown in Figure 10–4a and 10–4b was obtained at the Newtonian focus of the 61-inch telescope at Agassiz Station during the lunar eclipse, December 18–19, 1964. The telescope was stopped down to *f*/5.58 by an entrance stop in the pyrometer to ensure that the detector did not "see" anything but the mirror objective. The amount of precipitable water, measured from sounding balloons, was 1.4 mm for one air mass, an exceptionally low value for this observing site.

The post-detection time constant of the pyrometer was 0.2 seconds, and the size of the resolution element was 9″ × 9″ between half-power points. As the scan shows, the temperature anomalies in Copernicus, Milichius, Galilei, and an unnamed crater first become evident as a plus ΔT. The incremental temperatures ΔT shown in the scan are not corrected for the instrumental profile of the pyrometer. This correction will be important for Milichius, the unnamed crater, and Galilei. The power calibration (1.2 × 10⁻⁸ watts) of the record was obtained with a blackbody calibration.

Figure 10–5 shows a scan from Mare Crisium through Manilius, which was obtained a few hours before the lunar eclipse of June 24–25, 1964, at the Newtonian focus of the 74-inch telescope at the Radcliffe Observatory (Pretoria, Republic of South Africa). The amount of precipitable water for one air mass, measured from sounding balloons, was 2.5 mm.

The post-detection time constant of the pyrometer was 0.2 seconds and the size of the resolution element was 8″ × 8″ between half-power points.

To verify that the structure of the scan shown in Figure 10–5 has astrophysical meaning, we reverse the sense of scanning at the point indicated by the arrow. The second part of the scan clearly is almost a perfect mirror image of the first part; the small differences are due to the change in declination of the moon (0.114″ per second of time), which over a minute of time amounts to almost 80 per cent of the size of the resolution element and of the difference in scanning rate. The record shows very clearly the negative increment ΔT for the temperature anomalies in Proclus, Plinius, and Manilius.

Each one of the event marks indicated at the bottom of the scan shown in Figure 10–5 represents the time at which a picture was secured. The first mark from the left gives the values $\xi = +.879$ and

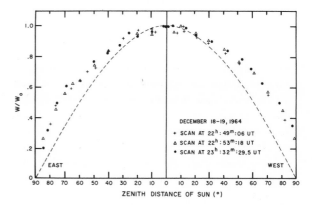

FIGURE 10–6: Relative radiant emittance obtained from scans across the lunar disk. The dashed curve gives the relative radiant emittance as a function of the solar zenith distance for an ideal smooth lunar surface.

$\eta = +.276$ as the orthographic co-ordinates of the center of the resolution element on the moon; at the second mark, $\xi = +.700$, $\eta = +.270$; at the third mark, $\xi = +.285$, $\eta = +.253$; at the fourth mark $\xi = +.065$, $\eta = +.251$. The root-mean square residual in position for this particular scan is 3.68″.

Figure 10–6 shows the relative blackbody radiant emittance of the lunar surface as a function of distance from the subsolar point. The observational data were obtained from three scans made a few hours before the lunar eclipse of December 18–19, 1964. The observing conditions for these scans are similar to the conditions for the scan shown in Figures 10–4a and 10–4b; the times indicated in Figure 10–6 are for mid-scan. The dashed curve has been plotted in Figure 10–6 to show the departure of the radiant emittance from the assumed cosine dependence.

The data indicated with crosses in Figure 10–6 were obtained from a scan that passed at 7°36′ from the subsolar point, the one indicated with triangles 5°42′ and the one given with dots only 18′. We did not try to fit a curve to these data points since the scans went through areas of completely different physical nature. In a future paper, we will analyze our temperature data as a function of the zenith distance of the sun for specific areas.

VI. LUNAR SURFACE MODELS BASED ON INFRARED MEASUREMENTS

Careful consideration of cooling curves of the craters Tycho and Copernicus, which we obtained during the lunar eclipse of December 18–19, 1964, and of infrared lunation data obtained by Murray and Wildey,[7] strongly suggests that the idealized one- and two-layer models of the lunar surface are not adequate for the unique interpretation of infrared eclipse and lunation measurements. These data suggest that the one- and two-layer models which assume temperature-independent thermal properties have not been properly applied to correspond with the actual conditions on the lunar surface, and that the effects of temperature-dependent properties should be more thoroughly investigated.

The thermal properties of various types of lunar-surface models that we will discuss may be characterized in terms of four criteria: (1) depth dependence, (2) temperature dependence, (3) horizontal surface variations, and (4) extent of the lunar surface over which averages are made.

Since infrared measurements usually resolve a region on the lunar surface over which the local solar zenith angle at a given time is nearly constant, several investigators (e.g., Wesselink,[8] Jaeger,[9] Jaeger and Harper[10]) have calculated surface temperatures which, strictly speaking, pertain only to one point, usually the subsolar point. Levin has noted that measurements of thermal radio emission obtained with very low spatial resolution on the lunar disk can be misinterpreted when compared with predictions of models which apply only to a point.[11] Piddington and Minnett, using an antenna which at 1.25 cm has a half-width of 23′, obtained a condition, based upon their phase lag measurements, which relates the temperature-independent thermal properties of the upper and lower layers of two-layer models to the depth of the upper layer.[12] Until this or some other condition is verified by radio observations of high spatial resolution, we cannot use it as a basis for discussion.

[7] B. C. Murray and R. L. Wildey, "Surface Temperature Variations during the Lunar Nighttime," *Astrophys. J.*, Vol. 139 (1964), pp. 734–750.

[8] A. J. Wesselink, "Heat Conductivity and Nature of the Lunar Surface Material," *Bull. Astron. Inst. Netherlands*, Vol. 10 (1948), pp. 351–363.

[9] J. C. Jaeger, "The Surface Temperature of the Moon," *Austral. J. Phys.*, Vol. 6 (1953), pp. 10–21.

[10] J. C. Jaeger and A. F. A. Harper, "Nature of the Surface of the Moon," *Nature*, Vol. 166 (1950), p. 1026.

[11] B. Yu. Levin, "Nature of the Lunar Surface Layer," *Astron. Zh.*, Vol. 40 (1963), pp. 1071–1075 [translated in *Soviet Astron.-AJ*, Vol. 7 (1964), pp. 818–821].

[12] J. H. Piddington and H. C. Minnett, "Microwave Thermal Radiation from the Moon," *Austral. J. Sci. Res.*, Ser. A, Vol 2 (1949), pp. 63–77.

Basic Equations and Computing Methods

Under the assumption of a plane-parallel lunar surface in which the thermal properties are only a function of depth x, the flux conducted outward in a solid material would be

$$F_c = K_o \left(\frac{\partial T}{\partial x} \right), \tag{27}$$

where K_o is the bulk thermal conductivity of the material. Between two planes each with radiant emissivity $\bar{\epsilon}_M$, separated by a distance s', the radiated flux is given by the equation

$$F_R = 4\epsilon' \sigma \bar{T}_3 s' \left(\frac{\partial T}{\partial x} \right), \tag{28}$$

where ϵ' is the effective emissivity, σ is the Stefan-Boltzmann constant and $\bar{T}$ the mean temperature between the two planes.[13] The very uppermost material on the lunar surface may be porous or composed of finely divided material. In this case the total heat flux in this material would consist in part of thermal conduction and in part of radiative transfer. If we consider a porous structure which contains holes of characteristic size, a, and if we further assume that the ratio, R, of radiated flux to conducted flux is not so large that significant departures of the temperature occur at any point from the mean temperature at that depth, the total upward flux will be given by the equation

$$F = [(1 - p) K_o + 4\epsilon' \sigma T^3 p a] \frac{\partial T}{\partial x}, \tag{29}$$

where p is the fraction of the structure occupied by the spaces. Since we have no prior knowledge of p, it is useful to rewrite this expression in terms of the effective conductivity K and effective spacing s, where

$$F = (K + 4\bar{\epsilon}_M \sigma T^3 s) \frac{\partial T}{\partial x}, \tag{30}$$

$$s = p a \frac{\epsilon'}{\epsilon_M}, \tag{31}$$

$$K = (1 - p) K_o. \tag{32}$$

The following arguments in no way depend upon an assumed porous structure, since only the thermal

[13] See also Wesselink, "Heat Conductivity and Nature of the Lunar Surface," and F. L. Whipple, "A Comet Model I. The Acceleration of Comet Encke," *Astrophys. J.*, Vol. 111 (1950), pp. 375–394.

properties can be ascertained by infrared measurement, and a fuller discussion of such structures would be pointless.

Before proceeding, let us consider whether radiative transfer could indeed play an important role. For a typical value of K, such as 5×10^{-6} cal cm^{-1}°K^{-1} considered below, and a temperature of 350°K, the conductive flux and radiative flux become equal for a separation $s \approx 200\ \mu$. This distance scale is not inconsistent with that measured by Wechsler and Glaser[14] for powdered rocks deposited under high vacuum, and may be appropriate for large regions of the lunar surface.

The heat conduction equation now assumes the form

$$\rho c \frac{\partial T}{\partial t} = \frac{\partial}{\partial x}\left[(K + 4\bar{\epsilon}_M \sigma T^3 s)\frac{\partial T}{\partial x}\right], \quad (33)$$

with ρ the density, c the specific heat, and t the time. At a point with orthographic co-ordinates (ξ, η) and bolometric albedo A_b, the flux into the surface is given by the difference between the absorbed insolation $(1 - A_b)\ I(\xi, \eta, t)$ and the energy radiated into space. Neglecting physical libration and the inclination of the moon's equator to the ecliptic, we find that the surface boundary condition becomes

$$(K + 4\bar{\epsilon}_M \sigma T_o^3 s)\left(\frac{\partial T}{\partial x}\right)_{x=0}$$
$$= \bar{\epsilon}_M \sigma T_o^2 - (1 - A_b)\ I(\xi, \eta, t), \quad (34)$$

where during the lunar day

$$I(\xi, \eta, t) = f(t)\ \sigma T_s^4\left[-\xi \sin\left(\frac{2\pi t}{P}\right)\right.$$
$$\left. + \sqrt{1 - \eta^2 - \xi^2}\cos\left(\frac{2\pi t}{P}\right)\right]; \quad (35)$$

P is the synodic period of revolution, T_o the surface temperature, and T_S the theoretical subsolar point temperature. With the value of 1.99 ± 0.02 cal cm^{-2} min^{-1} for the solar constant,[15] T_S is 395 ± 1°K. Wesselink has noted that the subsolar surface temperature for reasonable values of surface con-

ductivity never reaches this theoretical limit, since a small fraction of the incident flux is conducted inward. For example, for a homogeneous model with temperature-independent thermal properties and surface thermal parameter of $\gamma \equiv (K\rho c)^{-1/2}$ $= 1000$ in cal cgs units, which closely fits our data for the environs of Tycho, the calculated subsolar temperature is 393.2°K for $\bar{\epsilon}_M = 0.93$ and varies only by ± 0.1°K for ± 0.05 changes in $\bar{\epsilon}_M$.

During the penumbral eclipse, the insolation is reduced by a factor $f(t)$ determined by the portion of the solar disk occulted by the earth and by solar limb darkening. For this latter quantity the values at 6,000 Å have been used.

In the light of Jaeger and Harper's success with two-layer models, we consider a multilayered model, which will better approximate to a more realistic situation in which thermal parameters continuously vary with depth, if the data should warrant adding this complication. At the boundary of two layers, L and $L + 1$, the temperature is continuous,

$$T_L = T_{L+1} \quad (36)$$

and the heat conduction equation may be written as

$$\left(\frac{\rho_L c_L + \rho_{L+1} c_{L+1}}{2}\right)\left(\frac{\partial T_L}{\partial t}\right)$$
$$= \lim_{\substack{\Delta x_L \to 0 \\ \Delta x_{L+1} \to 0}}\left[\frac{(K_{L+1} + 4\bar{\epsilon}_M \sigma T_L^3 s_{L+1})\ (\partial T/\partial x)_{L+1}}{\Delta x_{L+1}}\right.$$
$$\left. - \frac{(K_L + 4\bar{\epsilon}_M \sigma T_L^3 s_L)\ (\partial T/\partial x)_L}{\Delta x_L}\right] \quad (37)$$

At the present time we are interested in a method of solving equations (33) and (37), both for an eclipse and for a lunation, which will readily allow computation with an arbitrary form of temperature dependencies of the thermal properties. Fourier techniques, despite their elegance, are clearly unsuited for such a nonperiodic phenomenon as an eclipse. Solutions utilizing Laplace transform techniques have been attempted, but for mathematical simplicity one must often oversimplify or even ignore the relevant temperature dependencies of each parameter. For these reasons and for its ready adaptability to machine computation, we have adopted the difference-equation approach throughout.

[14] A. E. Wechsler and P. E. Glaser, "Thermal Properties of Postulated Lunar Surface Materials," *The Lunar Surface Layer; Materials and Characteristics*, ed. by J. W. Salisbury and P. E. Glaser, New York: Academic Press (1964), pp. 389–410.

[15] C. W. Allen, *Astrophysical Quantities*, 2d ed., London: Univ. of London, Athlone Press (1963).

In each of our calculations, an initial temperature distribution has been assumed for a given position on the lunar surface. The deepest point in the moon is chosen so far in that it does not affect the surface temperature at any time and is held at a constant value, usually 230°K when not far from the subsolar point. After the integration of a complete lunation is performed, the temperature distribution is accurate, as calculations for several lunations show, to within 1°K to depths in excess of that to which the thermal wave can propagate during an eclipse. Using this initial temperature distribution and computed values of $f(t)$, we compute the temperature distributions during the penumbral and umbral stages of eclipse.

Unless otherwise stated, our cooling curves are computed for the crater Tycho ($\xi = -0.685$, $\eta = -0.140$), and on the assumption that $\varepsilon_M = 0.93$. In computing each model we have specified the diffusivity $\alpha = K/\rho c$, rather than the more familiar thermal parameter $\gamma = (K\rho c)^{-1/2}$, since α is the relevant parameter in the heat conduction equation (33), if one wishes to specify the distance scale of layers in dimensional terms.[16] For comparison with previous work the values of α were chosen to give convenient values of γ when c is assumed to be 0.20 cal gm^{-1}°K^{-1} and ρ to be 1 gm cm^{-3}, typical values from Wechsler and Glaser.

For direct comparison with observed temperatures, all of the calculated surface temperatures have been reduced to brightness temperatures by considering the decrease in apparent blackbody temperature that results from a change in irradiance by a factor of ε_M through our wide-bandpass filter. We have plotted these cooling curves as a ratio of temperatures to initial temperatures, $T_M/T_{M,max}$, as a function of t/t_o, with t_o the duration of the penumbral phase (56 minutes for the crater Tycho in the December 18–19, 1964, eclipse).

Discussion of Two-Layer Temperature-Independent Models

The homogeneous model was originally investigated by Wesselink and Jaeger. This model pro-

vides a convenient reference with which to compare observations and suggest the manner in which the model should be improved. We have computed such a family of cooling curves (presented in Figure 10–7), using time intervals Δt between 120 and 360 seconds, and 30 points beneath the surface ranging from a maximum depth of 25 cm for $\gamma = 1500$ to 70 cm for $\gamma = 250$. Figure 10–7 includes the computed pre-eclipse brightness temperatures $T_{M,max}$ as an indication of the manner in which they depend upon the models. Enough significant figures are given to show the dependence of these temperatures on the model.

The observed temperatures inside Tycho are also presented in Figure 10–7, along with an average of the observed temperatures 30″ east and west of the crater. Our deduced relative temperatures are systematically 3.5 to 4 per cent higher than those obtained by Sinton.[17] In part this may result from the use of different methods of determining the atmospheric transmittance, but more probably it results from his larger resolution element of 27.9″ and the concomitant smearing of significant detail.

As previous investigations have shown, the completely homogeneous model produces cooling curves which decrease more rapidly during the umbral phase of eclipse for anomalous craters than is observed. Our data for Tycho and its environs tend to corroborate this generalization although a completely homogeneous model for the environs of Tycho is not ruled out.

As a simple refinement upon these completely homogeneous models, we relax the condition of horizontal homogeneity. Such models as that in Figure 10–7, for example, show that when we consider a fraction of the lunar surface composed of "bare rock" of higher diffusivity, the slope of the cooling curve throughout totality is changed only very slightly. Therefore simple models that are homogeneous in depth but postulate a variety of surface materials will not fit our data for Tycho or its environs.

Jaeger and Harper have computed a family of cooling curves using two-layer models. One of these comes close to fitting the data obtained by Pettit[18]

---

[16] In a discussion of homogeneous models, where the depth variable is arbitrary, it is advantageous to rewrite the above equation in terms of a new depth variable $y = x/\ell_T$, where $\ell_T = \sqrt{4\pi K_p/\rho c}$ is the wavelength of the first harmonic in a Fourier expansion of the thermal wave in the lunar surface. Wesselink has shown that when this transformation is made, the relationships among the relevant parameters are homologous.

[17] W. M. Sinton, "Eclipse Temperatures of the Lunar Crater Tycho," *The Moon; Proceedings of the 14th International Astronomical Union Symposium, Leningrad, 1960*, ed. by Z. Kopal and Z. K. Mikhailov, London: Academic Press (1962), pp. 469–471.

[18] E. Pettit, "Radiation Measurements on the Eclipsed Moon," *Astrophys. J.*, Vol. 91 (1940), pp. 408–420.

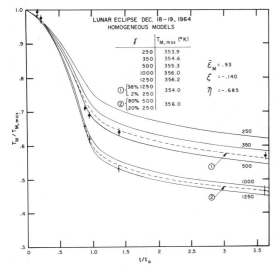

FIGURE 10–7: Observed brightness temperatures inside the crater Tycho (•), and the average of temperatures 30″ east and west of Tycho (+) are compared with a family of cooling curves for homogeneous models with temperature-independent thermal properties. The curves are computed for the orthographic co-ordinates of the center of Tycho and for the indicated radiant emissivity. The resulting theoretical pre-eclipse temperatures $T_{M,max}$ given here in tabular form are given to much higher accuracy than is warranted by the measured values of absorbed insolation, in order to show the dependence of $T_{M,max}$ upon the particular model. Estimated time errors are less than 1.5 minutes.

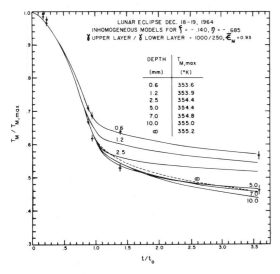

FIGURE 10–8: Observed temperature ratios for the crater Tycho and its environs; (•) inside the crater, (+) environs of crater. Theoretical cooling curves are computed for two-layer models with temperature-independent thermal properties. The thermal parameters γ, radiant emissivity $\bar{\epsilon}_M$, and upper-layer depths are indicated.

for an upland area on the edge of Mare Vaporum ($\xi = 0.0$, $\eta = 0.17$).

The addition of a substrate of higher diffusivity affects the eclipsed surface temperature in two ways. First, its higher thermal inertia and consequently greater thermal phase lag with respect to insolation increases the phase lag at the base of the top layer. For regions of the lunar surface in which the interface is heating up, that is, primarily west[19] of the subsolar meridian, the interface will be colder than at a similar depth for a homogeneous model consisting of upper-layer material. Since the radiating boundary demands a specified outward flux which, with the heat capacities, determines the thermal gradient in both layers, the surface temperature must be colder for two-layer models in which the source of heat is the internal energy of the upper layer. This effect is readily apparent in Figure 10–8 for those models with thick upper layers, and especially in Figure 10–9 where the temperature distribution beneath the surface is plotted for two models at three times: at the beginning and at the end of penumbral eclipse, and near the end of umbral eclipse.

On the other hand, when the advancing thermal wave reaches the interface, heat is withdrawn from a region of high thermal conductivity and the surface temperature falls more slowly. For thin upper layers this manifests itself in an abrupt leveling out of the cooling curves at the end of penumbral eclipse, while for a very thick upper layer (a centimeter or more deep) such an effect cannot manifest itself during an eclipse but is observable during a lunation.

The existence of three independent parameters (the diffusivities α_U and α_L of the upper and lower layers, and the thickness of the upper layer) introduces the possibility that a great many models can exhibit very similar eclipse cooling curves. It is also possible that the curves may be insensitive to values of one or more of the parameters. For the case of a moderately thick upper layer, that is, one for which the cooling curve deviates from the homogeneous case mainly during the umbral phase of eclipse, allowed values of α_U are bounded by the two conditions that the cooling curve for a homogeneous surface of the same ther-

[19] According to the astronomical convention, west is defined as the direction perpendicular to the lunar axis of rotation from the lunar sub-Earth point toward Mare Crisium. Thus the sun rises in the West on the moon.

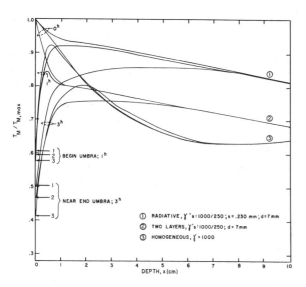

FIGURE 10–9: A comparison of the predicted internal temperature distribution for three different models, during a lunar eclipse. The temperature distributions are given as ratios to the pre-eclipse surface temperatures. The thermal properties are taken to be temperature-independent except that a radiative component to the conductivity is included in the upper layer of the radiative model. These models are computed for the lunar co-ordinates of Tycho and for $\bar{\epsilon}_M = 0.93$. The temperature distributions are given at $t = 0.00t_o$, $1.07t_o$, and $3.22t_o$ in order of decreasing surface temperature.

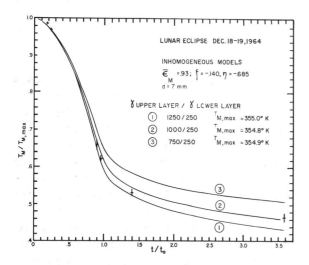

FIGURE 10–10: Effect of varying the upper-layer thermal parameter γ upon the two-layer theoretical eclipse cooling curves computed for the co-ordinates of the center of Tycho. Data obtained during the lunar eclipse of December 18–19, 1964, are plotted for the environs of Tycho (+).

mal properties must be the same as or slightly above the observed temperature at the beginning of umbral eclipse, but significantly below it at the end of total eclipse. The data we have obtained for the environs of Tycho may be described in terms of such a model with the thermal parameter of the upper layer between 1,000 and 1,250. Figure 10–8 shows curves for two-layer models, which assume thermal parameters of 1,000 and 250 for the upper and lower layers respectively, but assume variation in the thickness d of the upper layer. The data for Tycho itself may be described in terms of a model of the same materials but with a much thinner upper layer.

To investigate how the theoretical surface temperatures depend upon modifications of the model, we may take as standards the two models which agree well with the observational temperatures of Tycho and its environs, and systematically study the effect of changing each of the relevant parameters one at a time.

Variations of α_U manifest themselves in a significant manner for models of moderate thickness and for completely homogeneous models, since it is only α_U which prevents a rapid decrease to very low temperature at the end of the penumbral phase of eclipse. The cooling curves for such models, shown in Figure 10–10, are parallel but are significantly displaced during totality with respect to one another, as a result of different thermal gradients in the upper layers required to support roughly the same conducted outward flux. Very thin layers exhibit this effect, but it is decreased in magnitude almost proportionally to the depth of the layer.

Models with the same α_U but different α_L are very interesting, in that the qualitative behavior shown in Figure 10–11 depends upon the upper-layer thickness, d. If the layer is so thick that the thermal wave does not advance much farther into the surface than the interface, the effect of the lower-layer thermal inertia upon the material immediately above is important. Consequently, the curves for larger values of α lie below those for lower values. When the upper layer is so thin that the major source for radiance during total eclipse is the lower-layer material, the curves for greater α_L lie above. In addition, for models with small d, α_L manifests itself directly in the slopes of the umbral cooling curves; the slope decreases as α_L increases, and the lower layer approaches the condition of a constant-temperature heat source. Thus

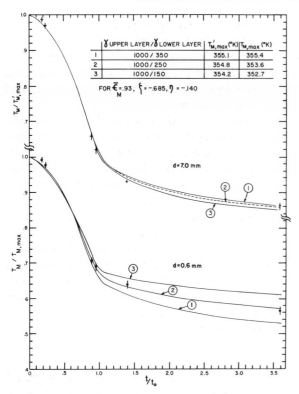

	γ UPPER LAYER / γ LOWER LAYER	$T'_{M,max}$ (°K)	$T'_{M,max}$ (°K)
1	1000 / 350	355.1	355.4
2	1000 / 250	354.8	353.6
3	1000 / 150	354.2	352.7

FOR $\bar{\epsilon}_M = .93$, $\xi = -.685$, $\eta = -.140$

FIGURE 10–11: Theoretical eclipse cooling curves for lower layers with different thermal parameters γ and for two different upper layer depths d. The plotted data pertain to Tycho (•) and its environs (+) and were obtained during the lunar eclipse of December 18–19, 1964.

for such anomalous regions as Tycho, with sufficiently accurate data one may be able to determine the diffusivity of the lower layer directly from relative measurements alone.

From these considerations, we conclude that the surface structure both in and around Tycho may be adequately, but not of necessity uniquely, described in terms of two different two-layer models, both of which consist of the same two materials (thermal parameters of 1,000 and 250), and which differ only in the thickness of the upper layer. As is apparent from Figure 10–8, the data for the environs of Tycho are consistent with upper-layer depths between 5 and 10 mm. Further calculations show that only a limited range of α_U and depth can fit our data for the environs of Tycho, for example the values $\gamma_U = 1150$ and $d = 4$ mm also give a reasonable fit.

In order to apply these models to other features on the lunar surface, we must consider how changes in the initial temperature distribution and surface temperatures before eclipse affect the

surface temperatures during eclipse. Since the loss of energy at the surface by radiation decreases rapidly with temperature, the relative umbral phase temperatures increase to the east and west of the subsolar point. Thus we would expect that the crater Aristarchus, at 48°E selenographic longitude ($\xi = -.685$), would exhibit relatively higher umbral brightness temperatures than a crater of similar structure near the subsolar point. As we have already seen, significant changes in the cooling curve of two-layer models can result from changes in the interface temperature. The cooling curves for similar models east and west of the subsolar point should not be precisely the same. These effects, demonstrated in Figure 10–12, apparently have not been considered in previous studies, a fact that may in part explain the inability of Saari and Shorthill to interpret eclipse temperatures inside Aristarchus.[20]

A similar relative increase in umbral-phase surface temperatures occurs toward high latitudes (see Figure 10–13). This results from an initial temperature distribution which approaches a constant value independent of depth, and from the less efficient radiating boundary. Thus, for comparison of infrared eclipse data with theoretical predictions, one must compute cooling curves for the particular lunar region investigated. The very slight differences in many cases between umbral-phase temperatures of radically different models do not permit the use of one set of models computed for the subsolar point for comparison with data obtained elsewhere.

We have computed cooling curves for another region of the lunar surface, the crater Copernicus. These curves, together with the measured temperature of Copernicus and its environs, are presented in Figure 10–14. Unfortunately, no measurements were made far into totality, where a more precise distinction between different models is possible.

In investigating the degree to which absolute as opposed to relative temperature measurements are necessary for obtaining the thermal properties of the lunar surface, we must distinguish among three effects. A positive error in the assumed temperature of the subsolar point, resulting from inaccurate

[20] J. M. Saari and R. W. Shorthill, *Infrared Mapping of Lunar Craters During the Full Moon and the Total Eclipse of September 5, 1960*, Seattle, Wash.: Boeing Scientific Research Laboratories, Geoastrophysics Laboratory, DI–82–1076 (1963.)

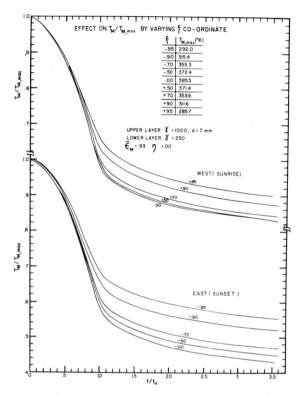

FIGURE 10–12: Theoretical eclipse cooling curves for a two-layer model are given for a range of lunar orthographic co-ordinates ξ. Note that umbral temperatures east of the subsolar point are significantly higher than those west.

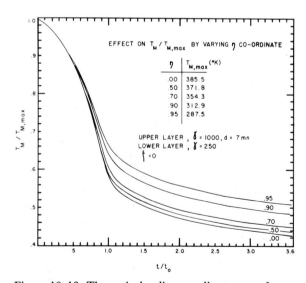

Figure 10–13: Theoretical eclipse cooling curves for a two-layer model are given for a range of lunar orthographic co-ordinates η.

values of the absorbed solar flux, is analogous to the effect of displacing the observed lunar feature in η toward the equator. A negative error has the reverse effect. As we have seen, a decrease in the absolute value of η displaces the relative umbral temperature downwards.

Secondly, systematic errors in the measured irradiance do not appear in a one-to-one manner in the derived temperatures, but are more significant at 350°K than at 200°K. For example, a relatively large error in the subsolar point temperature, reducing it from 395°K to 380°K, which is a 13.3 per cent error in assumed incident solar flux, increases the relative temperature from 0.463 to 0.477 at $t = 3.22 \, t_o$ for our standard two-layer model. A similar systematic error in measured lunar irradiance, with the same model, increases this relative temperature from 0.463 to 0.469, about half as much. The discrepancy is about 3°K or 15 per cent in lunar irradiance during the umbral phase.

While the incident solar flux may be accurate to about 1 per cent, the bolometric albedo for typical and anomalous regions of the lunar surface is not known with this accuracy. A further complication is that the change in radiant emissivity with observation angle, as investigated by Geoffrion, Korner, and Sinton for the subsolar point,[21] leads to an 11 per cent increase in radiant emittance for this region at full moon relative to the mean radiant emittance. Thus, an error in theoretical temperature for the subsolar point of 10 or 15°K is not inconceivable, and the resulting inaccuracies in lunar thermal properties should be taken into account.

The third effect is that for the same surface temperature but different mean emissivity, the rate of thermal emission varies. A 0.05 change in this emissivity causes a change of 0.0083 in relative temperatures at the end of umbral eclipse for the standard model.

Temperature-Dependent Properties

Since one- and two-layer temperature-independent models have apparently been successful in accounting for eclipse and radio observations, and since a multitude of plausible mechanisms have

[21] A. R. Geoffrion, M. Korner, and W. M. Sinton, "Isothermal Contours of the Moon," *Lowell Observ. Bull.*, Vol. 5 (1960), pp. 1–15.

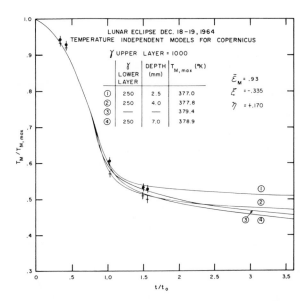

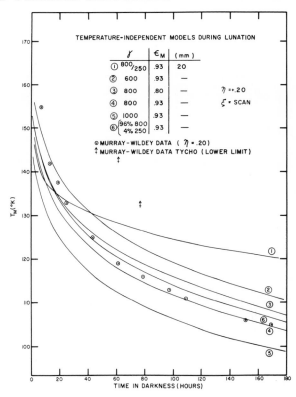

FIGURE 10-14: Data obtained during the lunar eclipse of December 18-19, 1964, for the crater Copernicus (•) and its environs (+), with theoretical eclipse cooling curves for homogeneous and two-layer models, both with temperature-independent properties.

FIGURE 10-15: Computed surface brightness-temperatures for models with temperature-independent thermal properties are compared with the data of Murray and Wildey. These temperatures are computed for a point at the mean orthographic co-ordinate of their scans, $\eta = +.20$, in terms of the number of hours after the passage of the evening terminator. In accordance with the suggestion of Wildey, these data points have been shifted five hours closer to the terminator. The brightness temperatures for the crater Tycho are lower limits because of uncertainties in position determination on the lunar surface.

been postulated to account for the formation of an upper "dust" layer of variable thickness across the lunar surface, there has been little motivation for introducing the mathematical complications of temperature-dependent thermal properties.

Murray and Wildey have recorded brightness temperature down to their instrumental noise level of 105°K, in the course of many scans up to eight days into lunar darkness. They are unable to fit the observed rate of decrease of temperature with time, the quantity most nearly independent of systematic errors, to any completely homogeneous models they have calculated. They speculate instead that horizontal variations in the conductivity such as "bare outcrop of boulders on the surface" might give a better fit. As is apparent in their Figure 8 and from our calculations in Figure 10-15, such completely homogeneous models have cooling curve slopes which are slightly less steep than is consistent with the data. If a few per cent of the surface material is of higher diffusivity (see Figure 10-15), the slope of the nighttime cooling curve decreases further. In addition, two-layer models in which the depth of the upper layer is shallow enough to be apparent during a lunation, that is, less than 6 cm, also have slopes which are less steep than single-layer models, and thus are inconsistent with the data.

Alternative explanations for the observed steepness of lunation cooling curves should be considered. With Murray and Wildey's spatial resolution of 26″ and a probable positioning error of two or three times this resolution limit, their data for regions within 12 hours of the sunset line should not be given much weight. Wildey suggests that their determination of the terminator is probably in error by 5 hours in the sense of reducing the observed brightness temperature at the terminator.[22] When such a time correction is applied to their data, it approximates better the predictions of homogeneous models.

Their final temperatures were measured very close to the limb and should be revised upward if

[22] Personal communication.

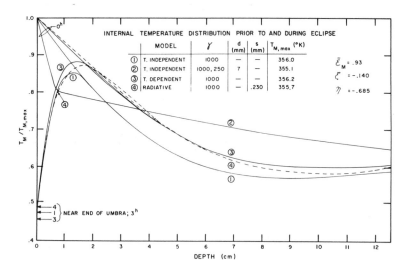

FIGURE 10–16: A comparison of the predicted internal temperature distribution for four models prior to and near the end of the umbral eclipse at the crater Tycho. Note that the differences of temperature beneath the surface are larger than those at the surface, and that a high internal temperature and a low surface temperature are not inconsistent.

limb darkening is important. An upper limit to this revision can be estimated from the observed 30 per cent reduction (Geoffrion, Korner, and Sinton) in the measured radiant emittance from the subsolar point when it is at the limb as compared to its mean value. At 105°K and for our wide-bandpass filter, this corresponds to about a 3°K surface temperature increase at the end of their scans. The data would then be brought into agreement with two-layer temperature-independent models that have upper and lower thermal parameters γ of roughly 800 and 250 and a depth of 4 cm. However, the effect of limb-darkening is probably much less than 3°K near the limb. The effect of small-scale surface roughnesses upon the insolation is probably the major cause of limb-darkening at the subsolar point, and a rough unilluminated surface should approximate a blackbody. It should also be noted that Murray and Wildey's scans covered both upland and *mare* regions, but they find no large-scale selenographic variations to suggest that their data should be interpreted other than in terms of the "time variation of temperatures at a single point."

Assuming that none of these postulated mechanisms is able to explain away the rapid decrease of temperature that occurs between 12 and 160 hours into lunar night, we are forced to construct models with temperature-dependent thermal properties. Radiative conductivity is one mechanism that would account for a decrease in conductivity at low temperatures, which in turn would produce the more rapid decrease in temperature during lunar night. At the same time, however, such a

model would have high conductivity during the lunar day, and thus would have temperatures between 2 and 10 cm below the surface that are significantly higher than if this heat transfer mechanism were unimportant. This is shown in Figures 10–9 and 10–16. This excess internal heat is almost entirely lost just before sunset and so plays no role in explaining nighttime observations, but it produces high surface temperatures near the sunset terminator, which should be observable.

In Figure 10–17 we have plotted the surface temperatures for several models with different K, and R, the ratio of radiative flux to conducted flux at 350°K. Each model exhibits the requisite slope; thus if radiative conduction is important for lunar surface materials, it is difficult to separate the effects of the conductivity from those of the effective spacing, s. An increase in s mimics a decrease in K. As in the temperature-independent models, the introduction of a lower layer of higher diffusivity decreases the rate of surface cooling when heat begins to be withdrawn from the lower layer itself. Thus we can say that if such a lower layer exists, it is probably more than 6 cm down. Considering γ to be defined in terms of the thermal conductivity excluding its radiative component, we find that values of γ between 900 with s between 300 and 600 μ, and 1,000 with s about 1,000 μ, could be considered consistent with the Murray and Wildey data. Measurements of the thermal conductivity of crushed basalt powder, cited by Buettner,[23] support the assumption of a T^3 term in

[23] K. J. K. Buettner, "The Moon's First Decimeter," *Planet. and Space Sci,*. Vol. 11 (1963), pp. 135–148.

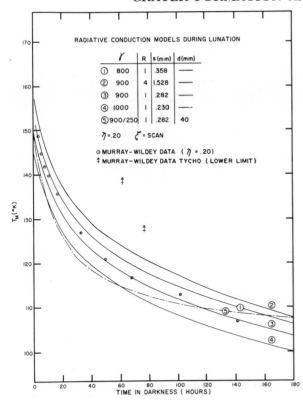

FIGURE 10–17: Computed surface brightness-temperatures for models including a radiative component to the conductivity. R is the ratio of radiative to conductive flux at 350°K, s is the effective mean spacing for radiative transfer beneath the surface, and d is the upper layer depth for a two-layer model. The value of γ characterizes the thermal parameters if there is no radiative transfer.

the conductivity with a distance scale $s = 1{,}000\ \mu$. With our assumed values of ρ and c, his results suggest values of γ near 750.

Eclipse cooling curves (shown in Figure 10–18), computed using homogeneous radiative models, fit the data for Tycho and its environs about as well as do homogeneous temperature-independent models. The internal temperature distributions, shown in Figure 10–16, for these two models are significantly different, however. In order to match precisely the observed cooling curve slopes, again two-layer models are necessary.

To our knowledge the only attempt in the literature to investigate the effects of temperature-dependent thermal properties other than radiative conductivity is the work of Muncey.[24]

[24] R. W. Muncey, "Calculation of Lunar Temperature," *Nature*, Vol. 181 (1958), pp. 1458–1459; "Properties of the Lunar Surface as Revealed by Thermal Radiation," *Austral. J. Phys.*, Vol. 16 (1963), pp. 24–31.

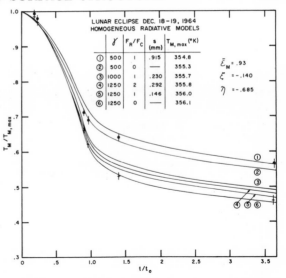

FIGURE 10–18: Computed ratios of surface brightness-temperatures to their pre-eclipse values for models including a radiative component to the conductivity. The meaning of the parameters is the same as in Figure 10–16.

In order to investigate the qualitative effects of temperature-dependent properties we have assumed, as has Muncey, the specific heat and the conductivity to be of the forms

$$c = c_o T, \quad K = k_o T. \tag{38}$$

As before, we cannot predict explicitly what effects the temperature dependence of the thermal properties will have upon the shape of eclipse and lunation cooling curves. For a given outward flux when the surface is cooling, one expects that a larger temperature gradient would be required near the surface than farther in where the temperature is higher. This is apparent if we compare the temperature distribution immediately below the surface during an eclipse for a temperature-dependent model with that for a temperature-independent model (Figure 10–16). The exact effect this variable temperature gradient has upon the surface temperature at any time, however, can be found only by obtaining the solution of the heat conduction equation in the material, subject to the appropriate boundary conditions.

The results of such computations (see Figure 10–19) show that homogeneous models characterized by the parameter $\gamma_{350°K}$ at 350°K, exhibit the same rate of cooling as do the lunation data. We have assumed that $\rho = 1$ gm cm^{-3} and $c = 0.20$

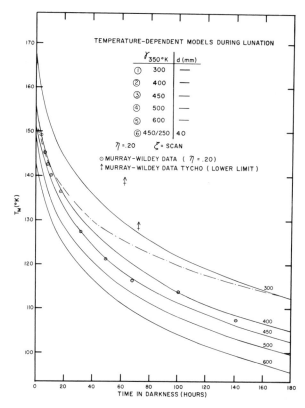

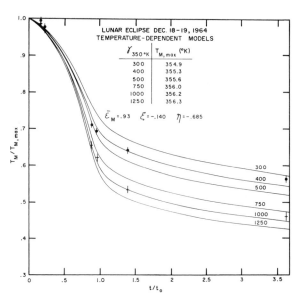

FIGURE 10–19: Computed surface brightness-temperatures for models assuming linear temperature-dependent
thermal properties, $K = K_o T$, and $c = c_o T$. The value
of γ characterizes these parameters at a temperature
of 350°K. Changes in the data of Murray and Wildey
are the same in Figure 10–15.

cal gm^{-3}°K^{-1} at 350°K. A model with γ_{350}°K = 450
appears to be about the best fit. Clearly, two-layer
models with depths less than about 6 cm are ruled
out, as are horizontally inhomogeneous models.

Homogeneous temperature-dependent models
come close to fitting the eclipse data for Tycho and
its environs, as shown in Figure 10–20. A typical
two-layer model (see Figure 10–21), such as that
having an upper layer of material 7 mm thick with
γ_{350}°K = 800 over a substrate with γ_{350}°K = 250 is
consistent with the data for the environs of Tycho.
A model with the same lower-layer material but an
upper layer characterized by γ_{350}°K = 400 and a
depth of roughly 6 mm would be appropriate for
the crater itself. However, any of a number of such
models could fit the data.

From the above discussion it should be apparent
that radically different models of the lunar surface
can exhibit similar, and in some cases indistinguishable, cooling curves during an eclipse and a

FIGURE 10–20: Computed ratios of surface brightness-
temperatures to their pre-eclipse values for homogeneous
models assuming linearly temperature-dependent thermal properties.

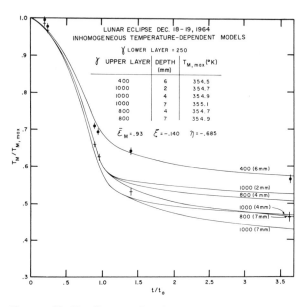

FIGURE 10–21: Computed ratios of surface brightness-
temperatures to their pre-eclipse values for two-layer
models assuming linearly temperature-dependent thermal properties.

lunation, and at the same time can have significantly different internal temperature distributions at millimeter and centimeter depths. Comparison of mm-wave radio observations of small regions of the lunar surface, with predictions based upon each of these models, may be able to distinguish among them.

The anomalous crater Tycho can be readily explained in terms of simple two-layer temperature-independent models, but several other models agree equally well with the data. In particular, the environs of Tycho can be explained by many different kinds of models; the large region scanned by Murray and Wildey can be accounted for by temperature-dependent or by radiative-conductivity models. In the light of our present inability to decide uniquely which of several plausible models applies even to any of the regions of the lunar surface we have studied, any detailed description of small-scale lunar surface structure, uncritically based upon any one kind of model yet devised, may be physically meaningless.

Part III.

Physics and Chemistry of the Lunar Surface

11.

Recent Discovery of Hot Spots on the Lunar Surface: A Brief Report of Infrared Measurements on the Eclipsed Moon*

R. W. Shorthill and J. M. Saari

Boeing Scientific Research Laboratory, Seattle, Washington

During the total lunar eclipse of December 19, 1964, measurements were made of lunar thermal radiation in the 10- to 12-μ region. The instrumentation has been described elsewhere.[1] The method of scanning the moon is shown in Figure 11–1. With two hundred line scans the lunar disk was covered in 16 minutes with a spatial resolution of 10 seconds of arc. On the illuminated moon, simultaneous measurements were made in the visible region using a photomultiplier at the same resolution. An example of the photometric data (phase angle 2 degrees 16 minutes) is shown in Figure 11–2 for the area of Mare Tranquillitatis. The contour interval is 1 per cent of the maximum photoelectric signal. The simultaneous full moon infrared data are shown in Figure 11–3, where the line-scan separation is indicated by the tick marks along the left margin. Data similar to that shown in Figures 11–2 and 11–3 have been obtained on the illuminated disk for twenty different phases from −125 degrees to +135 degrees.

The purpose of scanning the moon during the eclipse was to map the nonuniform cooling of the lunar surface, since we had previously determined that some of the ray craters cool less rapidly than their surroundings. The disk was scanned four times during the first penumbral phase, three times during totality, and once during the second penumbral phase. As expected, we found that the major ray craters show anomalous cooling as indicated in Figure 11–4. Unexpectedly, however, hundreds of localized areas were found that also show anomalous cooling. Some of these "hot spots" are indicated in Figure 11–5. Here the dots indicate the most obvious hot spots taken from the Sanborn chart recordings and the lines indicate areas of extended enhancements. Approximately 65 per cent of these hot spots appear in the seas, and 35 per cent appear in the uplands. Notice that Mare Tranquillitatis has more hot spots than any other mare.

Table 11–1 is a preliminary classification of 330 anomalies shown in Figure 11–5.

TABLE 11–1: Preliminary Classification of Thermal Anomalies on Eclipsed Moon

Ray craters	19.4%	
Craters with bright interior at full moon	41.8%	84.5%
Craters with bright rims at full moon	23.3%	
Craters not bright at full moon	0.6%	
Bright areas with much smaller craters	3.6%	
Bright areas associated with features like ridges	3.9%	8.7%
Bright areas not associated with any feature	1.2%	
Position unidentified or questionable	6.3%	

To date, almost 400 anomalies have been identified; it appears that an equal number of smaller hot spots in the data remain to be identified. Many hot spots showed signal profiles indicating that the anomaly was smaller than the sensor resolution

* The authors wish to thank Dr. A. H. Samaha, Director of the Helwan Observatory, Cairo, Egypt, for observing time and Dr. M. K. Aly, Associate Director, for collaboration on the 74-inch telescope. Transport of equipment under contract number AF19(628)-4371 was made possible by Dr. J. W. Salisbury, Chief of Lunar Planetary Research, United States Air Force Cambridge Research Laboratories.
[1] R. W. Shorthill and J. M. Saari, "Radiometric and Photometric Mapping of the Moon Through a Lunation," *Ann. N. Y. Acad. Sci.*, Vol. 123 (1965), p. 722.

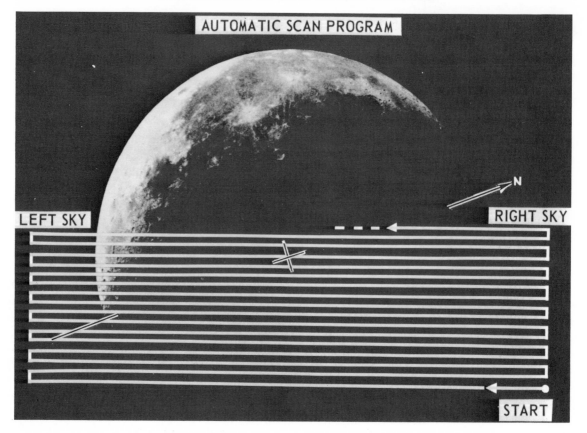

FIGURE 11-1: Automatic scan program. The scanning engine was mounted at the Newtonian focus.

(1/200 the lunar diameter); work is presently underway to make a correction for areal smoothing and assemble a catalogue of the more prominent anomalies.[2]

Contours of brightness temperatures during totality are shown in Figure 11-6 for the same area as in Figures 11-2 and 11-3. The last scan in totality started at 0255.0 UT and the time of the line scan through Mädler was 0302.8 UT December 19, 1964.

In addition to the hot spots, which for the most part are confined to craters, large areas of thermal enhancements were found. Mare Humorum in Figure 11-7 is an example of such an enhancement, the entire mare area being about 10°K above the surrounding uplands. Figure 11-8 shows more of the region around Mare Humorum extending to the southeast limb of the moon. Mare Tranquillitatis, Mare Foecunditatis, Mare Crisium, and Mare Serenitatis are also areas of extended en-

hancements. Only parts of Oceanus Procellarum and Mare Imbrium are enhanced (see Figure 11-5). Even though the region around Tycho is covered with many craters, only a few hot spots other than Tycho are observed, such as Heinsius A and the bright spot in the Hell plain ($\xi = -064$, $\eta = -543$) as shown in Figure 11-9. The radar data of Pettengill,[3] reported previously, show a high radar return from part of the rim of Tycho while our data show the anomaly, approximately centered on the crater, has four maxima and extends beyong the rim by about a half a crater diameter.

The only hot spot photographed with high resolution by the Ranger series was from Ranger VII. Figure 11-10 shows contours near the impact point; the closest hot spot was Bonpland-H, photographs of which have not as yet revealed any obvious explanation for this anomaly.

Since the data were recorded on magnetic tape

---

[2] Shorthill and Saari, "The Non Uniform Cooling of the Eclipsed Moon: A Listing of Thirty Prominent Anomalies," *Science*, Vol. 150 (1965), p. 210.

[3] G. H. Pettengill and J. C. Henry, "Enhancement of Radar Reflectivity Associated with the Lunar Crater Tycho," *J. Geophys. Res.*, Vol. 67 (1962), p. 4881.

it was possible to reconstruct an image on an oscilloscope and photograph it. Figure 11–11 shows the image reconstructed from the photomultiplier signal of a scan made at full moon (phase 2 degrees 16 minutes). The same procedure was used on the infrared data taken at full moon (phase 1 degree 58 minutes) shown in Figure 11–12. Perhaps the most interesting infrared image produced is that from the scan made near the end of totality and is shown in Figure 11–13. This photograph has not been retouched and the bright area in the lower right-hand corner is radiation from the side of the telescope. The segmented line along the left margin is due to the retrace circuits of the oscilloscope. The last figure, Figure 11–14, shows the full moon (visible) and the totally eclipsed moon (infrared) approximately the same size and orientation. Several features have been indicated to aid the reader in identification of the hot spots and extended enhancements.

MARE TRANQUILLITATIS

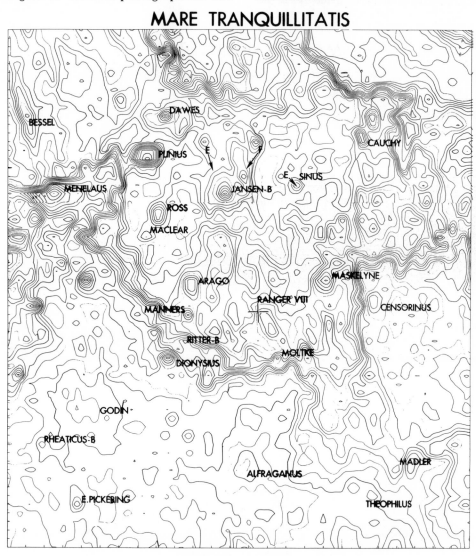

14.93 DAYS   12/18/64

ISOPHOTES ON THE FULL MOON

FIGURE 11–2: Isophotes on the full moon (phase angle −2 degrees, 16 minutes). These contours were drawn by a U.D.M. Orthomat XY plotter.

DISCUSSION

VOICE: I would like to ask if, in this last photograph of the moon, the high contrast film was used, or was there a full value placed in the electronics of data-handling for each of the oscilloscopes to give a threshold to the light values given?

SHORTHILL: We used a Polaroid camera with 3000 speed/type 47 film to photograph the oscilloscope face. The *f* number of the camera and the oscilloscope intensity were adjusted until the desired image was obtained. We plan to use certain enhancing techniques in order to bring out more detail in the images. No attempt to calibrate the images has been made. The images are useful in studying the gross thermal features; however, the thermal contour charts will be used for our detailed analysis.

VOICE: What about cooling curves?

SHORTHILL: Essentially we have reduced the third scan in totality. We have four scans in

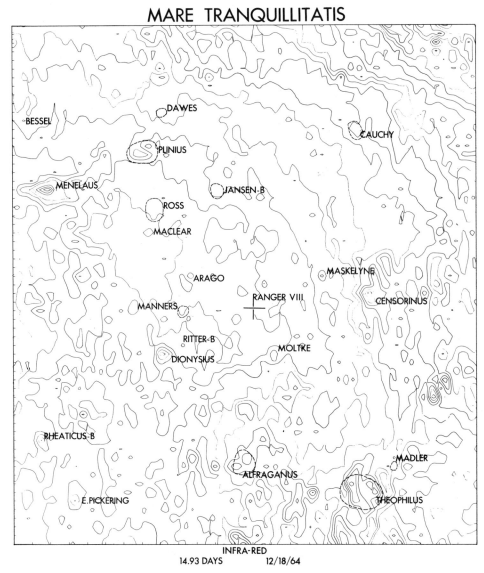

MARE TRANQUILLITATIS

INFRA-RED
14.93 DAYS   12/18/64

ISOTHERMAL CONTOURS ON THE FULL MOON

FIGURE 11–3: Isothermal contours on the full moon (phase angle −2 degrees, 16 minutes). Each tick mark on the bottom margin is separated by ten data values.

penumbra and three scans in totality and one scan in the last penumbral phase. We have worked on only the third scan in totality and we really don't have cooling curves as yet. It takes a long time to get all the points. We are mainly working on location of these points, so that when we go to the scans in the penumbra, we can locate these points, because it is hard to find them as they start to cool.

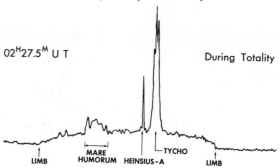

Figure 11–4: Infrared line scan across the moon through the crater Tycho. Note the structure in the crater Tycho and the enhancement of Mare Humorum.

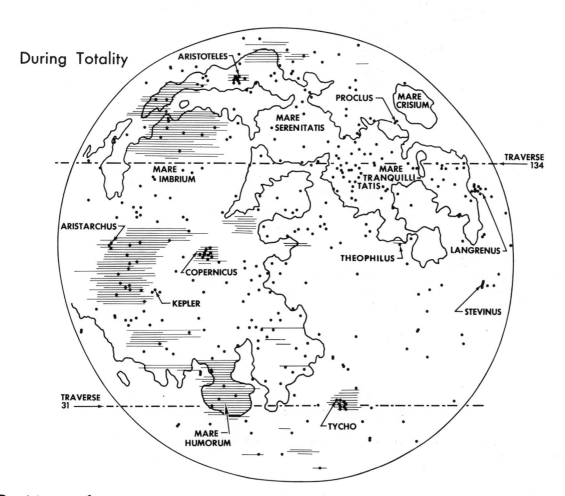

Position of Anomalies During the Dec. 19, 1964, Lunar Eclipse

Dots Indicate "Hot Spots"   Lines Indicate Thermal Enhancements

Figure 11–5: Position of hot spots on the lunar disk. Lines indicating thermal enhancements also show scan line separation.

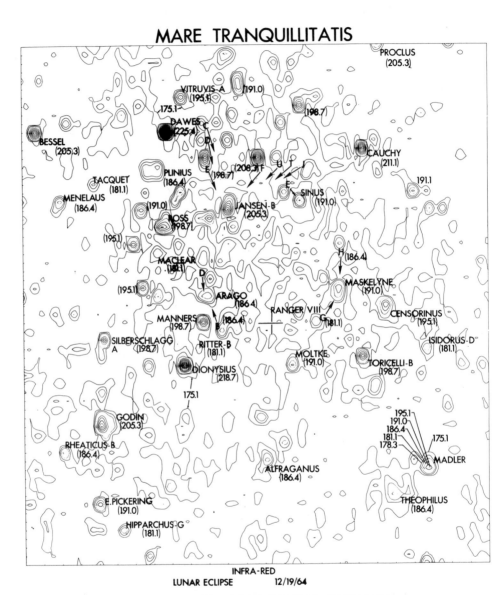

FIGURE 11–6: Isothermal contours on the totally eclipsed moon. Brightness temperatures are indicated in degrees Kelvin.

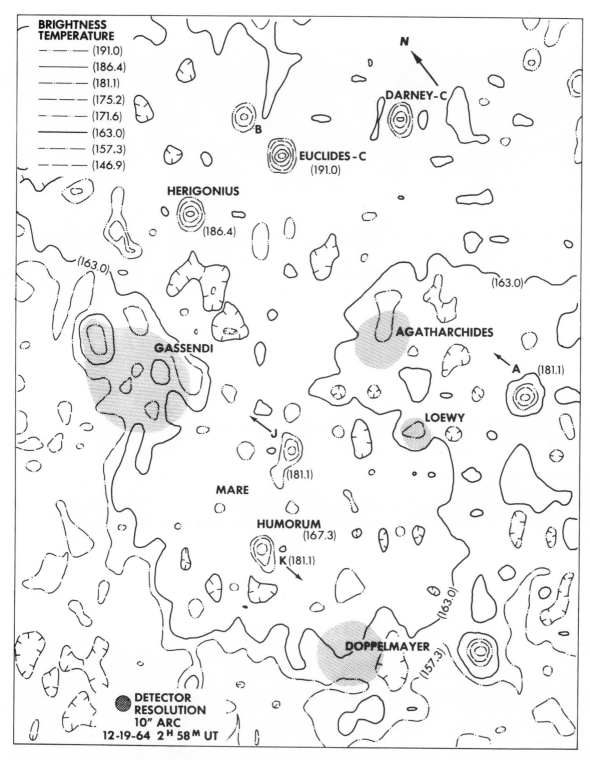

FIGURE 11–7: Automatically contoured region of Mare Humorum showing brightness temperatures. The general area of Humorum was found to be 10°K above its surroundings during totality.

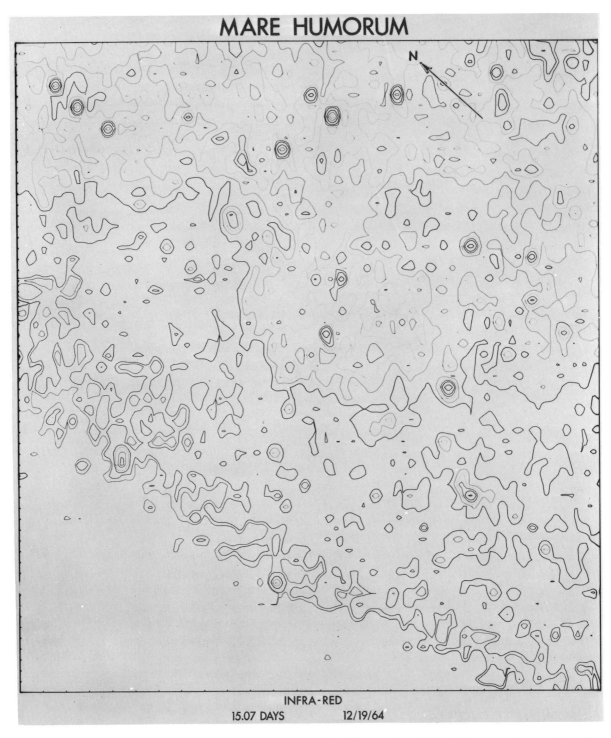

FIGURE 11–8: Isothermal contours of the Mare Humorum region. The southeast limb of the moon is near the noise equivalent temperature of our system.

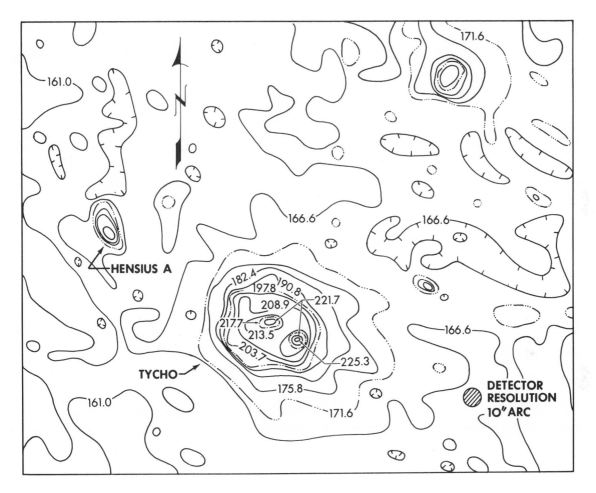

Brightness Temperatures in °K

2^H 19^M U T During Totality of the Dec. 19, 1964, Lunar Eclipse

FIGURE 11–9: Isothermal contours of the crater Tycho. The white spot in the Hell plain is shown in the upper right with a brightness temperature of 203.7°K.

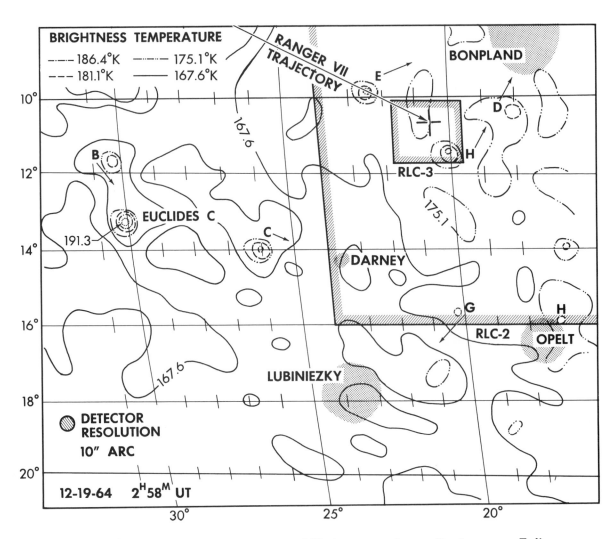

Thermal Contours Near Ranger VII Impact Area During an Eclipse

FIGURE 11–10: Isothermal contours near Ranger VII impact area. Bonpland-H is in the lower right corner of RLC-3.

FIGURE 11–11: Reconstructed visible full-moon image from the photomultiplier (S-11 response which peaks at 4450 Å).

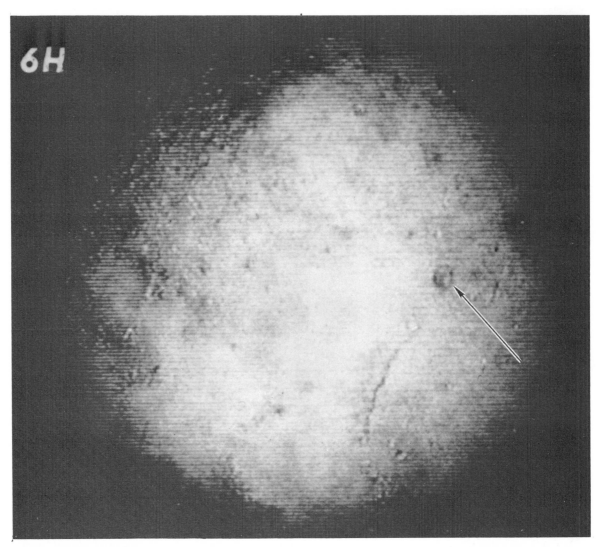

Figure 11–12: Reconstructed infrared full-moon image (10 to 12 μ). The crater Copernicus is indicated by the arrow. Dark tones on the photograph indicate cooler temperatures.

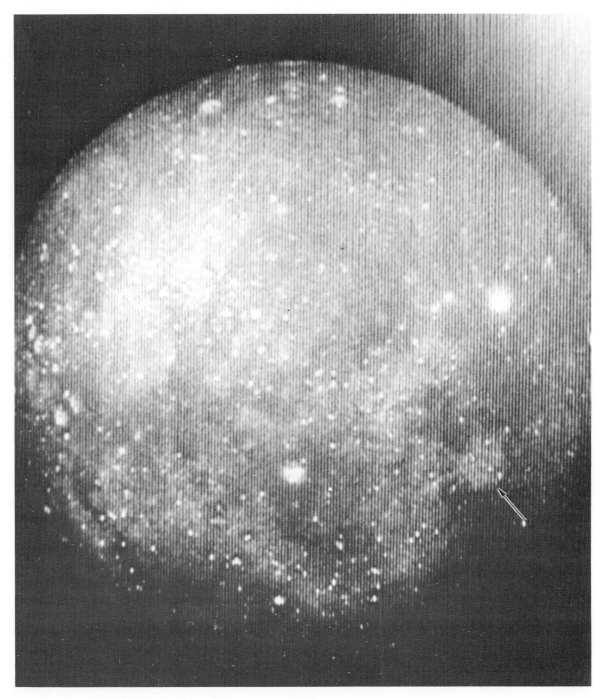

FIGURE 11–13: Reconstructed infrared image during totality. The enhancements and the hot spots are indicated by the bright patches and dots. Mare Humorum is shown by the arrow.

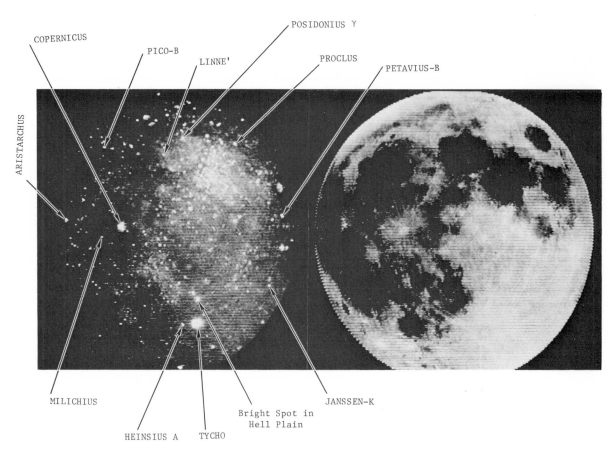

FIGURE 11–14: Reconstructed visible image of the full moon and infrared image of the eclipsed moon. The scales are about the same so thermal structure can be compared with visible features.

12.

Review of Radar
Observations of the Moon*

T. Hagfors

Lincoln Laboratory,† Massachusetts Institute of Technology, Lexington, Massachusetts

ABSTRACT

The methods available for the study of the moon by means of radar are briefly reviewed, a discussion is given of the information which can be derived from radar studies about the nature of the lunar surface, and results of radar observations of the moon are presented. From these results, it is concluded that the lunar surface is covered by a material having an effective dielectric constant of 2.6 and that the surface undulations on the moon have a mean slope of 11–12 degrees on the scale of about a meter. The enhanced reflectivity of young rayed craters has been interpreted as both a higher intrinsic reflectivity and a higher degree of roughness in these features.

I. INTRODUCTION

Before the first Ranger pictures became available, only radar observations could supply reliable direct evidence on the nature of structure sizes on the moon of the order of meters. There are still some questions of interpretation under debate in the radar astronomy of the moon, but there are many significant results about which there is considerable confidence. It is probably safe to assume that by the time the first man sets foot on the lunar surface, radar exploration of the moon will not have

been fully exploited. At that time, it will be possible to check directly the quality of the various interpretations made of the radar measurements, and the radar observations might then take on the task of examining wider areas of the moon than can easily be done by direct exploration by man. Also, it is very likely that the moon will serve as a reference target against which radar astronomical methods can be calibrated for use against the near planets Venus, Mercury, and Mars.

Section II describes in some detail the various methods of observation available to the radar astronomer in studying the moon. In the following section, a description is given of what can be learned about the lunar surface from these observations, and the reliability or accuracy which can be attached to the interpretation of the observational results is then discussed. In section IV, some of the more significant observational results available at the present time are summarized and interpreted in the light of the discussion in section III.

II. METHODS OF OBSERVATION

The first and most obvious observation that can be made is to measure the cross section. The power transmitted is used to compute the power density S_{in} at the target from known parameters of the transmitter antenna and the possible losses in the intervening medium. When the width of the beam is larger than the extent of the target, the illumination is approximately uniform. A certain amount at the incident power density $\sigma \times S_{in}$ is considered

* Indebtedness to Dr. J. V. Evans is gratefully acknowledged for a number of stimulating discussions on the material presented and also for making available some of the experimental results.

† Operated with support from the U.S. Air Force.

to be absorbed by the target and then reradiated omnidirectionally to give rise to the observed power density S_r at the receiver. The factor σ is the cross section of the target, and can, as we shall see, be related to physical properties of the target. The few cases where the antenna beam covers only a fraction of the lunar surface are still somewhat exceptional, and the definition of cross section in these cases must be carried out with some caution.

Range resolution in a radar system can be used to advantage to study the properties of the moon. The round trip delay time from the subterrestrial point to the limb of the moon is about 11.6 ms, and a range resolution of a few microseconds is quite readily obtainable in many radar systems. With a short pulse transmitted, the moon will, at any given time within the time interval during which the pulse passes the moon, be illuminated along a ring—or range ring—centered on the subterrestrial (or more correctly subradar) point. With a pulse length τ the surface area illuminated is

$$2\pi c\tau \cdot a = \text{area} \qquad (1)$$

where a is the radius of the moon and c the velocity of light, and the area is independent of the position of the range ring on the moon. For Earth-based radars, it is only the backscattering properties of the surface which are available for study. A more complete description of the scattering properties can be obtained only with the possibility of moving the transmitter and the receiver arbitrarily with respect to each other and to the surface.

Angular resolution cannot readily be obtained by making the antenna beams sufficiently narrow at ordinary radar frequencies and for this reason the so-called range-Doppler technique was developed.[1] In this technique, one co-ordinate is provided by the range rings just described. The other co-ordinate in the grid system can be derived from the following circumstance: partly due to the librations of the moon, but mainly due to the motion of the observer with the rotating earth, the moon will appear as if it is rotating with respect to the observer. The limb of the moon which is receding will therefore give a lower frequency of reflection than the center of the moon and the approaching limb a higher frequency. Lines of

[1] G. H. Pettengill and J. C. Henry, "The Enhancement of Radar Reflectivity Associated with the Lunar Crater Tycho," *J. Geophys. Res.*, Vol. 67 (1962), pp. 4881–4885.

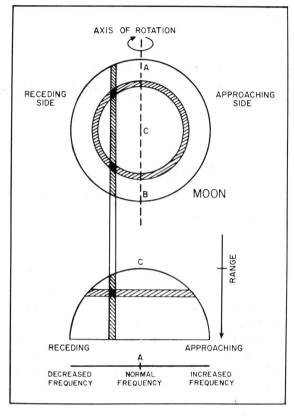

FIGURE 12–1: Range-Doppler co-ordinate system on the moon.

constant Doppler displacement will be straight lines across the lunar disk as shown in Figure 12–1. The distance measured out from the center of the disk along the instantaneous libration equator is proportional to the Doppler shift relative to the lunar center.

As can be seen, this provides a co-ordinate grid on the moon which is twofold ambiguous. This ambiguity can be resolved by observing over an extended period of time during which the direction of the Doppler axis is changing or by providing a beam sufficiently narrow to resolve a small fraction of the lunar surface. In the latter case, the range-Doppler technique provides a method to resolve to a small fraction of the beamwidth. The Cornell University antenna at Arecibo can resolve unambiguously areas of about 20×20 km, and even a finer resolution will probably be available with the Lincoln Laboratory Haystack antenna.

One further degree of freedom is available to the radar observer. The transmitted pulse will always have a definite state of polarization, linear, circular, or otherwise. The received wave can be observed with any set of orthogonal polarizations with properly designed equipment and the various components of the Stokes vector can be found as a function of the state of polarization of the transmitted wave.[2] This possibility does not appear as yet to have been fully exploited in radar astronomy.

III. DEDUCTIONS FROM MEASUREMENTS ABOUT LUNAR PROPERTIES

Let us discuss what deductions can be made about the moon on the basis of the set of measurements described in the previous section.

The cross section of a large dielectric smooth sphere can be shown to be:

$$\sigma = \pi a^2 \left(\frac{\sqrt{\epsilon} - 1}{\sqrt{\epsilon} + 2} \right)^2 = \pi a^2 \cdot R \qquad (2)$$

where ϵ is the dielectric constant. If R is known, the dielectric constant is found from:

$$\epsilon = \left(\frac{1 + \sqrt{R}}{1 - \sqrt{R}} \right)^2 \qquad (3)$$

The relative uncertainty in the determination of ϵ can be shown to be:

$$\frac{\Delta \epsilon}{\epsilon} = \frac{\Delta R}{R} \cdot \frac{2 \sqrt{R}}{1 - R}. \qquad (4)$$

In most radar observations reported,[3] the single measurement uncertainty in R is of the order of 3 db so that $\Delta R / R \approx 1/2$. For small reflection coefficients one finds that:

$$\frac{\Delta \epsilon}{\epsilon} \approx \sqrt{R}. \qquad (5)$$

The relatively large uncertainty in the determination of R stems from the fact that the radar parameters are very hard to determine accurately and it is not caused by variations of target reflectivity.

[2] M. H. Cohen, "Radio Astronomy Polarization Measurements," *Proc. of the IRE*, Vol 46 (1958), pp. 172–183.

[3] J. V. Evans and G. H. Pettengill, "Radar Studies of the Moon," *Radar Astronomy*, ed. by J. V. Evans and T. Hagfors, New York: McGraw-Hill (1965).

On the model consisting of a large smooth sphere made of dielectric material, the reflection coefficient should not vary with frequency unless ϵ does. If a frequency dependence were in fact to be found, this could be explained in one of several ways. It might be that the material is conducting and that the ratio of real and imaginary parts of the complex dielectric constant changes with frequency. This is held rather unlikely—the lunar material is usually considered to be essentially dielectric. The surface might also be nonhomogeneous, consisting perhaps of a dust layer supported by a firmer surface underneath. If the depth of the dust layer were thick at short wavelengths and thin at long wavelengths, there might be a marked variation of reflectivity. A thick dust layer of random depth could reduce the reflectivity observed with respect to the reflectivity of the supporting layer by as much as 50 per cent provided the dielectric constant of the dust layer were equal to the square root of that of the underlying layer. Yet another cause of a frequency-dependent reflection coefficient might arise from the presence of surface roughness. The surface might be mainly describable as an interface between vacuum and a dielectric medium at long wavelengths but not so to the same extent at shorter wavelengths.

A smooth sphere such as the one just described would give all the reflection from the Fresnel zone centered on the subterrestrial point. The size of this would be typically of the order of $\sqrt{a \cdot \lambda}$ corresponding to 1.3 km at $\lambda = 1$ m or 130 m at $\lambda = 1$ cm, and this is not what is observed in practice. The first refinement in the surface model is therefore to assume that the surface consists of plane facets and that the probability of having a normal $\vec{n}$ to one of these facets within a solid angle element $d\Omega$ is:

$$p(\vec{n}) \, d\Omega = \frac{1}{2\pi} p(\phi) \, d\Omega \qquad (6)$$

where ϕ is the angle between $\vec{n}$ and the mean normal to the surface. With this assumption and with the further assumption that geometric optics is valid we find that the power scattered back from unit area of the mean surface is proportional to:

$$P(\theta) \sim \frac{p(\theta)}{\cos \theta}. \qquad (7)$$

Considerable work has gone into the study of backscattering from an undulating surface from a physical-optics point of view.[4] In this type of approach the lunar surface is assumed to deviate from the ideal spherical shape by an amount ζ in the radial direction. ζ is regarded as a Gaussian random function of position on the mean surface and the nature of the height deviations is described by a variance and a correlation function relating the deviations at neighboring points. In this approach, the boundary fields are established by assuming the ordinary reflection and refraction laws to apply for the local geometric conditions. Since this condition implies that the local radius of curvature is much larger than the wavelength of the radar wave and since the depths of the undulations on the moon are also large compared with the wavelength, it turns out that the physical-optics and the geometric-optics approaches to lunar scattering are identical for all practical purposes.[5] With the plane facet model and on the assumption that the lunar surface is statistically homogeneous, we can reinterpret observations of power as a function of range directly as a distribution of surface slopes and we can, in particular, compute mean surface slope and rms surface slope as we may desire. The total cross section can be shown to remain largely uninfluenced by the presence of surface slopes. One can imagine that the surface slopes have only the effect of redistributing the first Fresnel zone over the surface.

Observation of orthogonal polarizations on reception has shown that not even the surface model described above is good enough—although there is still some debate on this problem. With the model just sketched, one should not expect appreciable depolarization of the transmitted wave since all reflections take place from facets normal to the line of sight. In other words, transmission of a right circular wave should result in a purely left circularly polarized wave, transmission of a linearly polarized wave should give rise to a linearly polarized wave in the same direction, etc. Since depolarization is actually observed, we must refine the surface model in order to explain this effect. There are several possibilities. It could be that there were, in addition to the largely smooth facets, some small-scale structure protruding from the surface, e.g., small rocks strewn over the surface, or alternatively, a substantial number of very small craters. Such objects would not necessarily scatter back in a polarization corresponding to that of the transmitted wave, and depolarization would hence be observable if an appreciable fraction of the scattered power were due to such a mechanism. A collection of small independent linear dipoles of random orientation has been shown to give rise to a depolarized component only 4 db below the expected polarization.[6] Suppose, as another possibility, that the surface is actually covered with a dustlike substance and that the reflection or scattering actually takes place from within the surface layer. For scattering at oblique incidence there would then be a systematic difference in the reflection coefficients for the two linearly polarized components in and across the local plane of incidence and again a depolarization could occur. With full control over the polarization of the transmitted wave and with the ability to observe any set of mutually orthogonal polarizations on reception, such a hypothesis can be tested. Some work in this direction is currently in progress at Lincoln Laboratory.

Finally, the availability of resolution of relatively small surface areas by means of the range-Doppler technique explained above makes some further conclusions about the nature of the surface possible. An increase in reflectivity from an area viewed at oblique incidence might mean that the area is rougher than its surroundings. It might also mean that the reflectivity—or the intrinsic dielectric constant—is higher in this area, or it might of course happen that both of these possibilities occur simultaneously. If a similar surface area is moved so that it is viewed at approximately normal incidence, however, the effect of increased roughness would be to decrease the reflectivity, whereas the presence of a material with a higher intrinsic reflectivity would tend to increase the reflectivity. Let us now turn to some observational results and discuss them in the light of the relationships just outlined.

[4] F. B. Daniels, "A Theory of Radar Reflection from the Moon and Planets," *J. Geophys. Res.*, Vol. 66 (1961), pp. 1781–1788; V. A. Hughes, "Diffraction Theory Applied to Surface," *Proc. Phys. Soc.*, Vol. 80 (1962), pp. 1117–1127; D. F. Winter, "A Theory of Radar Reflections From a Rough Moon," *J. Res. Nat. Bur. Stand.*, Vol. 66D (1962), pp. 215–226.

[5] T. Hagfors, "Backscattering from an Undulating Surface with Applications to Radar Returns from the Moon," *J. Geophys. Res.*, Vol. 69 (1964), pp. 3779–3784.

[6] Evans, "The Scattering of Radiowaves by the Moon," *Proc. Phys. Soc.*, Vol. 70B (1957), pp. 1105–1112.

IV. DISCUSSION OF OBSERVATIONAL RESULTS

Measurements of cross section have been carried out over a wavelength region from 8.6 mm to 20 m. The cross section expressed as a fraction of the physical cross section is given in Table 12–1 together with the source of the observation and the estimated uncertainty. Where no estimate of uncertainty was quoted by the source, the error was put equal to ±3 db. The same information is displayed in Figure 12–2 where cross section is plotted as a function of wavelength. Because of the large error bars it is difficult to establish a definite frequency dependence. Before the results of Davis and Rohlfs were published[7] it was customary to assume that there was no systematic frequency dependence. With these new results, however, this view might have to be revised in favor of a definite frequency dependence, the cross section being about twice as large at 10–20-m wavelengths as at shorter wavelengths. As will be discussed below,

[7] J. R. Davis and D. C. Rohlfs, "Lunar Radio-Reflection Properties at Decameter Wavelengths," *J. Geophys. Res.*, Vol. 69 (1964), pp. 3257–3262.

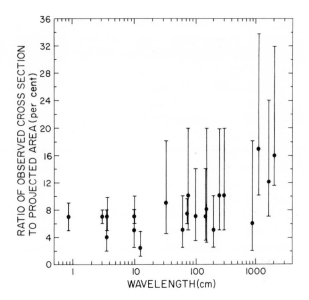

FIGURE 12–2: Measured cross sections as a function of wavelength.

TABLE 12–1: Values for the Radar Cross Section of the Moon as a Function of Wavelength Reported by Various Workers

Author	Year	Wavelength (cm)	$\sigma/\pi a^2$	Estimated Error db
Lynn, Sohigian, Crocker	1963	0.86	0.070	± 1.0
Kobrin	1963*	3.00	0.070	± 1.0
Morrow *et al.*	1963*	3.60	0.070	± 1.5
Evans and Pettengill	1963a	3.60	0.040	± 3.0
Kobrin	1963*	10.00	0.070	± 1.0
Hughes	1963*	10.00	0.050	± 3.0
Victor, Stevens, and Golomb	1961	12.50	0.022	± 3.0
Aarons	1959**	33.50	0.090	± 3.0
Blevis and Chapman	1960	61.00	0.050	± 3.0
Fricker, Ingalls, Mason, Stone, and Swift	1960	73.00	0.074	± 1.0
Leadabrand	1959**	75.00	0.100	± 3.0
Trexler	1958	100.00	0.070	± 4.0
Aarons	1959**	149.00	0.070	± 3.0
Trexler	1958	150.00	0.080	± 4.0
Webb	1959**	100.00	0.050	± 3.0
Evans	1957	250.00	0.100	± 3.0
Evans, Evans, and Thompson	1959	300.00	0.100	± 3.0
Evans and Ingalls	1962	784.00	0.060	± 5.0
Davis and Rohlfs	1964	1130.00	0.170	+ 3.0 − 2.0
Davis and Rohlfs	1964	1560.00	0.120	+ 3.0 − 2.0
Davis and Rohlfs	1964	1920.00	0.160	+ 3.0 − 2.0

* Revised value (privately communicated to Evans and Pettengill).
** Reported by T. B. A. Senior and K. M. Siegel, "A Theory of Radar Scattering by the Moon", *J. Res. Nat. Bur. Stand.*, Vol. 64D(1960), pp. 217–228.

this frequency dependence of the cross section, if proved real, will lead to a revision of the current view that radar reflections occur from a largely homogeneous surface. With an accurately calibrated reference target in satellite orbit, the different radars could be calibrated and the frequency dependence could be established quite accurately. A metallic sphere of cross section 1 m² will be put into orbit during 1965 and it is to be expected that a well-defined frequency law of the moon's radar cross section can be established then. This sphere is, however, too small for accurate calibration of the radar systems using the longest wavelengths, which, particularly in view of the latest results, is very unfortunate.

Taking the 7 per cent cross section and assuming a homogeneous surface for all depths of importance for the reflection of the wave, we obtain a dielectric constant slightly less than 3.0.

The reflection coefficient at normal incidence for a dielectric medium with $\epsilon = \epsilon_2$ covered by another dielectric of thickness d and with dielectric constant ϵ_1 is:

$$R = \frac{(1 - \sqrt{\epsilon_2})^2 - (1 - \epsilon_1)(1 - \epsilon_2) \sin^2 \alpha}{(1 + \sqrt{\epsilon_2})^2 - (1 - \epsilon_1)(1 - \epsilon_2) \sin^2 \alpha}. \quad (8)$$

As it is very unlikely that the moon is covered by a surface layer of constant thickness, it appears to be plausible to assume some probability density of the phase $\alpha = 2\pi/\lambda \, d \, \epsilon_1$. Provided the layer is thick measured in wavelengths, the probability density is approximately constant in the range of α from 0 to 2π. This average reflection coefficient becomes:

$$R_o = 1 - \frac{4\sqrt{\epsilon_1 \epsilon_2}}{(\sqrt{\epsilon_2} + 1)(\sqrt{\epsilon_2} + \epsilon_1)}. \quad (9)$$

Note that R_o has a minimum when $\epsilon_1 = \sqrt{\epsilon_2}$. The minimum value of R_o is approximately equal to half of that corresponding to a homogeneous layer with dielectric constant ϵ_2. A 7 per cent cross section could therefore also be obtained with a surface layer with $\epsilon_1 \approx 2$ and with an underlying layer with $\epsilon_2 \approx 5$. With this model the increase of reflectivity at long wavelengths could be explained by assuming that the thickness of the surface layer is typically of the order of several meters. The long wavelength radiation would then be reflected more as if directly from the underlying medium with dielectric constant ϵ_2.

TABLE 12–2: Values of Dielectric Constants for Some Solids (frequency range 10–10⁴ mc/sec)

Mineral	Dielectric constant ϵ	Reference
Quartz, sandstone	4.8	a
Quartz, fused	3.8	b
Dry sand	2.6	b
Olivine basalt	17.4	a
Olivine basalt, cellular	5.5	a
Rhyolite	4.0	a
Rhyolitic pumice	2.3	a
Ruby mica	5.4	b

[a] M. Brunschwig, *et al.*, "Estimation of the Physical Constants of the Lunar Surface," *University of Michigan Radiation Laboratory, Report 03544-1-F; AD 253 472, PB 162-020,* Ann Arbor, Mich. (1960).
[b] Von Hippel, *Dielectric Materials and Applications.*

The double-layer model appears attractive in another respect also. There has long been a discrepancy between the dielectric constants determined from studies of the polarization of thermal emission from the moon[8] and those determined from radar data.[9] The former have proved to be consistently low. With the double-layer model it appears at the present time as if it might be possible to reconcile the radar and the radiometric data to a certain extent at least.

We shall shortly present arguments to support an adjustment of the dielectric constant toward a somewhat lower value than 3.0 determined directly from the radar reflectivity. First, let us pause for a moment to discuss what might be the nature of a dielectric medium having a dielectric constant between 2 and 3. Table 12–2 gives the dielectric constant of some materials which might exist on the lunar surface. It is naturally impossible to draw any general conclusions from such a small sample. It appears, however, that the rather low dielectric constant observed cannot correspond to a solid rock surface, but must indicate that the surface material is in a less compacted form, possibly like sand, dust, or pumice. The compactness of the surface material is rather difficult to deduce even if we were to know

[8] N. S. Soboleva, "Measurement of the Polarization of Lunar Radio Emission on a Wavelength at 3.2 Cm," *Astron. Zh.,* Vol. 39 (1962), pp. 1124–1126 [in Russian]; C. E. Heiles and F. D. Drake, "The Polarization and Intensity of Thermal Radiation from a Planetary Surface," *Icarus,* Vol. 2 (1963), pp. 281–292.
[9] Evans and Pettengill, "The Scattering Behavior of the Moon at Wavelengths of 3.6, 68, and 784 Centimeters," *J. Geophys. Res.,* Vol. 68 (1963), pp. 423–447.

the type of surface material on the moon. There are several formulas relating effective dielectric constant to the fill factor of the material and it is not obvious which one to use. Twersky[10] gives:

$$\epsilon = 1 + \frac{3w_o \dfrac{\epsilon' - 1}{\epsilon' + 2}}{1 - w_o \dfrac{\epsilon' - 1}{\epsilon' + 2}} \qquad (10)$$

where ϵ' is the dielectric constant of the compacted material and where w_o is the packing factor. Odelevskii and Lenin, on the other hand, give the following:[11]

$$\epsilon = \epsilon' \left(1 - \frac{3(1 - w_o)}{\dfrac{2\epsilon' + 1}{\epsilon' - 1} + (1 - w_o)} \right). \qquad (11)$$

Both of these formulae are related to the Lorentz-Lorenz relation[12] but they give somewhat different answers. If we assume $\epsilon' = 8$ for the compacted material, (10) leads to $w_o = 0.55$ whereas (11) gives $w_o = 0.30$. If there is some very tenuous dust on the lunar surface like the fairy castle structure suggested by Gold[13] the radar data seem to indicate that this layer must be so thin that the radar cross section remains unaffected by it. That is, the dust layer is not likely to be more than about 1 cm deep. The material which gives rise to the radar reflectivity must have a fill factor of more like 50 per cent and is hence probably better characterized as a loosely packed sandy substance rather than dust.

The power returned as a function of range or delay has been extensively studied.[14] These data have been analyzed in terms of angle of incidence θ to the mean surface by the substitution:

$$1 - \cos \theta = \frac{c}{2a} \tau \qquad (12)$$

where τ is the delay measured with respect to the subterrestrial point. For a uniformly bright moon,

[10] V. Twersky, "On Scattering of Waves by Random Distributions, II. Two Space Scatterer Formalism," *J. Math. Phys.*, Vol. 3 (1962), pp. 724–734.

[11] V. D. Kroitikov and V. S. Troitsky, "Radiation Properties of the Moon at the Centimeter Wavelengths," *Astron. Zh.*, Vol. 39 (1962), pp. 1089–1093 [in Russian].

[12] A. R. von Hippel (ed.), *Dielectric Materials and Applications*, New York: John Wiley and Sons (1954).

[13] T. Gold, Paper presented at MIT Compass Seminar, Massachusetts Institute of Technology, Cambridge, Mass., January, 1964.

[14] Evans and Pettengill, "The Scattering Behavior."

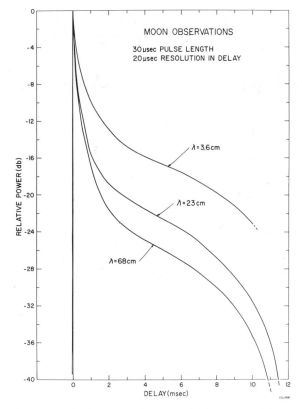

FIGURE 12–3: Power as a function of delay as observed at three different wavelengths.

such as observed at optical wavelengths, the power per unit area versus angle of incidence would be:

$$P(\theta) \sim \cos \theta \qquad (13)$$

whereas for a Lambert law of reradiation one would have:

$$P(\theta) \sim \cos^2 \theta . \qquad (14)$$

Figure 12–3 shows the power versus delay for three different wavelengths, viz $\lambda = 3.6$ cm, $\lambda = 23$ cm and $\lambda = 68$ cm. The trend toward a more pronounced highlight near the subterrestrial point with increasing wavelength continues also at longer wavelengths. When one approaches the millimeter wavelength region the highlight near the subterrestrial point has nearly vanished. This trend is illustrated in Figure 12–4 where the relative power is plotted against log 10 cos θ.

We conclude from this that the lunar surface is largely smooth locally at the longer wavelengths and that the model of the surface involving large specularly reflecting facets tilted with respect to the

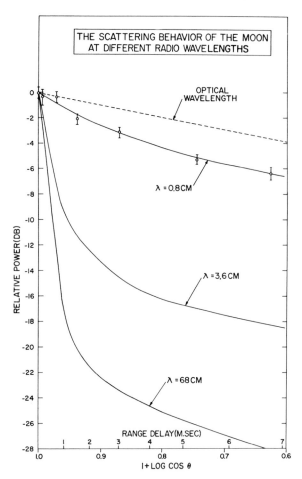

FIGURE 12–4: Plots of relative power as a function of log 10 cos θ.

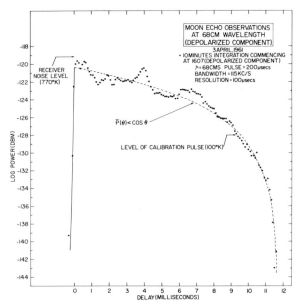

FIGURE 12–5: Depolarized (circular) power as a function of delay for λ = 68 cm.

mean surface is largely in agreement with the actual surface properties. However, at wavelengths less than one centimeter this picture is no longer a good approximation. At the short-wavelength end of the spectrum, structure of size comparable with the wavelength must be relatively much more abundant than in the long-wavelength region. Before attempting to make any deductions about the surface slopes we shall discuss the results of some measurements of depolarization caused by the surface because this has some bearing on the problem of surface slopes.

The two circularly polarized waves, resulting when a circularly polarized wave was transmitted, were observed as a function of time delay by Evans and Pettengill[15] at a wavelength of 68 cm. An example of one of their results is shown in Figure 12–5. The dotted curve in this diagram displays the

behavior of a uniformly bright moon, i.e., one where the power versus range varies as cos θ. It can be seen that the moon as viewed in depolarized radiation is indeed very closely uniformly bright. By comparing the polarized and the depolarized components the degree of polarization can be evaluated by

$$p = \frac{P_r - P_e}{P_r + P_e}. \tag{15}$$

The polarization factor p is plotted as a function of range in Figure 12–6.[16] As can be seen, the degree of polarization of the received signal at first declines rather rapidly out to about 3 ms and then more slowly out toward the limb where the polarization approaches about 30 per cent. Such depolarization could be caused by scattering from individual small objects. There could be a certain amount of multiple scattering involved, but this seems to be rather unlikely in view of the low intrinsic reflectivity of the lunar surface material. There could also be some penetration phenomenon which might account for at least part of the observed effect. Little has been done about the problem of depolarization at other wavelengths.

The presence of depolarization very strongly suggests that the surface model involving a surface which is largely smooth is inadequate. The correct

[15] *Ibid.*

[16] *Ibid.*

interpretation appears to be that there are two different scattering mechanisms present, one corresponding to a locally smooth surface to which we can ascribe a dielectric constant, and another component, diffuse, arising through the presence of microstructure on the surface. The dielectric constant which must be associated with the quasi-specular component will, therefore, be somewhat less than the one deduced by the use of the total cross section only. Before the dielectric constant is found, the diffuse component must be subtracted from the total return. For the 68-cm results, Rea, Hetherington, and Mifflin[17] have subtracted the diffuse portion of the polarized echo by assuming the power in this component to be between three and four times that in the depolarized echo. Employing this procedure, a dielectric constant of about 2.6 is obtained for the 68-cm data. At 68 cm, the quasi-specular component accounts for about 80 per cent of the reflected power, at .86 cm only 15 per cent can be ascribed to this component.

In deducing the value of the typical surface slope from the data, a similar procedure is followed. When the corrected form of the power-versus-range curve is found, the probability distribution of the angle between the mean normal and the actual normal is determined and the mean slope comes out to be about 11 degrees. Note that these slopes refer to a scale which must be of the order of one

[17] D. G. Rea, N. Hetherington, and R. Mifflin, "The Analysis of Radar Echoes from the Moon," *J. Geophys. Res.*, Vol. 69 (1964), pp. 5217–5223.

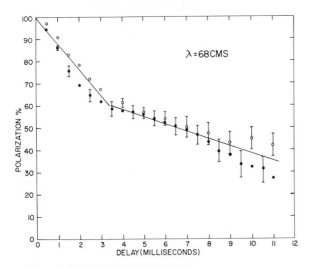

FIGURE 12–6: Depolarization factor p plotted as a function of delay for $\lambda = 68$ cm.

meter. At $\lambda = 3.6$ cm, where the subtraction of the diffuse component is more difficult, the mean slope comes out to be about 15 degrees. It would seem that the value of 11 degrees is the one which is significant in the design of vehicles for lunar landings.

Finally, let us briefly show some results of the first radar identification of a rayed crater, namely Tycho, on the moon by means of the range-Doppler technique.[18] Figure 12–7 shows four power spectra obtained at different ranges on the moon. The pronounced spike showing up in *b* and *c* has been identified with the crater Tycho. Many other rayed craters have since been identified, using the same technique, by the Cornell University Arecibo group. It has been shown that Tycho reflects about ten times as strongly as the surrounding area. It appears to be difficult to explain this number either as an increased intrinsic reflectivity or as an increase in roughness. It has been suggested that both explanations must be involved simultaneously, i.e., Tycho must be considerably rougher than its surroundings and it must also have a higher intrinsic reflectivity than the surrounding area.

V. CONCLUSIONS

The model of the lunar surface emerging from radar observations is therefore the following. On the scale of a meter the lunar surface appears to be locally largely smooth and to have mean slopes of about 11–12 degrees. On the scale of a few millimeters the surface appears to be rough, as if covered with material having this typical structure scale. This type of material seems to cover almost all areas on the moon. If the data are interpreted in terms of a homogeneous surface at all depths, the dielectric constant derived is about 2.6 and this could indicate a surface having the general character of coarse sand—the grains might not of course resemble those of sand encountered on the earth. If the surface has to be interpreted in terms of a double layer—a possibility which we consider unlikely but which nevertheless cannot be ruled out completely—the surface layer might have a dielectric constant as low as 2 and it might be supported by a denser material with a dielectric constant of, say, 5. Rayed craters seem to have a

[18] Pettengill and Henry, "The Enhancement of Radar Reflectivity."

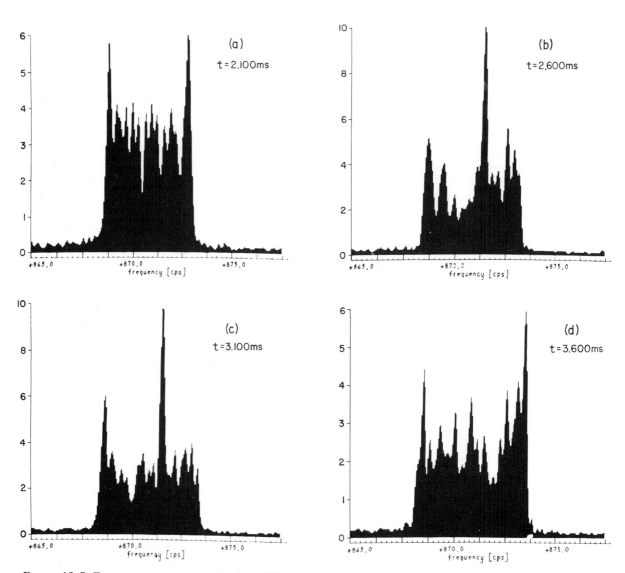

FIGURE 12–7: Frequency power spectra for four different ranges on the moon for $\lambda = 68$ cm.

higher intrinsic reflectivity than the mean surface and, in addition, appear to be rougher than their surroundings on the scale of meters.

DISCUSSION

UREY: What is the dielectric constant of some likely materials for the surface of the moon, in compact form?

HAGFORS: I would say 4 to 6, or thereabouts.

UREY: So that you could account for it by silicates, providing they are somewhat uncompacted?

HAGFORS: Uncompacted, yes.

VOICE: I am puzzled by the possibility of the two-layer model, because in it there is some absorption of the radar wave in the upper layer as it goes in and comes out. This absorption is a function of wavelength, so that as one would go to a shorter wavelength, one would see less and less of the underlying layer. We should therefore see a change in the reflection coefficient as a function of the wavelength.

The fact that there is none seems to be significant, because it indicates we are not reaching an interface, even on the longest wavelengths, which suggests that the surface is, in fact, uniform to a depth of 10 m.

HAGFORS: I am basing all my conclusions on the curve I showed here. We do not have any basis for saying whether there is a frequency dependence or whether there is none.

There is a slight indication that the cross section does go up at the longer wavelengths. But, in my opinion, it is not yet significant. So I would never state very strongly that it is really completely wavelength-independent.

13.

Terrestrial Calderas, Associated Pyroclastic Deposits, and Possible Lunar Applications

Robert L. Smith

U.S. Geological Survey, Washington, D.C.

Terrestrial calderas, generally speaking, are basins of subsidence larger than volcanic craters. They are formed when magmatic support is withdrawn from the roof of near-surface magma chambers causing the overlying rock to founder, usually along concentric faults or ring fractures.

Withdrawal of magma and subsidence may take place with or without surface volcanism at the site of the caldera. However most terrestrial calderas are caused by the rapid eruption of large volumes of ash and pumice from central vents or fissure systems within the area subsequently occupied by the caldera.

FIGURE 13–1: The summit area of Mauna Loa, Hawaii, illustrating a Kilauean type caldera.

FIGURE 13–2: An artist's conception of the beginning of a violent pyroclastic eruption from a large volcanic cone of andesitic composition, Oregon's Mt. Mazama.

FIGURE 13–3: An artist's reconstruction of the formation of Crater Lake, a Krakatoan type caldera.

I would like to discuss briefly three major caldera types: (1) Kilauean calderas, (2) Krakatoan calderas, and (3) resurgent cauldrons.

Figure 13–1 shows the summit area of Mauna Loa, Hawaii, and illustrates a Kilauean type caldera; calderas of this type form on the summit of some large basaltic shield volcanoes. They are flat-floored, vertical-walled shallow depressions probably caused by expansion of the volcanoes' summits due to vertical rise of magma into summit chambers followed by contraction due to lateral withdrawal of magma into rift zones on the volcanoes' flanks. They are rarely more than 6 miles in diameter.

Figure 13–2 shows the beginning of a violent pyroclastic eruption from a large composite vol-

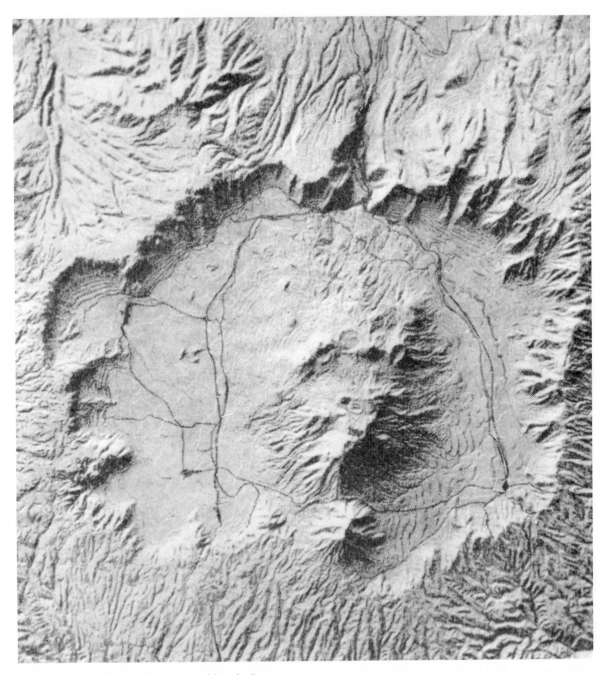

FIGURE 13–4: A relief model of Aso caldera in Japan.

canic cone of andesitic composition. Much of the early material rises high in the air and is deposited as direct fall. Most of the material, however, falls back on the volcano and forms gas-emitting, highly mobile avalanches or ash flows. While emitting gas these ash flows are analogous to fluidized systems, and hence flow off volcanic highlands and flood the lowlands much as water would. Some ash flows are known to have traveled 75 or 80 miles from their source.

The rapid eruption of more than a few cubic miles of such materials results in subsidence over the magma chamber causing Krakatoan type calderas as shown in Figure 13–3. This figure and

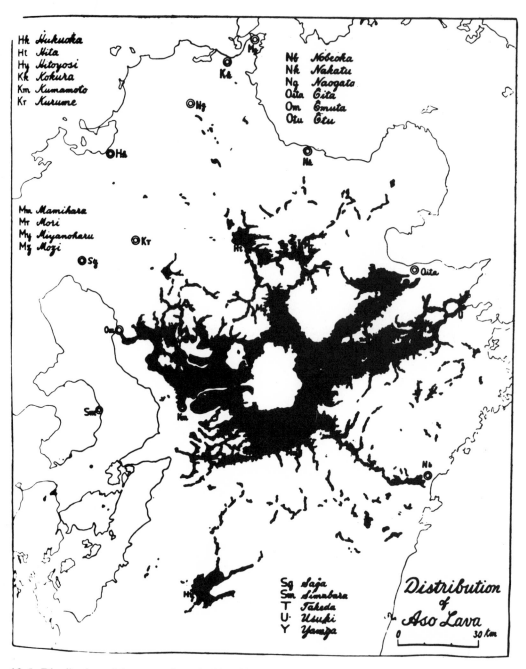

FIGURE 13–5: Distribution of the approximately 50 cubic miles of ash flows that surround Aso caldera.

Figure 13–2 are accurate artists' reconstructions of Oregon's Mt. Mazama and the formation of Crater Lake caldera.

Calderas of this type, formed by subsidence of the tops of large andesitic volcanoes are the most common type among the active volcanoes on the earth. These calderas, confined within the limits of volcanic cones, rarely exceed 8 miles in diameter. However, calderas involving subsidence of groups of cones and broad volcanic plateaus may be much larger and occur frequently among the prehistoric products of active volcanic terrains as well as throughout the older geologic section.

Figure 13–4 shows a relief model of Aso caldera in Japan. Aso is about 14 miles in diameter. Note the central cones with summit craters. Post subsidence volcanism is common to many calderas and the volcanoes may be randomly or symmetrically distributed. Figure 13–5 shows the distribution of the approximately 50 cubic miles of ash flows that surround Aso caldera. Some of these traveled at least 70 miles.

The Valles caldera in New Mexico is an archetypal resurgent cauldron. Figure 13–6 shows the Valles caldera and associated Bandelier ash-flow field superimposed on an older volcanic basement. The calderas that form at the earth's surface over resurgent cauldrons are distinguished from other calderas by the presence of a central mountain mass that was formed by domical uplift of the subsided calderá floor. This structural dome may or may not be associated with intracaldera volcanism.

Figure 13–7 is a geological sketch map of the Valles caldera showing the structural uplift and a ring of peripheral volcanoes. Our interpretation of the sequence of events (see Figure 13–8) is as

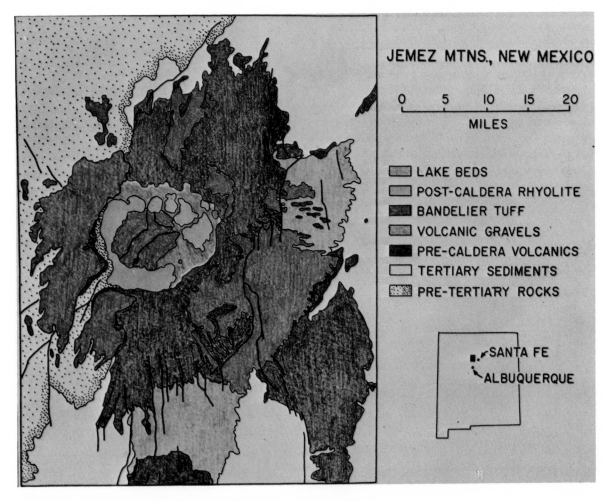

FIGURE 13–6: The Valles caldera in New Mexico, a resurgent cauldron.

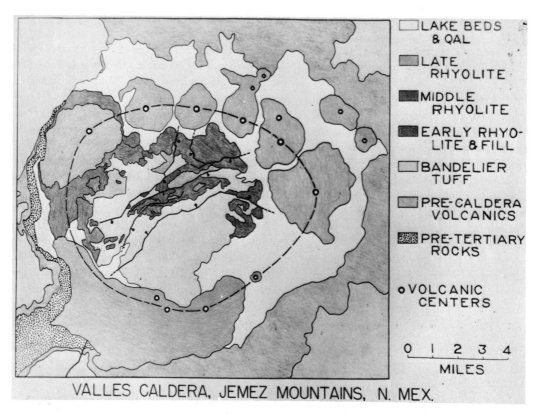

VALLES CALDERA, JEMEZ MOUNTAINS, N. MEX.

FIGURE 13–7: A geological sketch map of the Valles caldera.

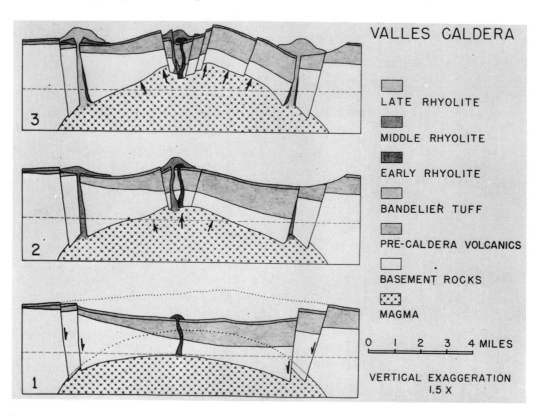

FIGURE 13–8: An overview of the sequence of events in the formation of the Valles caldera.

follows: (1) eruption of the Bandelier ash flows; (2) subsidence of a cylindrical block of crust about 10 miles in diameter to a depth of several thousand feet; (3) burial of the subsided block, to a depth of about 2,000 feet by volcanic ash, lavas, and erosional detritus including lake beds; (4) uplift and doming of the subsided block; (5) emplacement of ring volcanoes; and (6) revelation by erosion of the presence of the pre-subsidence surface of the Bandelier ash flows at a higher elevation than before collapse. I should mention that the central uplift is complexly faulted and transected by a graben-like rift zone over 3,000 feet deep.

In Figure 13–9 a shaded relief map shows the caldera, resurgent dome with transecting graben faults, and ring of volcanoes. Figure 13–10 shows sketches of three other resurgent calderas. Note the transecting fault fractures in the structural domes of each caldera and the chain of craters associated with the Long Valley caldera. Not shown is a greater chain of craters and volcanic domes north of this caldera.

Figure 13–11 is a short review of what I have said:

Kilauean calderas—primarily caused by subsidence without surface volcanism. A terrestrial one is not known to exceed about six miles in diameter.

Krakatoan calderas—exemplified by Crater Lake, Oregon, where pyroclastic eruptions cause subsidence resulting in random or symmetrically located central volcanoes. The maximum size of the central-vent-type ones is about 8 miles. Others are much larger, the maximum known circular ones being about 18 miles across and elongate ones being about 60 miles long. Much larger ones are suspected.

VALLES CALDERA

0   1   2   3   4   MILES

FIGURE 13–9: A shaded relief map of the Valles caldera.

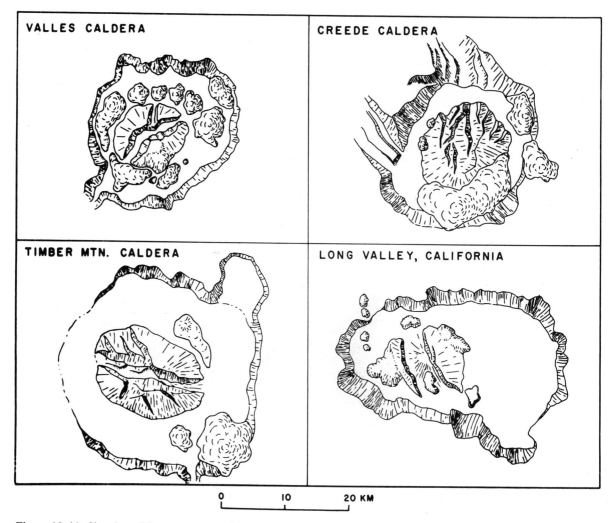

Figure 13–10: Sketches of four resurgent calderas.

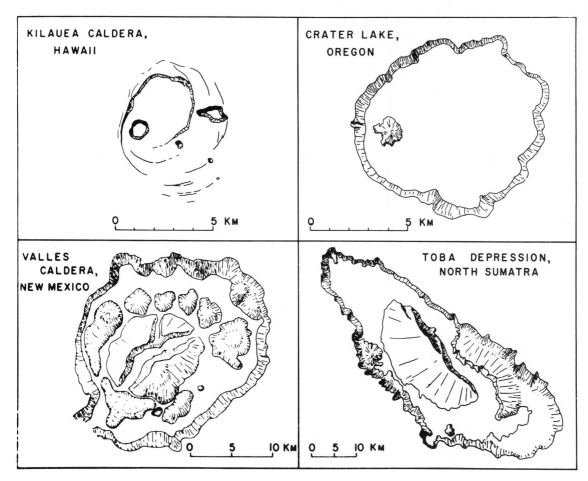

FIGURE 13–11: A review and comparison of the Kilauean caldera, the Krakatoan caldera, and two resurgent cauldrons.

Resurgent cauldrons—typified by a central structural mountain mass. The maximum known circular ones are about 16 miles across and the maximum elongate ones are about 60 miles long.

The Toba cauldron, a resurgent structure 60 miles long by 20 miles wide, formed following the eruption of nearly 500 cubic miles of ash flows. The ash-flow field surrounding Toba covers more than 10,000 square miles.

Terrestrial ring structure complexes are evidence for the existence of calderas in older volcanic terranes. The total ring structure problem is much too complex even for a summary. However, let us say that these structures are associated with plutonic and volcanic rocks of nearly all compositions and that they are now generally recognized as the subvolcanic equivalents of calderas. They give us additional information on caldera size and distribution. Figure 13–12 shows four granite ring complexes, all containing down-faulted volcanics intruded by a plutonic mass and bounded by ring dikes. These we interpret as resurgent cauldrons.

In Figure 13–13 a ring structure, 14 miles in diameter, is more probably related to Kilauean

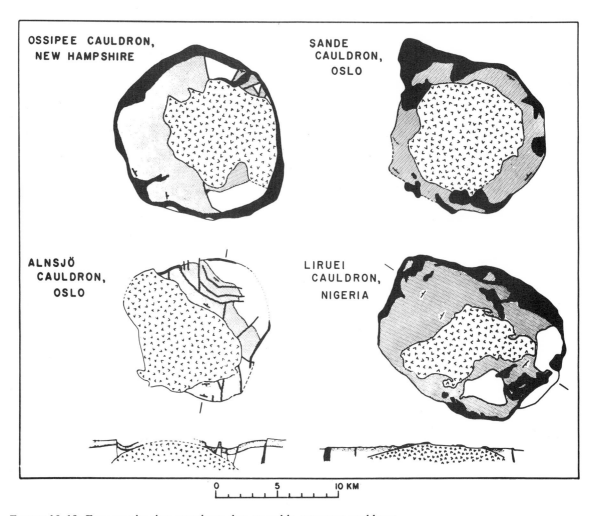

FIGURE 13–12: Four granite ring complexes that resemble resurgent cauldrons.

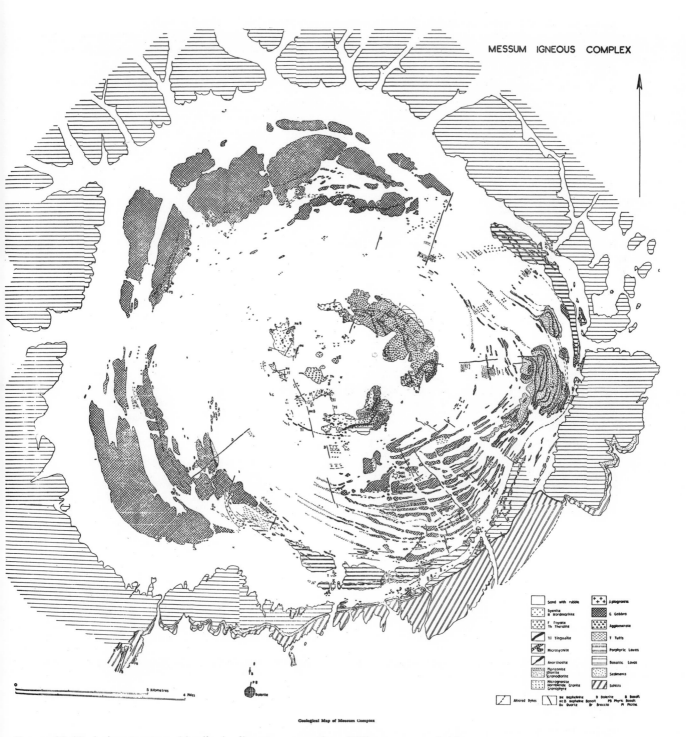

FIGURE 13–13: A ring structure, 14 miles in diameter, suggestive of Kilauean type subsidence.

type subsidence and this structure suggests that Kilauean calderas can be larger than those presently known.

Most terrestrial ring complexes and many calderas occur in groups related to major tectonic features of the crust but not necessarily to the great orogenic lineaments. Figure 13–14 shows such a group in Australia. Many of these are multiple structures, showing overlap in both time and space. The largest is about 70 miles long. The total area embraced is about 12,000 square miles.

Figure 13–15 shows a caldera group in Africa, also covering an area of about 12,000 square miles. These structures are associated with ash flows.

Figure 13–16 is a crude distribution map of known ash-flow deposits and calderas in the western

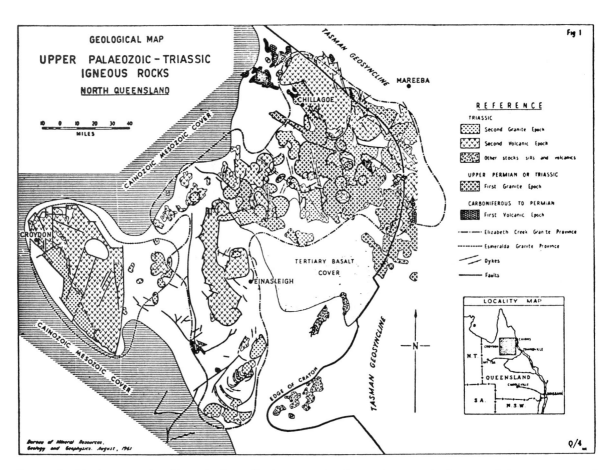

FIGURE 13–14: A group of cauldrons in Australia that are related to major tectonic features of the crust, some of which show overlap in both time and space.

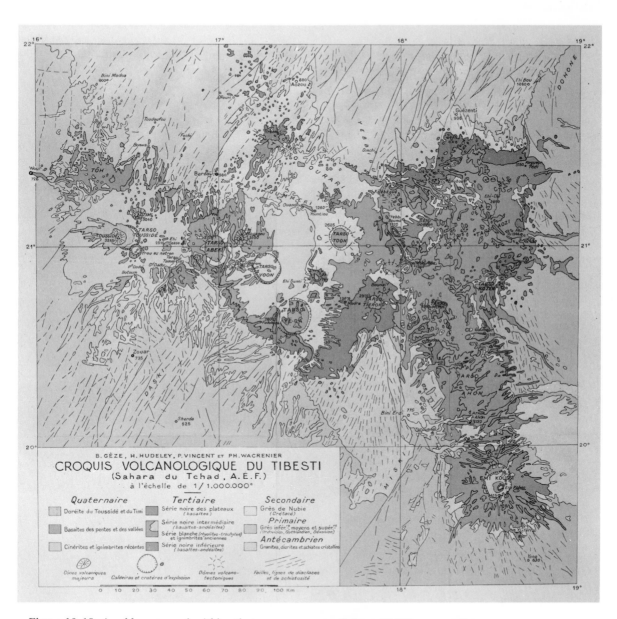

Figure 13–15: A caldera group in Africa that covers an area of about 12,000 square miles.

United States, exclusive of Alaska and Hawaii. This slide was made several years ago and a few more calderas have since been found. The small stippled areas are single localities that are probably parts of large ash-flow sheets. I think it safe to say that large areas in between existing outcrops were once covered with continuous ash-flow deposits. Six years ago only four calderas known in this area had been discussed in the literature; today the number is fifteen with more to come.

I have made some crude estimates based on volumes of ash-flow materials from known calderas, and maximum and minimum caldera sizes, and these suggest that a minimum of 250 and a maximum of 2,500 calderas will be required to account for the ash-flow deposits in the Tertiary of the western states. If Mexico is added, the number will be doubled. Some of the individual ash-flow sheets contain volumes of the order of 1,000 cubic miles. Their sources, although suspected, are unknown,

but certainly these must be caldera structures larger than any known to us at present. The larger these structures become, the more difficult it is to document them by field mapping, especially in structurally complex areas. My colleagues and I have spent over fifteen years on the Valles caldera alone.

The point I wish to make is that the deposits shown in Figure 13–16 span only the last 60 million years of the earth's history in this area. The ring complex and caldera groups just shown, probably spanned even less time. In the absence of erosion and extensive sedimentation, these and other areas of the earth's surface would be greatly pock-marked by very large holes of volcanic origin.

What all this means in terms of the moon is problematical. We have learned that there probably is no volcanism on the moon, that there may never have been any volcanism on the moon, that there may be some volcanism on the moon, and that there is a lot of volcanism on the moon. These

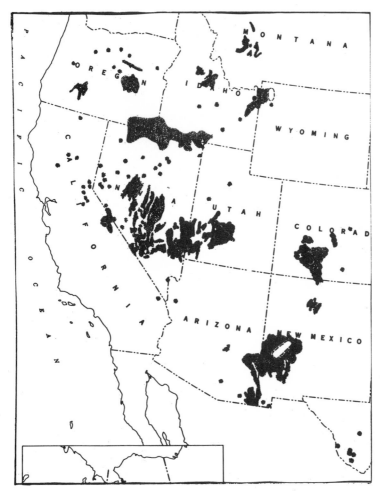

FIGURE 13–16: A crude distribution map of known ash-flow deposits and some calderas in the western United States, exclusive of Alaska and Hawaii.

deductions were based on a great amount of dedicated scientific effort by their proponents.

As we look at the central peak of Albategnius, in Figure 13–17, recall Aso caldera (Figure 13–4) with its central volcanoes with summit craters; I suggest that this is a comparable feature. However, the possibility that this summit crater is a chance hit by a meteoritic body raises doubt in my mind.

Alphonsus (Figure 13–18) shows some features that are typical of volcanic structures:

1. The transecting central fault. Nearly all calderas known to me are located on rift systems, fault intersections, or fault dislocations. These commonly show post-caldera movement.

2. The concentric rim faults. These are common to many of the lunar craters, and nearly all terrestrial calderas show them. They are related to the ring-fracture system along which subsidence took place.

3. The cracks on the floor. They are similar to cracks that develop on cooling lava lakes or on Kilauean caldera floors, but occur here on a much larger scale.

4. Maars or pit craters. These may develop on

FIGURE 13–17: An example of a summit crater on the moon, the central peak of Albategnius.

larger cracks; the dark-haloed craters suggest them.

5. The central peak. It could be a steep-sided volcanic dome; I have seen very similar domes rising above basaltic plateaus in Iceland with reliefs of the same order of magnitude that we see here.

As we look at Alpetragius I am tempted to suggest that this is a resurgent cauldron. The central dome even shows a suggestion of the transecting graben that I mentioned in connection with terrestrial resurgent structures. Other central mounds in lunar craters show this feature better. Other lunar craters seem to show structurally controlled smaller rim craters. This is another feature common to calderas.

We could go on in this way, singling out suggestive features and build a case for caldera volcanism on the moon. However, as I look at many of the lunar craters of larger type around which my discussion has centered, they all appear to have one feature in common that is incompatible with all terrestrial calderas known to me. I speak of the hilly, hummocky, and chaotic topography that forms their rims and immediate surroundings.

FIGURE 13–18: Typical volcanic features of Alphonsus.

I feel that most geologists having a wide knowledge of volcanic structures would interpret this feature as a debris ring deposited as a result of an explosion. The explosion concept and its documentation has been convincingly presented by Eugene Shoemaker. The only question I raise is, could some of these explosions have been volcanic rather than from impact? If so, they might then have had a more normal subsequent volcanic history. Volcanic explosions of such magnitude are inconceivable on today's earth. Perhaps it is naive to consider them on the moon.

DISCUSSION

CAMERON: I would like to know if anything on the order of shatter cones was found in and around the Valles.

SMITH: No.

GREEN: I noticed in Figure 13–1 that there are a number of interesting features and patterns on the Hawaiian volcano that you showed. I wonder if you would discuss briefly what sort of flow markings you would expect on lava flows and ash flows, what their scale would be, and whether we should observe these on the Ranger pictures.

SMITH: I thought quite a bit about this problem, and I see nothing on the Ranger photographs that suggests to me the presence of lava flows. But this could be due to the resolution.

VOICE: What kind of contours occur on ash flows? How level are they? Or how hilly? How flat does an ash flow spread out, from what kind of distances?

SMITH: Are you speaking of the surface of ash flows?

VOICE: Yes.

SMITH: Primary surface over most of the sheets, as far as we know—and most of them are older and their tops have been stripped off. There are only a few examples of historic ones on which we have any control.

According to the literature, these floors were level enough to have ridden a bicycle over. I quote Dr. Griggs, who first went into the Valley of Ten Thousand Smokes in Alaska.

The surface, however, is pock-marked with tiny funnel-shaped vents which are fumaroles. We find the same structures in older ash flows. Of course, I concede it a possibility that some of the small lunar pits could be fumarolic structures on ash-flow surfaces, but I really like the impact picture better.

VOICE: When you said "riding a bicycle over," that means flat, but not necessarily level.

SMITH: We can infer that ash flows are emplaced over a temperature range of something like perhaps 1,000 degrees, at least 500 or 600 degrees. The colder ones have a tendency to produce crescentic pressure ridges much as you see in lava flows. In other words, their viscosities are higher, they are probably thicker, and are moving at a slower rate. So they tend to mound. They would have a pattern of flow-bounded, crescentic pressure ridges as you have seen on lava flows.

As you go up in the temperature scale and get into the higher temperature gas-emitting type, they are more fluid, they are probably thinner, and I suspect that their near-vent surfaces are nearly level.

But as you approach the distal ends, where turbulent flow becomes laminar flow and the speed is slowing down, you also get mounding.

VOICE: In terrestrial impact craters you can find broken rock due to the pressure effects anywhere you drill in the crater. In Canada we have now drilled about six of these and find this to be universally so. I would not expect this to be so in a collapse feature. To what extent do the observations exist, and have these been drilled in many places?

SMITH: I didn't go into this problem because it gets rather complex. We are in the process of trying to differentiate calderas on the basis of whether the subsidence was chaotic or whether the subsidence was uniform.

I think in Kilauean type structures, you are simply dropping a piston bounded by concentric fractures. There may be brecciation around the fractures but not in the central block. This is also true of the very large structures, we think, like the resurgent cauldrons and the very large calderas.

You are dropping a circular block of crust along boundary fractures. You would get brecciation on those fractures. However, in the most common calderas on the earth, which are in the big andesitic volcanoes of the orogenic belts, we think that subsidence is chaotic and that this is probably the reason that you get random distribution of central volcanoes versus symmetrical distribution in the larger structures.

14.

Lunar Ash Flows

John A. O'Keefe

Goddard Space Flight Center, Greenbelt, Maryland

The main part of what I have to say will be addressed to a point raised by Dr. Gold, primarily concerning the softened appearance which we see on some of the lunar craters. Professor Gold drew attention to the fact that craters which have diameters on the order of 100 m generally have a smooth appearance, which is true neither of those which are larger nor of those which are smaller.

Mrs. Winifred Cameron made counts on Ranger VIII of the percentage of craters which show a shadow, which provide a rough, crude measure of the smoothness of the crater. If it has a shadow, it is evidently fairly steep-sided. If it lacks a shadow, it means the side is gentle. As we come toward smaller and smaller craters we get fewer and fewer shadows, but the curve then turns around and climbs back up somewhat. As in many cases where a count is made, the effect isn't as striking as the photograph, but it is there nevertheless. I really think that we can rely almost as much on Professor Gold's remarks as we can on this observation (Figure 14–1).

Consider the central peak of Alphonsus, which is a locus of a good deal of volcanism; in particular, this central peak is probably volcanic in nature. The reason for thinking so is the absence of craters on the sides of this peak. The absence is genuine. It is not simply due to the fact that the peak is at a steep angle, because the angle that the sun makes with the ground, from the base to the peak, on the average, can be shown by actual measurements of the height of the peak and the width of the base to be about the same as the angle that the sun makes on the Ranger VII photographs where the shadows are sufficient to outline the craters. There-fore, if there were ordinary, primary craters we would see them even on that slope. In the second place, at the south and north ends of this peak we see that the surface is nearly parallel to the sunlight. We would see craters in these regions if they were there. In fact, counts of craters made in a region at the south side of the peak show about one-third as many as in a typical region of the interior of Alphonsus. This feature genuinely has fewer craters than the rest of the surface of Alphonsus.

There are some very tiny craters on the slopes, which are evidence that this surface is capable of sustaining an ordinary impact. That is to say, an ordinary impact on it will leave a crater behind. It is not true that the surface is unable to form a typical crater wall. Neither can it be due to the proposition that this lump is something thrown out of Mare Imbrium which landed here after-ward. If we adopt the out-throw hypothesis, then we are led to suppose that the surface of Alphonsus may be, and probably is, older than Mare Im-brium. There are lots of reasons for thinking that. There are more craters here than elsewhere. But that is not the reason we fail to see tiny craters on the surface.

The argument is as follows: Dr. Shoemaker pointed out that for small craters there is a very significant deficiency below the number which we would have expected on the basis of our ordinary knowledge of sizes of meteorites and the length of time that the solar system has existed, which, as Professor Urey has brought to everyone's attention, must be something like $4.5 \cdot 10^9$ years.

There is a real deficiency. Whatever the explana-tion of this deficiency may be, it means that we do not see anywhere, even in the rich areas, all the craters produced during the last $4.5 \cdot 10^9$ years. Whatever the explanation of this degradation, we cannot hope to measure a difference like that between the age of Mare Imbrium and the age of the rest of Alphonsus by making crater counts of these very tiny craters. The difference between them must be due to something which happened much later. In other words, the lunar surface, at the small crater level (below the 300-m level) has forgotten what happened at $4.5 \cdot 10^9$ years ago. The larger craters may remember, but the smaller craters have certainly forgotten. Therefore, this difference between locations is due to the fact that the peak is a recent formation. It is a recent formation and clearly volcanic in nature, because it is very difficult to imagine upheaving anything like that in any way other than volcanism.

Northwest of the central peak we see another area in which there is a reduced number of craters, extending northward across Alphonsus. This area, I believe, is also volcanically produced. I suggest

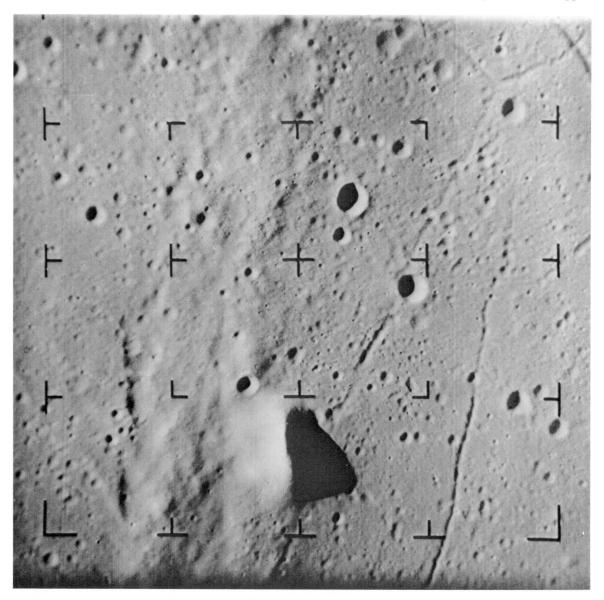

FIGURE 14–1: Television picture taken by Ranger IX prior to impact on March 24, 1965, at 06 08 20 PST. Spacecraft altitude above the moon 58 miles. Shows region of central peak of Alphonsus with rille running through its shadow toward upper right.

that the significant difference between the crater count in this region and in a typical region of the floor, such as the region to the northeast, is due to the fact that the northwest region is the result of recent volcanism and the other is something older.

The surprising thing that is found when making counts of craters in this northwest area, however, as compared to the counts of craters in the northeast area, is that there is nearly the same number of craters in each region. The difference is that if craters *with shadows* in the northwest region are counted, they are much fewer than those with shadows in the northeast. Roughly 9 per cent of the craters in this region have shadows, as opposed to 30 per cent in the other. In all, the northwest region has about half as many craters as the northeast. In other words, the number of sharp craters is greatly diminished in the northwest as compared with the northeast, whereas the other craters are not so much reduced.

Figure 14–1 shows a softening of the crater outlines. This is the phenomenon to which we have been alluding, and to which Gold and all of the Ranger discussants drew attention. It is a fact that larger craters, craters above a certain size, are softened in some way or another.

I propose that this softening is the result of the deposit of a blanket of ash over the moon's surface. This explanation will certainly work. Dr. Jaffe has produced photographs of the result obtained by placing a layer of powder over a typical lunar crater, and that explanation works. Moreover, if this is an erupted rock, it would be a viscous one. If it were a viscous rock, then we would expect it to be associated with ash because, in general, we find that the viscous lavas and the ash flows go together; tholoids and cumulo domes go with ash flows in this particular sense.

Let us consider why this ash flow, if it is one, leads to shallow craters. There are two possible mechanisms which can produce this. One is very well known in the ash-flow field. It is what is called differential compaction. When an ash flow is laid down, the surface is completely horizontal. Underneath it we suppose there is a feature similar to that shown in Figure 14–2. The ash flow then compacts and the amount of compaction is clearly going to be greatest where there is the greatest thickness of ash, and least where there is the least thickness. So the topography is like that outside the ash flow but with a reduced vertical scale. This is only a

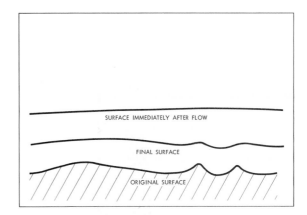

FIGURE 14–2: Differential compaction in ash flows.

possible explanation, but it is a standard phenomenon of ash flows.

Cook's monograph on the ash flows of eastern Nevada has formulas which calculate the original thickness of the ash flow on the basis of the aforementioned kind of differential compaction, utilizing figures on the height of the topography and the thickness of the flow in various places.

The second possible mechanism involves the principle that there are two conceivable types of ash flows. In the one which was discussed above, there is a dense flow, moving over the ground like a liquid; in the other there is a dilute flow, mostly a dusty gas, from which the material is deposited afterward. In the second case a blanket simply would have been laid over the ground, and like any other blanket, a blanket of snow for example, it may reveal what is underneath.

Figure 14–3 shows a dark-haloed crater. Note that there is again a softening of some of the craters in this region, as though a blanket had been deposited over it. I think there has been a deposit of some kind because it has plugged up the rille both ways, but we also have a softening of the contours of the material underneath. This effect probably rules out a basaltic flow, because if there had been a flow of basalt it would have been liquid and would have filled up these holes. When this basalt finally froze there would have been some contraction; however, it would only have amounted to about 2 per cent, and the 2 per cent shrinkage is not enough to account for what we see.

Figure 14–4 is a full-moon picture of the region around Alphonsus. The central peak and the dark-haloed crater are visible. This picture shows that the dark-haloed craters inside Alphonsus appear to be the same kind of material which we see on the mare. Let me make this point more clearly. In Figure 14–4 we see a typical piece of lunar terrain. There are the tiny craters, and the soft craters on the order of 100 to 200 or 300 m across. This is the last A frame on Ranger IX. The tiny craters are sharp, the larger craters are soft. This distinction between the soft, large craters and the hard, small craters is the basic point.

Here are the consequences of this idea in some

detail. This peculiar result, which makes the 10-m craters look more like 1-km craters than like 100-m ones is easily explained by the ash-flow mechanism. Suppose the flow is 20 m thick. It will obliterate craters 10 m across; it will greatly soften those which are 100 m across; and it will leave the 1-km craters with about the same shape as before. Accordingly, any small craters seen on the surface must be post-flow. In this case they will, of course, be sharp. Hence the large and small craters will

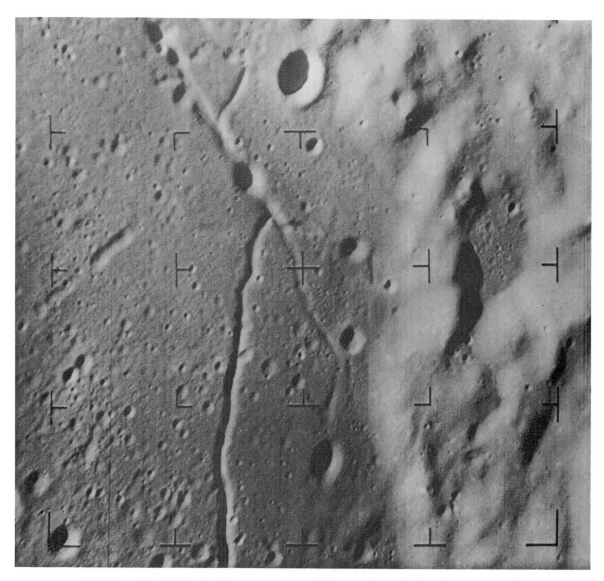

FIGURE 14–3: Television picture taken by Ranger IX prior to impact on March 24, 1965, at 06 08 20 PST. Spacecraft altitude above the moon 115 miles. Shows east edge of Alphonsus floor with part of surrounding wall in right 1/3 of picture. Crater floor is cut by prominent rilles which are lined with dark-halo type craters that have covered part of the rille. The crater walls have soft contours and are almost featureless.

resemble each other more than the craters in the middle.

An alternative explanation for this softening of the craters is erosion. However it is my contention that erosion is not the true cause. In the first place, Dr. Hapke spoke on the blackening of small fragments on the moon's surface. According to him, little pieces, about 10 μ thick, were blackened by proton bombardment in a period of something like 100,000 years. Let us be liberal and call this 100 μ in 100,000 years. Clearly, if the moon's surface is eroded at such a rate that this particle is removed in less than 100,000 years, the moon will never be blackened. Therefore to be consistent with the darkening data, the erosion rate can be, at most, something like 1 μ per thousand years, or 1 m per billion years, and 4.5 m in the age of the solar system. Four-and-a-half meters

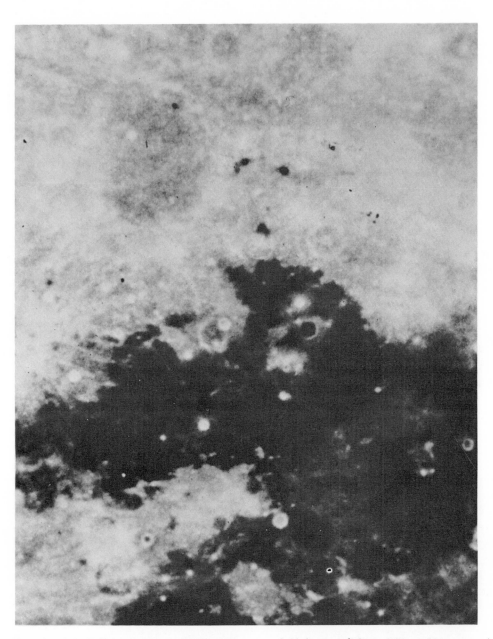

FIGURE 14–4: Full moon picture of the region around Alphonsus. (Mount Wilson Observatory.)

isn't enough erosion to alter in any serious way the kind of contours that we have seen. A crater 300 m across can't be softened by taking 4.5 m off some place on the outside.

In the second place, this number of 1 m per billion years is the sort found when studying the erosion of meteorites. Meteorites in space cannot be eroding at a rate very much higher than this because if they were, they would have been destroyed. There are meteorites, fist-sized objects, which show cosmic ray bombardment over periods, in the case of irons, of hundreds of millions of years, and in the case of stones, of tens of millions of years. If the rates of erosion are calculated from these, the above level is not reached, but you get within a factor of ten of it. The erosional theory has a very narrow escape, even from the meteoritic data, but it can't elude the data from the blackening.

Finally, I will comment on the remarks made by Dr. Smith about the calderas. This caldera remark is one more piece of evidence, in my opinion, that the ash-flow mechanism exists on the surface of the moon rather than the basaltic flow. The evidence is not completely clear cut. As Dr. Smith pointed out, calderas of the Hawaiian type, basaltic calderas, do exist, and some are quite large. But in general the very large, circular calderas go with ash flows rather than with basaltic lavas.

DISCUSSION

VOICE: To have the considerable amount of volcanism that you require on the moon, where are you going to get the energy? The moon's density is about 3.2, which means there will be a small, very deep core (if a core exists at all) in comparison with the earth, which has a major core and therefore the source of heat for major earth volcanism.

O'KEEFE: I didn't think the core of the earth had much to do with supplying the heat for volcanism.

VOICE: For the heat of the machine.

O'KEEFE: Most of the heat that supplies the energy for volcanism comes from the uranium, potassium, and thorium, and so forth, which are mostly in the mantle. There is very little radioactive material in the core.

VOICE: Put the heat in the mantle. There would still have to be a major core in the moon if there were that much heat in the mantle of the moon.

O'KEEFE: You mean if you had that much heat in the mantle, the moon would have to have a core if it had that much iron. Evidently it doesn't, because we know the density is wrong.

VOICE: It doesn't have much of a core at all.

O'KEEFE: That is what I think. At any rate, the core contributes very little to the total energy put out by the earth.

WHIPPLE: We have been working on erosion rates from the meteorites, the ordinary stones, which show from the cosmic ray exposure maximum ages on the order of 50 million years. Some people think there were no stone meteorites in this area of space 50 million years ago, and all of them have come since. I think erosion is a better explanation.

This leads to an erosion rate of about 50 angstroms per year, or a half a centimeter in 10^6 years. In 10^5 years you have 500 μ. That is for a hard stone. The evidence for the cometary meteoroid is that the rate is less certainly somewhere between 10 and 100 times greater than that, but indications are that it is much greater. If we put on, say, 20 times, that would be at a rate of 10 cm in 10^6 years. In terms of microns, it is a little higher than Dr. O'Keefe wanted. The point is, however, that most of this is rather small material. It is going at a fairly high velocity into this fairy castle structure. I am unclear in my own mind what it does when it gets there. Dr. Gault's pictures showed that in sort of fragile, weak material it just bored a hole. I would assume that the major effect then would be to bore holes in the very upper fairy-structure-type material. But in so doing, it tends to break it up. It may overturn it some. I can't be sure about the overturn.

It is very hard to change this number, or the larger one that may apply to weak structures, to say exactly what that will do to the surface material in terms of Hapke's calculation of the 100,000 years. I think this figure, for a regular stone, is a highly significant one, and it is going to stay within a factor of two indefinitely. The major uncertainty is the effect of the shielding of cosmic rays by the outer material and just how that operates. I think this figure of 50 angstroms per year stands, and the fairy-structure material is weaker, therefore there must be somewhat larger erosion rates.

O'KEEFE: How about Norton County?

WHIPPLE: Norton County has a greater age, but Norton County is not chemically like the ordinary chondrites that give this rather uniform value. I don't think that the exposure age is yet easily interpretable. Whether that old age means anything or not is another question.

O'KEEFE: Professor Whipple's figures give 5 mm in a million years. My figure was 1 mm in a million years. If we adopt Professor Whipple's rate, then in a time equal to the age of the solar system, we get 25 m, which on a crater 300 m across will have little effect. On a 100-m crater, it will tend to abolish it. The erosion rates from meteorites will just work, but not the darkening rates. The meteorites go out to the asteroid belt where there is perhaps more erosion.

VOICE: In fairness to Gold, I would like to point out that I think the erosion mechanism he has in mind is not one of removal of material from the surface, but mainly of redistribution. So that at all times, there is a mixture of dark material with a little light material, and the dark material slides down the slope or whatever the electrodynamics of the situation would require. In this way one could have much larger erosion rates than this. There is not a complete removal from the surface, but just a redistribution.

O'KEEFE: Yes, but it works out the same way. If, instead of saying that 10 μ are darkened in 100,000 years, you say 10,000 μ or 1 cm darkened, it is going to take 100 million years, because the chunks have to be darkened one by one. It doesn't make any difference concerning the over-all rate whether 1 cm is darkened in 100 million years or whether 10 μ are darkened in 100,000 years. The surface still can't be dug away fast enough and kept black.

VOICE: I am confused by your numbers. The numbers I understood Hapke and Gold to come out with were purely sizes, not depth of penetration of protons. Was there a number given as to how deep these numbers would go in darkening the surface? Was there some limit on this?

O'KEEFE: They just darken the tiny skin on the outside. Naturally you don't darken the particles at any depth.

VOICE: Isn't the argument, on the color basis, that the surface which is light has been exposed for less than this order of 100,000?

O'KEEFE: Even the light areas, the highlands, have to be darkened by proton bombardment to match the curve.

VOICE: I think we should point out that on the slope the erosion rate, or the effects of sputtering, can be much more pronounced because what can happen here is that the dust particles are sputtered loose and then under gravity fall down. I think this might explain why there is much less cratering on slopes, because it just tends to fill in these things faster.

O'KEEFE: I think it is very difficult to answer that problem. It is very difficult in general for Gold and his group to answer why we don't see bare rock on slopes.

DUBIN: I don't know whether your argument on the coloring holds if you look at it another way. Suppose the flow occurred and then the erosion rate was very slow, say within the last million years, extremely low erosion. This has been computed in terms of the micrometeoroid flux, the effect of which is one gram per square centimeter. There are four billion years for the integrated flux of meteorites and dust on the moon. If the erosion rate is very slow, then the upper surface, as Hapke and Wehner have shown, would be darkened in time on the order of 10^5 years. There would be plenty of time for darkening to occur, and this is observed. It does not discount the Gold argument.

O'KEEFE: I think that what you are saying is that micrometeorite flux might have been more rapid in ancient times, and this might save the situation. The difficulty with that is that when we go to craters smaller than 300 m, as Shoemaker pointed out, we are not looking at very ancient times. The surface has forgotten, in the region at which we are looking, what happened in very ancient times. It is looking at times not very far back.

KOPAL: I was going to try to answer the previous question. I think much point was made yesterday, and partly also today, about the gradients on the moon trying to help the dust to migrate. These gradients are on the whole very low.

On the belt, which has so far been mapped by the Aeronautical Chart and Information Center, we have some idea of distribution of slopes. It appears that less than 2 per cent of the entire lunar surface is inclined to the horizontal by more than 5 degrees.

VOICE: You talk about the central peak and low cratering that you observe in the central peak, and this central highland area, but you don't talk about the crater walls. Are you postulating that the central peak and the crater walls are formed by the same volcanic processes? You certainly see the same phenomena in terms of cratering; the same features are observed in both. I prefer the hard rock idea for that reason.

O'KEEFE: What happens is that toward the outer part of Alphonsus there is a zone in which the craters are very few, in which there is remarkable smoothness. Beyond that, and not very well seen in the Ranger IX photographs, there is a region in which the crater density picks up again. I am sorry that I don't have a slide that illustrates this well. It was well-illustrated on the first lunar slide that was shown by Dr. Smith. It is just an inner zone, just in the inner wall of the crater, that we see this remarkable smoothness. Mr. Whitaker has many times pointed out that in the inner walls of the craters there is a ring which is white and nearly crater-free.

I do follow your suggestion; I believe that this ring is a ring of eruption. I think it is a ring in which material has been erupted around the central peak. In other words, I regard Alphonsus as a ring complex, as a cauldron of subsidence, in which there has been a ring fracture and viscous materials moving up all around it.

VOICE: I think there is a suggestion in the Ranger VIII photographs that there is this same kind of softening in the highlands that border Mare Tranquillitatis. Unfortunately resolution isn't good enough, but at the boundaries of the mare there is a suggestion that those highland surfaces look somewhat comparable to the crater rims around Alphonsus.

There is another point that hasn't been brought out relative to these dark areas, and their inter-relationships and associations with the crater rims. That is that you get the darkening right up to the edge where you start up the slope. Assuming that these darkened areas are deposits on the surface, because of the softening and so forth, you would expect to get some sort of a uniform distribution of this material. From what we see both in the west and the east area, with the two dark areas, there is an interfering uplifted area with this rim, and with the material coming in, and yet you don't see dark material on those slopes.

It apparently has been deposited there, but then migrated down off those slopes. There is evidence for this migration off of the rim material, and an

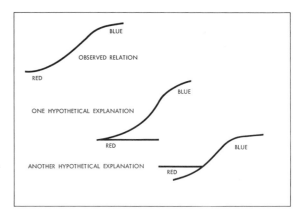

FIGURE 14–5: The relations of red and blue ash flows.

evidence that there is a stripping process going on here.

If you look at the photographs closely, you will see that those uplifted areas do definitely interfere with the pattern of that dark, deposited material.

O'KEEFE: That is very true and interesting. The point here is that when you mix gas and dust together, what you get, in effect, is material of very high effective molecular weight. The thing spreads out very flat as it moves. It does not become a great cloud. It becomes a thin, flat cloud. There are many pictures of a thin, flat cloud that moves; that is what I suspect is going on in the place you are talking about. Moreover, I suspect this is the explanation for the curious phenomenon that Mr. Whitaker pointed out, namely that there are many places where a slope comes down and a flow comes up and meets the slope and stops. If this is due to a gas of high effective molecular weight in the flow, it is easy to suppose that it ponds at the foot of that slope.

The critical question in the case of these flows on the moon is this one: Which is older? What you usually have is a blue material on the hill and red down below. If the blue is younger than the red, then, of course, you have a toe of a slope like this (Figure 14–5), which is what they have been suggesting.

If, on the other hand, the blue is older than the red, then you have this situation: red is ponding at the foot of the blue slope. That, I suspect, is the sort of thing we see in the area you are talking about, where the dark-haloed crater material meets the slopes at the wall.

15.

Thermal History of the Moon and the Development of Its Surface

B. J. Levin

O. Schmidt Institute of Physics of the Earth, Academy of Sciences, Moscow, U.S.S.R.

The macrostructure of the lunar surface was formed and modified, during the entire time of the moon's existence, by the bombardment of different cosmic bodies and also, during some period of time, under endogenic forces which have made possible lava effusions. I do not agree with Gold's hypothesis and, as do many other students in this field, I believe that lunar maria and lunar ring-plains with flat dark floors are filled with lava.

The nearly equal surface density of the "postmare" craters on the surfaces of all lunar maria indicates that their formation occurred during the same epoch. Therefore the first question which must be discussed is whether this idea of nearly simultaneous formation of the maria agrees with the calculations of the thermal history of the lunar interior.

The calculations show that with a content of radioactive elements similar to that in meteorites (U, $n \times 10^{-8}$ g/g, Th, $4n \times 10^{-8}$ g/g, K, 8×10^{-8} g/g, where n is of the order of 1 to 3) the interior of the moon had to be partially molten after 1.5 to 2 aeons after its formation.[1] Due to the small pressure inside the moon the melting temperature increases only a few hundred degrees from the surface to the center. Therefore the melting embraces the whole interior down to the center,

[1] B. J. Levin and S. V. Majeva, "Some Calculations of the Thermal History of the Moon," *Dokl. Akad. Nauk SSSR*, Vol. 133 (1960), pp. 44-47; *Soviet Physics Doklady*, Vol. 5 (1961), p. 643; Levin, "Thermal History of the Moon," *The Moon; Proceedings of the 14th International Astronomical Union Symposium, Leningrad, 1960*, ed. by Z. Kopal and Z. Mikhailov, London: Academic Press (1962), pp. 157–167; Maeva, "Some Calculations of the Thermal History of Mars and the Moon," *Dokl. Akad. Nauk SSSR*, Vol. 159 (1964), pp. 294–297.

excluding only the outer layer of 300–400 km thickness (Figures 15–1 through 15–3). The increase of the "initial" temperature from the surface to the center arose because the central part was formed some hundreds of millions of years earlier than the outer part and therefore had begun to accumulate radiogenic heat at a much earlier time. Because different minerals have different melting temperatures (in our calculations the range of melting temperatures is presumed to be 200 degrees, although it might be somewhat larger), the light melts, already in the stage of partial melting, were formed and rose upward to the base of the solid outer layer and through it to the surface. At that time the thickness of the solid outer layer must have decreased to 50–100 km. This period was favorable

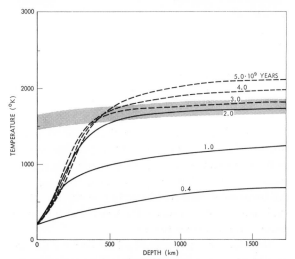

FIGURE 15–1: Temperature distribution with depth in the moon at different ages, for $n = 1$. The shading shows the range of melting temperatures.

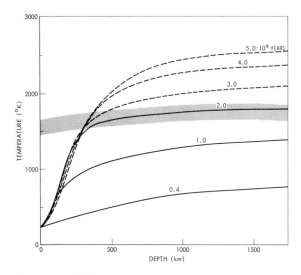

FIGURE 15–2: Temperature distribution with depth in the moon at different ages, for $n = 2$. The shading shows the range of melting temperatures.

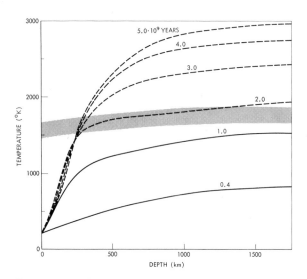

FIGURE 15–3: Temperature distribution with depth in the moon at different ages, for $n = 3$. The shading shows the range of melting temperatures.

for lava effusions. After the end of differentiation the situation became unfavorable for effusions. Due to the transport of radioactive elements toward the surface which inevitably occurred during the differentiation, and which facilitated the escape of radiogenic heat into space, the heating of the moon ceased, and it began to cool.

Therefore, contrary to MacDonald,[2] we do not ascribe any real meaning to the temperature distri-

[2] G. J. F. MacDonald, "On the Internal Constitution of the Inner Planets," *J. Geophys. Res.*, Vol. 67 (1962), pp. 2945–2974.

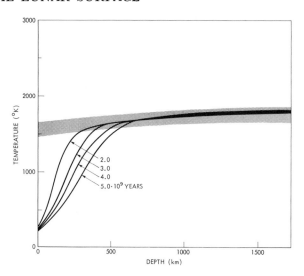

FIGURE 15–4: The course of temperature changes in the lunar interior during the stage of cooling at various ages. The shading shows the range of melting temperatures.

butions calculated for the post-melting times because the supposition of uniform distribution of radioactive elements used in the calculations is no longer valid. (Although in our calculations the latent heat of melting is taken into account, these temperature distributions have only a formal mathematical meaning and therefore are shown in Figures 15–1 through 15–3 by dotted lines.)

The period of differentiation of the lunar interior is not embraced by the present thermal calculations and the calculations of the cooling of an already differentiated moon start from some arbitrary "initial" temperature distribution. The initial moment of these calculations is placed at 3.5 aeons ago. It was assumed that in the deep interior the "initial" temperature corresponds to the beginning of melting and toward the surface it drops to 230°K (the constant surface temperature of the moon in the equatorial region). Variations of this "initial" temperature curve practically do not affect the calculated temperature distributions at the present time.

After differentiation, the lunar interior contained about the same concentration of radioactive elements as the terrestrial ultrabasic rock dunite. (According to different authors the U- and Th-content of dunites is characterized by $n_i = 0.2$–0.8 and the content of K is 2.5×10^{-4} g/g.) An example of the course of temperature changes in the lunar interior in the stage of cooling is shown in Figure 15–4. For different values of n_i the distributions of temperatures for the present time are very similar (Figure 15–5). (Different assumptions as to the total content of radioactive elements in the moon, which after differentiation are manifested in different thicknesses of the "lunar crust" enriched in radioactive elements, have almost no influence on the present-day temperature of the interior.) According to present calculations the outer part of the moon down to a depth of 500–800 km, i.e., embracing 60–80 per cent of the whole mass, (depending on n_i and the role of radiative heat transfer) is solid, while the central part is semi-molten.

The comparison of the surface heat flow corresponding to the release of all heat generated inside with the heat flows calculated for the stages of heating or cooling of the moon shows very clearly that without differentiation the moon would continue to be heated for a long time but that the differentiation resulted in the substantial excess of lost heat flow over the "equilibrium" one (Figure 15–6). For different values of the parameters involved in the calculations (ϵ is the opacity in cm^{-1}) Majeva[3] obtained the following values of the lunar heat flow q in 10^{-6} cal/cm^2 sec:

n_i	0.2				0.8				"Equilibrium" flow	
ϵ	10		40		10		40			
n	1	2	1	2	1	2	1	2	1	2
q	0.37	0.46	0.35	0.45	0.34	0.44	0.32	0.41	0.23	0.33

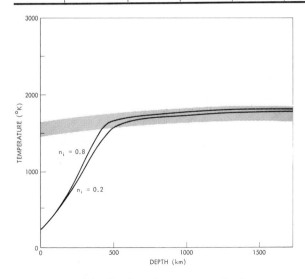

FIGURE 15–5: Distribution of temperature in the present moon for $n_i = 0.8$ and 0.2.

The decrease of lunar surface temperatures from the equator toward the poles has a marked influence on the thermal conditions in the interior. A surface temperature of only 90°K in the polar regions gives a thickness of the solid outer layer 50–100 km greater than in the equatorial region (Figure 15–7). The solid matter being denser than the semi-molten, this difference in thickness, combined with the spherical outer surface, would be a

nonequilibrium one. For isostatic equilibrium a small flatness of the figure of the moon is necessary and according to the author's hypothesis[4] this is the cause of the deviation of the moon from hydrostatic equilibrium.

The above calculations of the thermal history of the moon led to the conclusion that a period

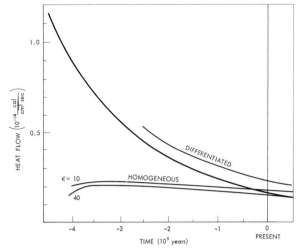

FIGURE 15–6: Heat flow through the lunar surface for differentiated and homogeneous models.

[3] Maeva, "Some Calculations."
[4] Levin, "On the cause of the Nonequilibrium Figures of the Moon," *Astron. Zirc.* Vol. 285 (1964) [in Russian].

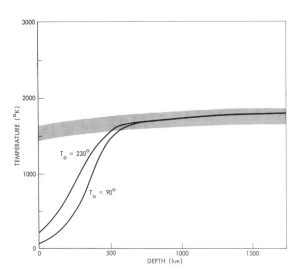

FIGURE 15-7: Effect of surface temperature T_0 on the temperature distribution in the moon. The shading shows the range of melting temperatures.

favorable for lava effusions occurred 2.5–3.0 aeons ago (depending on the content of radioactive elements in the moon). At that time spontaneous breakthrough and effusions of lava were probably possible. However the circular form of lunar maria, which are themselves encircled by mountain ranges, indicates that their formation was connected with impacts of very large bodies, bodies of greater size than those which formed the largest "postmare" craters. Therefore a second question which arises is whether the thermal history of the moon is compatible with the history of the bombardment of its surface.

My positive answer to this question[5] is based on the idea of the common origin of the moon and the earth. As was noted by O. Schmidt,[6] in the course of the accumulation of planets inelastic collisions of bodies occurred in their vicinity, leading to a decrease in velocity and subsequent capture of the bodies. This would lead to the formation around the growing planet of a swarm of bodies from which one or several satellites could be formed. As was shown by Ruskol[7] the mass of the circumterrestrial swarm formed in this way must be adequate for the formation of the moon if maximum dimensions of bodies from which the earth accumulated were of the order of tens or hundreds of kilometers. This just corresponds to the dimensions of planetesimals evaluated by other means. (For smaller dimensions of bodies, a greater frequency of collisions would lead to an excessive mass of the circumterrestrial swarm.) Due to a sharp increase of density in the swarm in the direction of the earth, the moon was formed at a distance of 5–10 earth radii and then gradually receded from the earth in the course of the tidal evolution of the earth-moon system. Simultaneously with the accumulation of the moon, several other bodies must also have been accumulated in the circumterrestrial swarm, particularly in its outer regions. Shifting away from the earth, the moon swept up these bodies or, in other words, these bodies bombarded the moon. The formation of the lunar maria must be connected with the sweeping up by the receding moon of the outermost ring of large bodies, as was suggested by Kuiper.[8]

The calculations of the time scale of tidal evolution of the earth-moon system show that with the present-day value of the "tidal-friction coefficient," the recession of the moon would take only about 1½ aeons. However, it is natural to suspect that at the time when the earth was in its early stages, when its interior was not yet heated and softened and the oceans did not yet exist on its surface, the "tidal-friction coefficient" was much smaller than it is now. (The arguments by MacDonald,[9] who believes that in the past the tidal friction was greater than now, do not seem convincing to me.) Assuming a smaller tidal friction in the past, it is possible to reconcile the epoch of the sweeping up of the

[5] Levin, "A Modern Form of the Meteorite Hypotheses of Moon Relief Formation," *Bulletin Vsesoyuznogo Astronomogeodezicheskogo Obschestva*, No. 30 (1962), pp. 6–19 [in Russian]; "The Modern Form of the Impact Hypothesis of ,Lunar Relief Formation," *Proceedings of the 13th International Astronautical Congress, Varna, Bulgaria, September 1962*, Vol. 1, ed. by N. Boneff and I. Hersey, Vienna: Springer-Verlag (1964), pp. 11–20.

[6] J. Schmidt, "Genesis of Planets and Satellites," *Izv. Akad. Nauk SSSR, Ser. Fizich*, Vol. 14 (1950), pp. 29–45 [in Russian].

[7] E. L. Ruskol, "The Origin of the Moon. I. Formation of a Swarm of Bodies Around the Earth," *Astron. Zh.*, Vol. 37 (1960), pp. 690–702 [translated in *Soviet Astron. -AJ*, Vol. 4 (1961), pp. 657–668]; "The Origin of the Moon," *The Moon; Proceedings of the 14th International Astronomical Union Symposium, Leningrad, 1960*, ed. by Z. Kopal and Z. Mikhailov, London: Academic Press (1962), pp. 149–155; "On the Origin of the Moon. II. The Growth of the Moon in the Circumterrestrial Swarm of Satellites," *Astron. Zh.*, Vol. 40 (1963), p. 40 [translated in *Soviet Astron. -AJ*, Vol. 7 (1963), p. 221].

[8] G. P. Kuiper, "The Moon," *J. Geophys. Res.*, Vol. 64 (1959), pp. 1713–1719.

[9] MacDonald, "Tidal Friction," *Rev. Geophys.*, Vol. 2 (1964), pp. 467–541.

outer parts of the circumterrestrial swarm with the epoch of maximum melting of the lunar interior.[10]

During its accumulation and during its recession through the circumterrestrial swarm the moon was bombarded mainly by bodies of this swarm which had small relative velocities while the bombardment by bodies moving around the sun and therefore having large velocities relative to the moon was of secondary importance. When the moon had emerged from the circumterrestrial swarm, i.e., after the formation of lunar maria and lava-filled ring-plains, the formation of primary craters by low-velocity impacts practically stopped and the main role in the formation of the primary "post-mare" craters was played by high-velocity impacts of asteroids and cometary nuclei. This was accompanied by intensification of secondary crater formation.

Finally, I would like to offer some critical remarks concerning the ideas presented by Shoemaker[11] on the history of the formation of the Mare Imbrium region. He writes, "The general flooding by mare material of nearly all significant depressions present in the late Imbrian time indicates the occurrence at one time of widespread melting inside the moon. The melting took place long after most of the depressions filled by mare material were formed." On the series of figures given in Shoemaker's paper the following sequence of events is depicted for the Mare Imbrium area: formation of a huge crater which became the Inner Imbrium Basin; a slumping of the outer basin perhaps triggered by the event that formed the Sinus Iridum; the formation in the Archimedian epoch of craters in the area of the outer basin; the filling of lower-lying areas by the mare material; the formation of "post-mare" craters in the Eratosthenian and Copernican periods.

On the one hand it is difficult to understand what was the cause of the slumping of the large outer basin of Mare Imbrium and where the underlying material was shifted. On the other hand, it is difficult to understand how a huge lava effusion could occur without the slumping of large surface regions to fill the place vacated by effused lava.

I suggest that the lava effusion from the inner basin and the slumping of the outer basin must have occurred simultaneously. This process was helped by the existence of the radial fissures as well as the circular ones (similar to those observed at present around the Mare Humorum), and by the instability of the outer solid layer lying on the semi-molten matter of lower density. There are no objections, from the point of view of physics, against such simultaneity of effusion and slumping, and, in addition, I can see no objections from the point of view of lunar stratigraphy.

DISCUSSION

KOPAL: I should like to call to Professor Levin's attention one well-known fact. The observations of the lunar surface disclose that the present lunar surface is capable of sustaining altitude differences of between 4 and 5 kilometers. These are not only local but regional. In order to sustain such a difference in level, stress differences must be generated in the interior amounting to 10^9 dynes per square centimeter.

I submit that if you melt the moon, or if you melt any appreciable or large fraction of the lunar interior, up to a half or more, you will not be able to sustain these differences. They will subside and the moon will be more spherical.

Since these differences are an indubitable fact, I cannot but conclude that large-scale melting has not occurred—that large-scale accumulation of radiogenic heat in the interior has not occurred because the value of n which you have taken was much too large. In the early days, the shape of the moon alone disclosed that the moon can't have as much radioactive material inside as your value of n would indicate.

LEVIN: It seems to me that even if one adopts the smallest value obtained for meteorites, n about 1, the melting of the interior of the moon is inevitable. After the differentiation, there will be a cooling of the moon, and at present we must have a thick, solid layer which I hope will sustain large stresses.

KOPAL: The meteorites are not a fair sample of the bulk of the lunar matter as far as radioactivity is concerned.

LEVIN: You think meteorites are not representative of the radioactive content of the moon?

KOPAL: Yes.

[10] Ruskol, "The Tidal Evolution of the Earth-Moon System," *Izv. Akad. Nauk SSSR, Ser. Geofiz.*, Vol. 2 (1963) [translated in *Bull. Acad. Sci. U.S.S.R., Geophys. Ser.*, Vol. 2 (1963), pp. 129–133].

[11] E. M. Shoemaker, "The Geology of the Moon," *Sci. Amer.*, Vol. 211 (1964), pp. 38–47.

VOICE: I think that there is rather good evidence now that meteorites are not a good sample of the earth, and that particularly the potassium content of the earth as a whole may be five times lower than the number that you have used for the moon, and if we reject the composition of meteorites, or radioactive content of meteorites, as representing the earth, why not also reject it for the moon?

LEVIN: I see no important evidence to reject. Different evidences exist. The evidence of radio astronomers is of a far larger content of radioactive elements because they obtain a heat flow larger than is given by our calculation of the thermal history of the moon. You on the contrary propose to adopt a lower content of radioactive elements.

It seems to me that the range of values of n which we have used in our calculation is reasonable. Perhaps it is the lowest value that will be correct, perhaps the larger. But the range is approximately correct.

VOICE: May I say that this subject was reviewed in a paper by Wasserburg, MacDonald, Fowler, and Hoyle in *Science* about a year ago and also later reviewed by MacDonald, and it isn't the uranium and thorium content which is the important thing here. It is the potassium content. If the potassium content between achondritic meteorites and chondritic meteorites varies by more than a factor of five, there are some evidences, some lines of argument which we can't review here, which are very convincing, that the potassium content of the earth is not 8×10^{-4} g/g.

LEVIN: I understand, of course, that the estimated probable content of radioactive elements in the earth, in the moon, and in the meteorites also, has been changed very often during the last year. I am not a chemist and I cannot predict what changes will be proposed by the chemists in future years. (In the discussion about the content of radioactive elements in the moon, it remained unmentioned that Wasserburg *et al.* [1964] propose to decrease the potassium content in the earth, as compared with meteorites, and to increase the uranium and thorium content. So the total generation of radiogenic heat remains about the same or even increases. If we are to adopt Wasserburg's values, the general character of the thermal history of the moon remains the same, as was shown by Phinney and Anderson [1965] and Anderson and Kovach [1965].)

O'KEEFE: Is it possible to reconcile your ideas with those of Dr. Kopal by referring to the idea of partial melting? It seemed to me that whenever you spoke about it, you were speaking not about total melting but about partial melting.

LEVIN: Yes.

O'KEEFE: We know that inside the earth we have partial melting and still we have the kind of stresses which are found in the interior of the moon. Perhaps that is the difference.

LEVIN: Perhaps so, yes. It seems to me to be desirable to clarify your remarks about partial melting. Indeed, I suppose that the redistribution of radioactive elements which began after the partial melting of the lunar interior prevented its total melting. But the partially melted silicates, as well as the totally melted silicates, are both unable to sustain stresses. It is important that only the central part of the moon is (partially) molten, while a thick outer layer is solid and therefore can sustain stresses.

WHIPPLE: On this subject, there is quite a bit of evidence now that the meteorites as such heated very rapidly after they were formed, possibly going through the cycle for 200-km bodies in the order of 300 million years. There is quite a school advocating that there was indeed short-lived radioactivity, like aluminum 26 and others, that added even more heat in the early stages rather than less heat. This is a very controversial subject but a lot of evidence is piling up with the meteorites that the heating was very rapid. If the moon had contained as much as 10 per cent of the typical meteorite matter, formed at the same time as the meteorites, your cycle would have taken place even more rapidly.

LEVIN: Yes.

WHIPPLE: The other point I wanted to make is a question I can't answer offhand. Perhaps it is answered in the literature. If you indeed have the surface material with somewhat lower density than the mean material of the moon underneath, after the other material is heated, and you throw out a very large crater wall in the original Mare Imbrium, not just 5 km or whatever it is, but quite a bit taller than that, then perhaps you do indeed have a bit of isostasy working so that you do not have to support the whole thing by pure stresses, but you can support quite a bit of the present load, assuming a great deal of slumping.

In fact, most all of the Mare Imbrium wall did slump, so that you don't have to support it all by stress. You can support it partially if not totally by isostasy. I am not really convinced that the situation is very bad. What I am trying to say is that I think these arguments support Dr. Levin, who always seems to see eye to eye with me, or I with him, on these matters.

UREY: Mr. Chairman, I would like to remark that the opinion about the short-lived radioactive heating is not unanimous.

VOICE: Dr. Levin, relative to your last point, where you disagree with Dr. Shoemaker, I agree with you in your disagreement, but in agreeing to this, I disagree with both of you. I think you are bringing out a very important point, namely the timing of the flooding of the basins. And I am reminded of the work by Dr. Khabakov, of Leningrad, who pointed out to me that there was a long time period represented in the formation of Mare Nectaris which he proved by the long time span represented by the craters between the shores of Mare Nectaris and the concentric Altai Scarp.

The point is that at least one of the lunar maria took a long, long time to subside, and that it was concurrently filled as it subsided with lava flows, according to Khabakov. This, of course, is counter to an impact-generated basin that would also be concurrently filled.

LEVIN: When I speak about the simultaneous lava effusion and the subsidence, it does not mean that the lava effusion was a very quick process. It could take a few millions of years, but it was short, as compared with the age of the moon.

VOICE: But the key point is that we have a difference of opinion as to the time it took for the basin to form.

LEVIN: I must say that I disagree with many ideas by Khabakov, because he does not take into account the development of the lunar interior. He suggests that its development was entirely similar to that of the earth, but the difference in dimensions between the earth and the moon, and the corresponding difference in pressures and the melting temperatures, led to an enormous difference in thermal history.

LOWMAN: I would like to say I agree with Dr. Levin's calculations on the present thermal balance of the moon. I did a little work myself based on Gordon MacDonald's paper and found that at the present time, with the most conservative assumptions, there should be partial melting in the interior of the moon at about 500-km depth now.

I would like to know if you took into account additional heat sources like adiabatic compression, chemical combinations between perhaps free radicals, the kinetic energy of in-falling bodies, and finally, of course, the short-lived isotopes.

LEVIN: We have taken into account only the radiogenic heat produced by uranium, thorium, and potassium. The part of this heat generated during the formation of the moon is taken into account in the form of initial temperature distribution. The part of the kinetic energy of infalling bodies retained by the moon is probably substantial and must increase the initial temperature. The role of compression, although substantial for the earth, is negligibly small for the moon. According to our views, the accumulation of the moon occurred during the last stage of the accumulation of the earth when the short-lived isotopes had already decayed. At the time of the accumulation of the moon the volatiles were already lost from planetesimals, and therefore the presence in the moon of substantial amounts of free radicals seems to me to be most improbable.

HAPKE: I would like to remark that the large heat flows near the surface of the moon and the large temperature gradients which Troitskiy deduces from his radiothermal measurements are far from being generally accepted.

LEVIN: I also don't accept them.

VOICE: I would like to say that Dr. Levin's mechanism and the impact theory are both compatible with the time interval between the formation of the mare basins and their filling. If you calculate what kind of heat anomaly should be caused by an impact of the size of Mare Imbrium and how long this heat anomaly would take to decay, the heat anomaly should persist there for something in the region of 200 to 500 million years, during which time the isotherms at best will rise. This is assuming a certain homogeneity for the materials involved and certain velocities for the impacting body. This will considerably affect the local heat flow around the Imbrium basin, and the elevation of the isotherms could cause melting at higher elevations in the crater much later, a long time after the formation of the basin, and then lead to volcanism.

LEVIN: We have not made such calculations; therefore I cannot judge about the role of this heating.

16.

Electronic Polarimetric Images of the Moon

V. P. Dzhapiashvili and L. V. Xanfomaliti

Abastumani Observatory, Abastumani, Georgia, U.S.S.R.

Polarimetric observations of the moon and planets have been carried out at the Abastumani Astrophysical Observatory (Georgia, U.S.S.R.). The polarimetry is based on an electron polarimeter, the latest model of which gives a reading of the degree of plane polarization, polarization angle, and stellar magnitude of the object (down to the 13.5 magnitude using the 40-cm refractor). Interesting results have been obtained with this device.

Since 1962 observations of polarimetric images of the moon have been carried out with a special device, an electron polarovisor, which is a scanning polarimeter. Polarimetric observations of the moon in different phases permitted the detection of a class of objects on the lunar surface similar in shape to ring-plains and craters, but not always spatially identified with them. The nature of these objects (for which the conventional names "polaro-ring-plains" and "polaro-craters" are proposed) is perhaps connected with the history of the moon and results from the strained state of hypothetical glasslike mass accumulations or from gas outflows on the moon's surface.

As early as 1960 the authors found that the polarimetry of the lunar surface near full moon with the small polarimeter aperture (1.5 seconds) shows the existence of small objects (less than 2.5 km) with very different physical properties. At the beginning of 1963 the second model of the polarimeter was put in operation. With this device the effect of inversion of the polarization of seas and continents near full moon was found. At present the third model of the electron polarimeter is in operation. This polarimeter permits improved resolution of the separate details on polarimetric images of the moon.

17.

Radio Measurements of the Moon

F. Drake

Cornell University, Ithaca, New York

I want to discuss briefly some radio measurements of the moon at 3-mm wavelength, which seem to develop points relevant to some of the previous discussions, particularly the infrared measurements of Shorthill, Saari, and Ingrao. The work I want to call to your attention was done principally by Bruce Gary at the Jet Propulsion Laboratory and Joseph Stacey at Aerospace Corporation, using the 15-foot antenna of Aerospace Corporation which is shown in Figure 17–1.

With this antenna, lunar maps with a resolution of one-tenth of the lunar diameter have been made at a number of epochs in a lunation as shown in Figure 17–2. In making the maps, a sensitive radiometer and digital techniques were used.

The next three figures show maps that are typical examples of the results obtained. The brightness-temperature contours are in degrees Kelvin. Figure 17–3 is nearly at the time of the new moon. The optical appearance of the moon is shown in the upper right-hand portion of these

FIGURE 17–1: The 15-foot antenna of Aerospace Corporation.

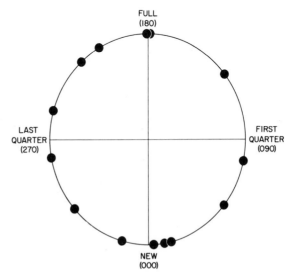

FIGURE 17–2: Lunar phases at which maps of brightness-temperature have been made.

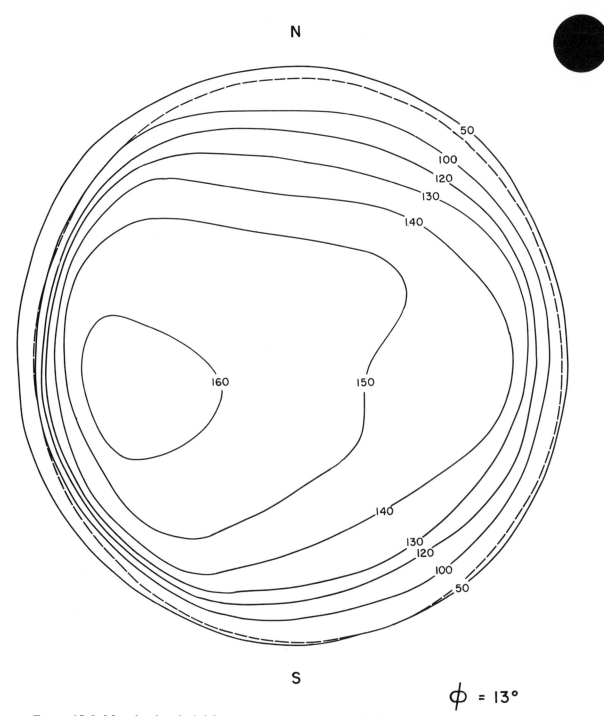

FIGURE 17–3: Map showing the brightness-temperature in degrees Kelvin, at the time of the new moon.

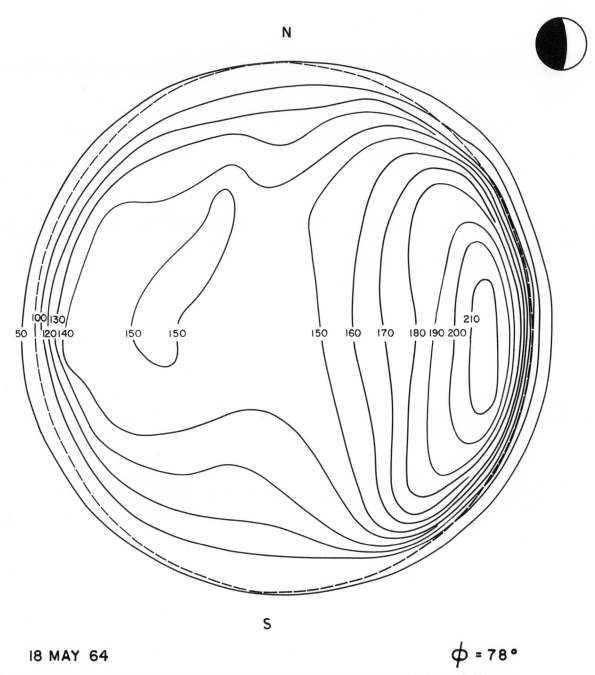

N

S

18 MAY 64

$\phi = 78°$

FIGURE 17–4: Map showing the brightness-temperature in degrees Kelvin, at the time of half-moon.

figures. We note that the level from which the radiation is coming is about 3 cm or so below the surface of the moon, very near the surface. The minimum temperature is about 150°K.

Figure 17–4 shows the half-illuminated moon. It can be seen that where the sunlight is illuminating the moon, it has become much warmer. The minimum temperature is still 150°K in the portion of the moon that has been dark longest.

Figure 17–5 shows the situation at full moon. Notice the asymmetrical displacement of the contours, which is the well-known phenomenon of the thermal wave lagging the insolation.

Looking at Figure 17–6 we see that the maximum temperature is about 280°K, which means that even at depths of only a few inches, slightly below the level to which we are observing, the temperature never rises above the freezing point of water.

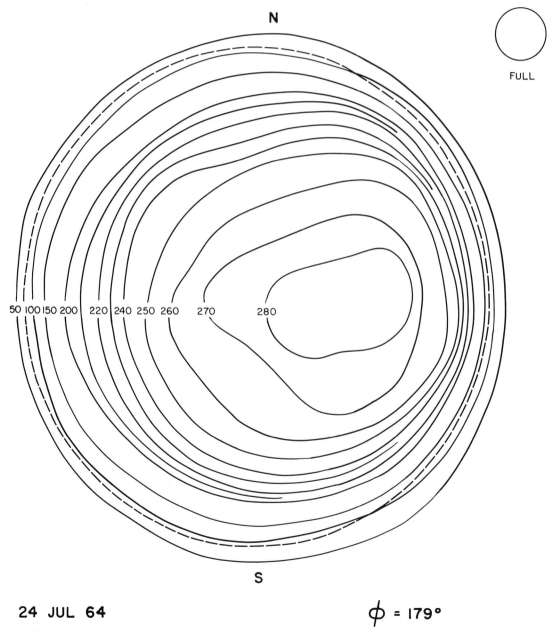

24 JUL 64 $\phi = 179°$

FIGURE 17–5: Map showing the brightness-temperature in degrees Kelvin, at the time of the full moon.

Another important point which is not apparent from these figures is that when we follow a point of constant phase, we observe essentially the same temperature, suggesting that the radioemissivity of the moon at this wavelength is very nearly isotropic. This requires structures everywhere on the surface of the moon that are a fraction of a millimeter in size or larger, which is consistent with remarks made previously by many speakers here.

These data can be used to produce a lunar phase curve, which is shown in Figure 17–6. This curve is for the equatorial region. The mean temperature is 206°K, somewhat less than that observed at longer wavelengths. However, when one takes the errors into account, there is not yet a significant difference in the mean temperature from, say, 21 cm to this wavelength. This is contrary to the conclusions of Troitskiy.

An interesting point in this curve is that the phase lag is about 22 degrees in general, as it should be, indicating we have a substance of very low thermal conductivity. However, there is an abrupt rise in the curve at phase 90 degrees when the sunlight first appears. Also, there is a point of inflection at about 280 degrees phase. This means that the simple theory for the lunar radio emission cannot fit these phase curves. We appear to have principally material of very low thermal con-

ductivity, but to explain the abrupt rise at 90 degrees and the point of inflection, there must be a second component, not in depth, but on the surface. This is, of course, consistent with the irregularities shown by the radar and by the infrared measurements.

Figure 17–7 shows phase curves for many latitudes. These can be used to compare measurements at specific points on the moon with the average lunar behavior to find thermal anomalies in the radioemission.

This part of the reduction is still in process, but Figure 17–8 shows the results one gets when one compares simply the mare ground against the highland ground. We find systematic temperature differences which average out to 3.0 ± 0.3°K.

This 3°K difference is more than can be explained by the albedo difference effect on the solar radiation absorbed. That albedo difference, using the optical albedo, would give a 1.5°K difference. There is another 1.5°K that is unexplained. We think a simple explanation of this is that the infrared emissivity of the mare ground is slightly less, about 4 per cent, than with the highland ground.

Figure 17–9 shows an attempt to go a bit further. Here we have a detailed map of the differences in

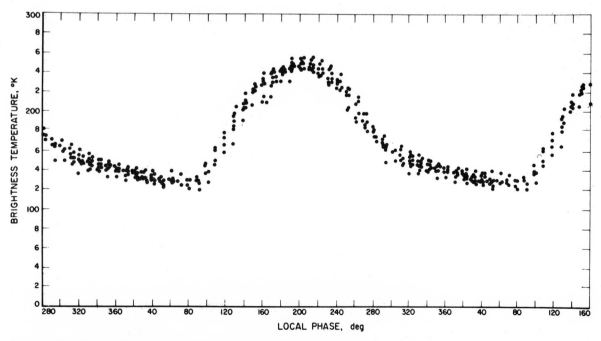

FIGURE 17–6: Lunar phase curve for the equatorial region.

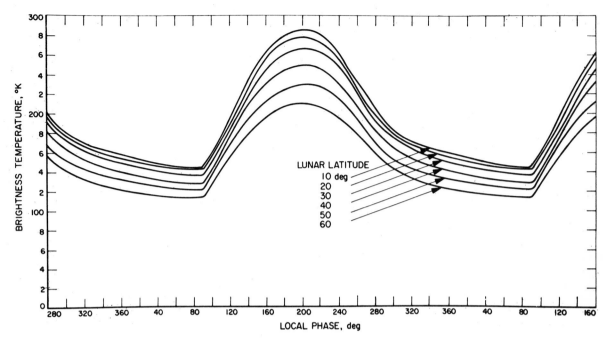

FIGURE 17–7: Lunar phase curve for many latitudes.

temperature from the expected means, super-imposed on the map of Shorthill and Saari. The gross structure of these anomalies simply correlates with the arrangement of highland and mare regions indicating that it is simply the albedo difference that is causing most of the observed effect. In addition there is some correlation with the anom-alies indicated by Shorthill and Saari, for instance, in the Mare Humorum. Our coldest region is where they see the fewest thermal anomalies, also. This correlation is again consistent with a low infrared emissivity, because such emissivity will also explain the thermal anomaly seen in the infrared at the time of a lunar eclipse.

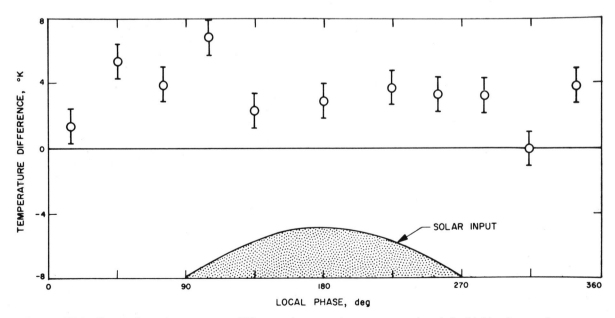

FIGURE 17–8: Comparison of temperature differences between the mare ground and the highland ground.

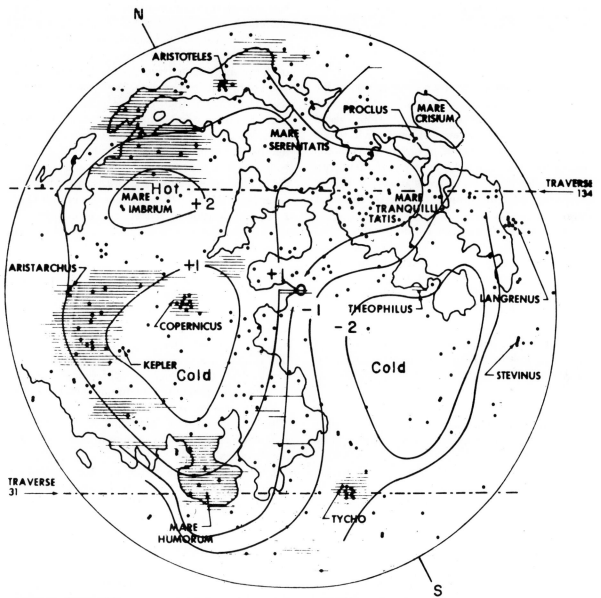

FIGURE 17–9: Detailed map of the differences in temperature from the expected mean superimposed on the map of Shorthill and Saari.

The points I want to stress are the apparent correlation with the infrared anomalies; the fact that we need 1-mm structures all over the surface; the fact that at least a two-component model seems required; and lastly that the phase curve itself supports the old idea that the moon is a very poor thermal conductor in its upper layers.

DISCUSSION

SAGAN: I would like to comment about the problem of relating measured brightness temperatures with temperatures deduced from visual albedoes. We have measured the reflectivities as a function of wavelength of a variety of common materials. There is in almost every case a tendency for the reflectivity to increase markedly longward of about 7,000 angstroms, that is in the near infrared.

There is a perfectly good reason for this—electronic transitions occur in the ultraviolet and visible. The infrared fundamentals occur in the three- to ten-μ region. Around 1 μ we are stuck with the second or third overtone, which is down many orders of magnitude in absorption coefficient.

Therefore, a photon incident on a granular material at 1 μ can make many reflections and re-emerge without having been absorbed, which is not the case in further infrared or shorter wavelengths.

It follows that the infrared reflectivity of the moon should be much larger than it is in the visible, and the integrated or bolometric albedo should be much larger.

I would think that a discrepancy of 1 or 1.5 degrees between the predicted and computed temperatures for the maria and highlands can be attributed to a difference in the infrared part of the reflectivities, or at least a major fraction of it, and we can also probably squeeze something out of the emissivities.

DRAKE: Yes. I am in agreement. Let me clarify what I said. It is simply that there is a difference, evidently in the infrared emissivity, between the highlands and mare, and this we take as evidence that there are different materials there. By that I am trying to confirm statements made previously by other people.

INGRAO: I would like to make the following remarks: It frightens me that we are discussing models and talking about differences in temperature in absolute measurements of only a few degrees.

I made a very simple calculation that I wanted to discuss yesterday—it is to compute the coefficient of the propagation of errors in the parameter that you use to reduce your data. When you do that, it will frighten you that you are using parameters that are not so reliable to reduce your data. You are assuming emissivities for the reflectance of the middle that you don't find there always.

We worked out two-layer models, and the infrared measurements are very insensitive when the outer layer is bigger than 2.5 mm. So I don't think I am in a position to accept or to give validity to infrared measurements that give a result of dust layer thickness or measurements of the order of 10 or 7 mm, because all of the lines go together. There is a very small difference between the cooling curve for 7 mm, 10 mm, and infinity. The answer will come only from people like Dr. Drake in the radio measurements.

I have a curve that gives, at different times of the eclipse, the variation in temperature with depth. You can detect the difference that Dr. Drake is talking about in models at 3 cm, but in the infrared measurement we cannot do anything to differentiate between some models because the systematic errors are big enough to mask a distinction between models.

I think, as Dr. Drake did, that the selection of models will come from measurements. We can complement with the surface measurements. Moreover, in our paper (Chapter 10) we make an analysis computing the efficiency of propagation of errors, and we bracket the errors that you should expect for high temperatures of this solar point and for low temperatures. To take subsolar temperatures, you will have to find very, very carefully what technique has been used in order to find out if there is any systematic error hidden there.

Part IV.

Conclusions

18.

Summing-up of the Conference

Ernst Öpik

Department of Physics and Astronomy, University of Maryland, College Park, Maryland

In a brief summary only a few impressions and highlights can be given, so I may not be quite accurate in referring to everything mentioned. The real hero and center of the discussions was Ranger, appropriately introduced by a movie. The photographs were 100 per cent successful.

Important results from this splendid enterprise are already forthcoming. They are interpreted in the light of ground-based observations and knowledge, which supplement the Ranger results. The first provide an over-all view, the woods; and the second allow us to see the close-range details, the trees and flowers.

The present conference is an impressive show of concerted scientific effort to this effect. The subject of the moon is no longer only a privilege of astronomers, but has become a matter of concern to a wide range of scientists. An astronomical photograph may yield the cream of its information, at first-hand inspection, or in a day. But after that, an astronomer sits down and spends, on the average, several months of hard work extracting the significant features and then interpreting and publishing them. This may take years. The thousands of Ranger pictures are awaiting this process which, because of their importance, is happily accelerated as much as possible, so that the astronomical tradition may not be applied to them. One thing cannot be accelerated: scientific thinking.

The moon itself was the second hero: at the outset of the conference Professor Urey gave an exposé of the moon's possible origin and other points, for instance, its thermal history as of a cold body. In a typical theoretician's approach, there are usually only two possibilities: (1) either, (2) or. Professor Urey pointed out the two ways our satellite could have arrived at its present location: (1) either as a direct offspring of the earth, the result of a tidal catastrophe or something similar, or (2) captured by the earth when the moon, ready-made, approached the earth from more distant portions of the solar system, probably from the opposite point of the earth's orbit.

Professor Urey found the first possibility uninteresting, as this undesirable offspring would then have lost all traces of its previous existence or existences in the catastrophe. The second possibility would be much more inspiring. It would allow us to decipher mysteries of a world in the making, thus rendering the study of, and visits to, the moon more attractive.

Later on, Professor Levin returned to the same problem. He had a different preference: that the moon was formed somewhere near the earth at some five to six radii of the earth, and then, after becoming one solid body, by its tidal action put itself farther out, where it is now. Professor Levin calculated the thermal history of the moon on a different basis and found that it is difficult to escape the conclusion that the moon was melting and that at present a solid crust of several hundred, say, 800 km, conceals a molten core, the same kind as on earth, except that it could be of different composition and density. This seems somewhat academic, if you talk about Ranger, but one of the results of Professor Levin's calculations was to give figures showing the thermal flow or expected thermal flow on the surface of the moon.

The surface of the moon is becoming more familiar to us—we find that thermal conductivity determines the gradient, or the rate of rise of temperature inward, which has very practical applications in connection with Ranger and the current Apollo projects. This even suggests how deep we should dig in the lunar surface to find the desired degree of cold, heat, or mechanical support; thus the origin of the moon, this purely theoretical cosmogonical problem, turns out to be intimately linked with the urgent practical problems of space.

Observations of radio emission from beneath the lunar surface are as yet inconclusive, and the theoretical conclusion about the heat flow, of course, is subject to a degree of uncertainty. However, it will instigate the continuance of infrared, radio, and radar observations that have been reported at this meeting to study the actual distribution of temperature, conductivity, and density beneath the lunar surface.

The Ranger pictures give a close view of the lunar surface. In brief it could be said that there is little difference in principle between its large-scale and small-scale appearance. Craters of all sizes are found on the lunar surface. There is little doubt that most of them are of impact origin. The size of craters is directly related to the size of the impacting projectiles; a general cosmogonical law tells us that when there are no specific limitations, the number of small bodies is always greater than the number of large bodies and therefore the number of small craters on the lunar surface is much greater than that of the larger craters.

Then we heard of comparisons of lunar conditions of cratering with terrestrial experiments, including nuclear explosions and direct laboratory experiments with low velocity impact, which to an unexpected degree could imitate many features, and not just one feature, of the lunar craters. One could get a different kind of crater according to the softness or hardness of the surface into which the projectiles were heading.

In the distribution of crater sizes it was first found that the craters are either direct-hit (primary) craters or secondary craters. Secondary craters are caused by ejecta from the primary craters. As pointed out recently, in the first stage of the making of the moon from relatively low-velocity projectiles, the ejecta would be less numerous than at present when the moon is bombarded by fast interplanetary particles, meteorites, comet

nuclei, and so-called asteroids of the Apollo group, which may be the same thing. With this fast bombardment, the number of ejected boulders would be greater.

We heard a calculation of the probable frequency distribution of small lunar craters in the Ranger pictures. The distribution was compared with the meteoritical and experimental terrestrial craters, assuming the number of large primary craters to be given and the number of ejecta to be determined experimentally from a nuclear explosion. The numbers generally agreed down to a diameter of 300 m; below this diameter there was a progressive deficiency of small craters, which is reasonably explained as a result of age and erosion. The smaller the craters are, the more quickly they disappear. If a 300-m crater had been eroded in 4 billion years, one of 1 m would be eroded in 15 million years.

These figures seem to agree with other estimates of erosion. Dr. Whipple stated that about 25 m eroded in 4.5 billion years in space by meteorites. A crater of 300-m diameter would have a rim elevation of about 20 or 25 m, or close to the expected erosion limit; an erosion limit over the age of the moon of about 30 m has been estimated by myself from an inspection of the best ground-based photographs. There seems to be agreement on these points. The number of craters and the distribution of the small craters seem to agree well for different regions. So Ranger VII in Mare Cognitum and Ranger VIII in Mare Tranquillitatis give the same general rate and distribution. In Alphonsus, Ranger IX, the crater density is considerably greater but the number of small craters seems to agree again.

There is a very important point to be cleared up. The peak of Alphonsus has attracted attention for various reasons, and today we heard that it would be of volcanic origin because there are too few small craters on it. Now, there are various circumstances which should be cleared before such a statement is made. First, the illuminated portion of the peak is inclined by some 25 or 30 degrees to the sun rays, whereas the floor of Alphonsus is 12 to 15 degrees. So the shadows would be missing in the peak craters.

Second, there is much more erosion on the mountain slopes, causing small craters to disappear sooner than in the valleys. The slopes are of harder bedrock, and smaller craters are produced there by equal projectiles. So the statistics are subject to various limitations. In general, the picture of a steady-state distribution of small craters which can be eroded in less than 4 billion years seems to be a good working hypothesis, the number of new craters formed being in equilibrium with those disappearing from erosion.

The greater number of larger craters on the floor of Alphonsus need not be puzzling because of the many ways of explaining them. One explanation is that we have two kinds of crater populations, one in the highlands, the other in the maria. In the highlands we see the last impacts of the projectiles, or planetesimals, which made the moon. They are much more numerous, about 50 times more craters per unit area having been produced during this final bombardment (which may have lasted only for 100 years) than in the maria which received the continuing impacts of interplanetary stray bodies during the subsequent 4 billion years until today.

If the floor of crater Alphonsus was made 100 years earlier before the end of the last intense bombardment and the formation of the maria, this would be the explanation even though the age of the maria could be but 200 years later. The time here is only relative. It is not to be measured in years when you look at crater density, which was determined by the rapidly changing properties of surrounding space. A greater crater density in Alphonsus as compared to the maria would be quite natural for such not-exactly-contemporaneous formations. They are all problems for which Ranger pictures should be scrutinized especially.

Now the important point is the surface structure of the moon from the practical standpoint of lunar landings. The Ranger pictures give us a close view of the geometry of the outer surface of the moon, but we have to decide or guess from ground-based observations what is inside, at present, before we land. We have to make a special effort in this direction because it is a prerequisite to landing.

The lunar surface, with very few exceptions, is completely covered with dust. A thin layer of dust, of course, is present everywhere as shown by the thermal data for lunar eclipses, although it may be very thin on the mountain slopes and especially in young craters like Tycho. There it may be half a centimeter thick, covering hard rock, or it could be much thicker.

A uniform model of the lunar surface does not exist as yet. The properties of the material must change with depth and must differ from the topmost dust layer on the lunar surface. Purely thermal observations have been made during lunar eclipses, when variations of emission in the infrared window, between 8- and 14-μ wavelength, were recorded.

Some evidence indicated that the low conductivity of the surface layer was increasing inward, but the thermal observations, being confined to a thin surface layer, could not be deciphered easily; radio emission and radar observations must solve the problem. Radar would tell the density and dielectric properties of the material, and radio emission would tell the run of temperature with depth.

At present there is not complete agreement on the interpretation. It seems that the outer portion, perhaps a few millimeters, of the lunar surface dust layer everywhere on the slopes and on the maria is a very porous, little-conducting material, but that the conductivity and density increase inward, a direct consequence of mechanical laws. The overlying layer by its weight increases the contact surfaces of the dust particles, so that in addition to radiative conductivity, more and more direct bulk conductivity transmission of heat by contact from one grain to another takes place with the increasing depth. Thermal data have been interpreted by two-layer models, dust overlying a denser conducting material, but the conclusions are not convincing; a continuous-transition model is required.

Radar data indicate that the dielectric constant is usually less than that of solid rock, and that not only the upper layer but all the accessible layers are of a highly porous structure. It is not a dense material—maybe 50 per cent of the density of rock, and 50 per cent porous volume. But it doesn't mean that it is all dust; it may be compacted in a sponge-like fashion. Rather, a layer of dust of equal volume density would more or less give a similar effective dielectric constant. At the same time it would show similar thermal and optical properties, perhaps for a specific whisker-type shape of the dustgrains.

One of the spectacular conclusions from continuous radar scans is the high radar reflectivity of the new or young craters, Tycho, Copernicus, and many other spots of the lunar surface observed as hot spots during eclipses. These hot spots are colder in thermal emission in daytime and warmer at night. This indicates that the conductivity is higher, or that the bedrock is close to the surface. The hot spots give a high dielectric constant comparable to solid rock and a reflectivity for radar that is almost ten times that of the average. In these regions, such as in Tycho, we expect solid rock to be found quite near the surface, and the question arises where to land first on the moon. It seems that these young craters are the most likely places. Also, one could be sure that nowhere on the moon would well-supported space vehicles sink deep into the dust layer. A layer which is very fluid would not support the small craters which we see in the Ranger pictures. They would sink under their own weight and would not allow the dust to stick to their slopes, as it actually does.

Clever composite photographs of the moon, infrared positives and ultraviolet negatives superimposed, enhance the very delicate color and show regions of different coloration sharply delineated on the surface of the maria. This can be explained only by some flow phenomenon, either of the overlying dust, or of the underlying, presently solid, bedrock which could be solidified lava. The meteorite impacts would affect the color of the overlying dust, by mixing. These photographs indicate extended regions of inhomogeneity which can be explained only on the basis of flow, either a liquid or volcanic ashes. The most probable, of course, is liquid lava flow. There has been much talk about whether lava is possible or not on the moon. Even on the pure impact model, some melting would be caused around the impacting projectile and lava could spread out from there. But we can't be certain.

The question of volcanoes on the moon has now taken second or third place. The original theories of lunar craters were based mostly on the volcanic theory. Other explanations, like Gilbert's, who proposed a meteorite theory in 1893, were simply ignored for quite a long while. In 1916, for example, I was not aware of Gilbert's work. I thought I had discovered the theory myself—and published it. There seems to be a good case for certain volcanic phenomena on the moon.

That question is about the same as the one concerning the age of lunar formations: What do we consider recent on the moon? The maria may be just a few thousand years younger than the old highland formations, yet by the small number of craters on them they may seem billions of years younger. The same may be true of the volcanic structures. Whatever is suspected of volcanic character on the moon is connected either with the

maria or with the level marialike crater floors, like the floor of Alphonsus where flow of a liquid is suspected. These small-scale suspected volcanic structures could be even a consequence, not the cause, like a geyser from the molten lava which could have been produced by the meteorite impact. This has not been said here, but I add it now.

Alphonsus has been mentioned many times in connection with possible volcanic activity. At present, the chief concern is with the so-called dark spots of Alphonsus which are very prominent and which attracted amateurs of astronomy decades ago because of the suggestion that there was vegetation on the moon. These dark spots have craters in their midst and it is probable that the dark material is connected with the craters. There are still two possibilities. The crater could either be a volcano, a kind of volcanic eruption, or it could be the product of a splash of impact into an underlying dark substance. These dark and light substances, as we heard, were changing color like chameleons.

Until quite recently there were hopes of identifying lunar surface materials from their photometric properties, as compared with those of terrestrial materials. These hopes have practically vanished. What we have heard indicates that the photometric properties of the lunar materials do not depend so much on the chemical composition of the lattice structure as on the treatment before.

Possibly the lunar material could be many kinds of rock in a highly broken up state, damaged by radiation, and subjected to proton and alpha ray bombardment for ages, which has blackened them, changed the polarization properties, and partly cemented the grains together. The lighter areas on the moon evidently are less subject to this bombardment. The bright portions are still dark enough to account probably for some degree of blackening. Because there is more denudation on the slopes from meteorite impacts, and because the dust has a tendency to flow downhill (on the highlands light-colored material is thrown out also by the bright young craters), the thickness of the protective dust layer on the slopes is small, which leads also to intense erosion of the bedrock. Therefore in the highlands the color is brighter because it contains more fresh bedrock substance and less blackened material.

From our terrestrial standpoint we can already explain the photometric and polarimetric properties of the moon without needing to explain what it is, but the landings on the moon must be concerned with it. That is a problem which cannot be solved so simply from here and which requires some samples of the lunar material which must be processed chemically, mineralogically, and petrographically before direct landings are made on the moon.

19.

Panel Discussion

WHIPPLE (presiding): I know that all of you are going to have questions of the speakers since the major purpose of our present endeavor is not only to find out about the moon but to ascertain whether it is a safe place on which to land. This subject is of great interest to us and NASA and, some of us suspect, for the USSR.

I think the first question to which we might direct our attention is the bearing strength of the surface of the moon. One question, which I thought was a very interesting one, was raised by Don Gault. He asked whether, in the case of the fumaroles or the ash deposits around the dark craters in Alphonsus (the gas presumably carries dust up and gives the dark material), there would be time for degassing, and whether that material might not be cold-welded, as we suspect most of the dusted surface is. Perhaps Dr. Wehner would have some opinion as to whether material thrown quickly out into a vacuum and landing would stick together as well as material that had been in space a long time before this happened.

WEHNER: I really don't know what to answer in this case. With respect to these black spots I would only make one small point: I don't think it is carbon or graphite, because they have a very high sputtering rate and this is not a physical sputtering—it is a chemical reaction. Some carbon compound with hydrogen is formed and these sputtering yields can be by two orders of magnitude higher than those of other materials.

With respect to the sticking together, it would probably be cold-welded together, given a long enough time, at least 10^6 years.

WHIPPLE: So it might still be safe to land in that area?

WEHNER: Yes.

WHIPPLE: Would anyone care to speak now about the question of the bearing strength?

GREEN: One of the points about bearing strength which is overlooked is the very important effect of grain shape. If material is thrown out from a volcanic center, very commonly the grains are highly vesiculated and the adhesion effects are strictly second-order effects, as are vacuum effects, relative to the much more important factor of grain shape. In the evolution of volcanic fragments, there is a high degree of vesiculation and in vane penetrometer tests in Paricutin ash the shear strength was anomalously high because of the highly vesiculated and embayed nature of the grains. Shear strengths of up to 50 pounds per square inch were obtained in this material, in or out of vacuum, when vibrated at a resonant frequency of about 25 cycles per second to achieve maximum compaction. I think grain shape relative to shear strength has been neglected in many of these works.

JAFFE: I think it is possible to get direct evidence from the Ranger photographs as to the strength of the surface, at least as far as the lower bound is concerned, because it can be concluded from the appearances of a number of smaller craters that they are covered with some form of overlay. In that case the overlay must have sufficient strength to maintain its position against lunar gravity. I have been measuring these slopes and taking the resulting data, assuming a lower limit for the density of 10 per cent of solid, and coming up with numbers for what must be the minimum strength. Going through the normal equations of soil mechanics, to provide a bearing strength for this, the lower limits I get are about 10 grams per square centimeter for a bearing width of 10 centimeters, and of the order of 100 grams per square centimeter for a bearing width of 1 meter. Now, these are lower bounds. All the assumptions along the way were made to make sure these were low values. The actual strengths may be considerably higher. There is no way of telling that with this particular technique because it may well be that the strengths of the overlay are considerably higher than the minimum required just to hold the overlay in position against lunar gravity.

O'KEEFE: I would just like to say a word about that black haloed crater that is, in fact, deposited volcanic ash. It must be erupted hot, because otherwise it couldn't be detached from its surroundings. This is not strictly true, but normally new volcanic ash is hot when it is detached. Now, a cloud like that cannot lose heat in any simple way. It can only lose it either by radiation to space or by heat transfer to the ground. A simple calculation using the values of kpc, which are so often measured, shows that the heat transferred to the ground is about 500 calories per square centimeter in a period of about 17 days. Heat transferred to space is about two calories per square centimeter per second. Neither the radiation to space nor heat transfer to the ground will cool the cloud quickly enough to avoid sintering or welding it. Therefore, when it arrives it will still weld. There is not likely to be a cold ash deposit in space.

WHIPPLE: I see. Now, does anyone else want to pursue this question of the bearing strength or are we going to agree it is somewhere between 100 kilograms per square meter and the other considerably larger figure?

UREY: May I make a remark? A couple of years ago Gordon MacDonald made a very similar calculation to that of Jaffe, using the shape of craters as seen from the earth, and he came up with very much the same figure, as I recall.

GOLD: I want to say something about the question of deriving a bearing strength from angles of repose. We can quite easily show you powders with vertical angles of repose and extremely small bearing strength and all manners of combinations. So I don't feel that that kind of a calculation has any very great power behind it.

WHIPPLE: The assumption of density is there.

GOLD: Yes, but there is also the angle at which it can recline to be considered and that depends on the internal friction of the material in a different way, for that collapses with additional weight on the surface.

JAFFE: This is not purely an angle of repose calculation. This considers also the height of the slopes. So actually there are two unknowns, the cohesion and the friction. What I did was to consider the entire range of possible friction angles from that of the observed slopes all the way down to 10 degrees and solved the bearing strength equations with these two unknowns and picked the minimum. The minimum is the numbers which I quoted. This does not assume any particular value of cohesion. As a matter of fact, it turns out that the lowest values are obtained if the material is essentially cohesionless. Now, I do not believe that it is cohesionless. It seems to me that this would be unlikely in a vacuum. If there is appreciable cohesion, then the minimum strengths obtained by this technique are considerably higher, so again I state them only as lower bounds. If cohesion is present the numbers will be higher.

VOICE: What is the value of the specific gravity of the material you are using?

JAFFE: I assume 0.3. The result is directly proportional to whatever is assumed for the density.

WHIPPLE: I thought one or two of the most interesting observational results presented were Whitaker's two-color pictures and Shorthill's of the temperature distribution at the middle of the eclipse. I thought it might be interesting to show those again to start the discussion on the problem of the nature of the maria and some of the associated problems.

WHITAKER: In Figure 19–1 (Figure 3–2 of my paper) we can see the South Pole at the top, and then Mare Imbrium, Copernicus, the rays, Pytheas, Leverrier, and Helicon. Also visible are Sinus Iridum, Plato, Mare Frigoris, and the highlands just west, astronautically, of Sinus Iridum; and we can see Aristarchus, the very well-known yellowish diamond, and Sinus Roris. The rays nearly always appear dark, that is redder, and yet where they come to the bluer material they appear to leave off and then start again on the other side. This happens quite frequently.

WHIPPLE: Now we will see Figure 11–5 from Shorthill's paper, turned upside down for comparison (Figure 19–2). Do we find any correlation in this?

SHORTHILL: There is a hot spot in this photograph that appears right on the edge of Sinus Iridum in the first one.

WHIPPLE: There was a considerable difference among the maria in Figure 19–1 and 19–2. Dr. Shoemaker, would you arrange those maria in terms of age?

SHOEMAKER: I am in disagreement with some of my colleagues trying to do this. I think that most of them are about the same age.

WHIPPLE: Which is one, two, three? We agree

that all of them might have been developed very closely together.

SHOEMAKER: Are you speaking of the basins or the maria material?

WHIPPLE: The basins. Is this Mare Imbrium?

SHOEMAKER: Yes.

WHIPPLE: And Mare Tranquillitatis up here?

SHOEMAKER: Yes.

WHIPPLE: Now, this is hotter than that.

SHOEMAKER: No.

WHIPPLE: No? Now, which is older?

SHOEMAKER: Mare Imbrium is younger. The basin is younger than Mare Serenitatis.

WHIPPLE: Now, this is the youngest and the darkest and therefore has less conductivity. We agree that these are older and they have more

conductivity. I have a feeling there is something in this, but I don't know exactly what it proves. Where does Mare Humorum stand in your age calibration.

SHOEMAKER: We don't have a relative point for Humorum relative to these others because it is so far separated. It is very difficult to work out the pre-Imbrium sequence, but roughly the sequence is thought, by Donald Eggleston, who has worked on this problem, to be (going back in time): Mare Imbrium, Mare Serenitatis, Mare Tranquillitatis, and then Mare Nectaris. That would be approximately the sequence. Have you any feeling where Mare Humorum may fit in this pattern?

GOLD: Why should the time of the sequence of the formation of the basins be of consequence to

FIGURE 19–1: Composite UV-IR photograph of Mare Imbrium region. Darker is redder.

the properties of the filling material now?

SHOEMAKER: I don't know.

GOLD: Do we need to discuss the sequence of the filling, which might be quite different?

WHIPPLE: I don't know.

GOLD: Indeed, I would deny that there was any sequence in that because I believe in their close propinquity, but if they were filled suddenly then the sequence of the filling is not likely to have been the same as the sequence of the making of the basins.

WHIPPLE: That is quite possible, of course, but I got the impression that there was a real difference

among these maria basins and it must have great significance in terms of the material that is there. Now, it might not be in age, it may be something else; but there must be some reason why there is this large difference here.

GOLD: If the mare basins were entirely separate from each other then you might believe that the time of formation and the time of flooding go close together in each case and therefore the sequence is the same for the two processes. However, there is clearly a chain of adjoining and overlapping basins; the filling cannot have been a separate one for each because then there would be big borderlines

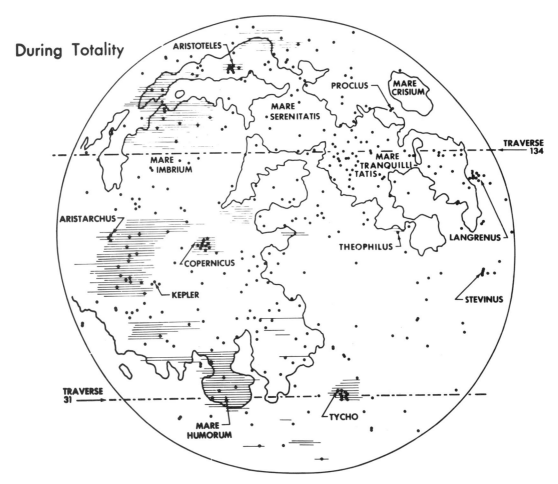

Position of Anomalies During the Dec. 19, 1964 Lunar Eclipse

Dots Indicate "Hot Spots"   Lines Indicate Thermal Enhancements

FIGURE 19–2: Position of hot spots on the lunar disk. Lines indicating thermal enhancements also show scan line separation.

between the overlapping basins. It is clear that whatever flooding occurred, occurred jointly over a number of basins together, so therefore there is no reason for believing that the sequence of the flooding is the same as the sequence for the formation of the basins.

WHIPPLE: My impression was that if indeed Mare Imbrium were last and flooded over, say, Serenitatis, then there might be less conductivity through Serenitatis than there is through Imbrium. But this seems to show the reverse, for example.

GOLD: And the other point we spoke about was that we are concerned with such a minute layer for the thermal measurements—only the top inch or so, and that top inch has, after all, many times been treated with ray materials from all the highlands and all the craters everywhere that have been sputtered over it; so it is very hard to see why the other flooding which occurred is really determining the present thermal properties.

JAFFE: Since, in general, during eclipses, we don't see the thermal wave propagating very far into the moon, it is quite likely that the dust is sufficiently thick over most of the moon so that we are not seeing differences in thicknesses of the layer here. I'd like to suggest that the differences in this picture are differences in the intrinsic conductivity of the material and that they represent compositional differences from place to place.

WHIPPLE: I would be quite happy with that. Does anybody else want to comment on these photographs? Does anyone want to account for the fluids over those areas and would anyone choose a fluid other than lava?

O'KEEFE: Well, I would like to suggest that, beginning with Aristarchus, we are perhaps seeing material which has not yet had time to darken. Aristarchus has been very active in historic times. Perhaps around it we see brown—a brown layer which is of an age of the order of 100,000 or more years. Now this would imply that in general the brown maria are newer than the blue maria. I know that this is not accepted everywhere. However, there are a couple of brown places that look a little bit like volcanos. There are other places on the moon, apart from the rays, of course, which are always yellow, and which look as though they were brown and new. One of the difficulties—and I can't trace this out accurately—is in Mare Serenitatis, where we have an inner brown area and an outer blue ring, which could be accounted

for on the supposition that Serenitatis had bent down and formed a bowl under the first layer of ash and later had received a second coating which came in the center where the bowing had already occurred. It is a fact that around the edge the thing is much more rilled, and it is also a fact that Ranger VIII came down in a blue mare which has slightly more craters than that of Ranger VII, which came down on a chocolate brown mare.

SHOEMAKER: Did you want to speak about correlation?

KENKNIGHT: I would generally like to support the view that O'Keefe takes in this regard. Perhaps not with regard to ash, but our data is in support of the findings of Hapke in that the materials which are quite red or brown, relative to these other materials which are gray, are indeed materials that are high in silica. We have seen enough cases of typical lunar materials so that we can say with confidence that the reason the high grounds are red, or light in color, is that these grounds are high in silica. These low-lying areas must be high in basaltic materials. I find this to be in support of the contentions this morning of Levin. The exact thermal history that has to go along with this is a very complicated business in which we have to talk about at least three time scales. The time scale which is of the greatest importance at the moment is only about a hundred thousand years or so and is a competition between the bombardment of small particles and the bombardment by protons, and so on. But this type of processing cannot obliterate the kind of thing one sees here. It can only obliterate dimensions on the order of 10 or 100 meters or on that order of magnitude. This is interesting enough to the persons who have to put their feet down in it, but we had better not confuse that with the broader scale phenomena which involve longer time scales. There is surely an additional time scale that is associated with the long time, with uranium atoms decaying.

And there is another time scale which must have come even before that, which is associated with the cosmic radiation—aluminum 26. The very fact that we can say with reasonable confidence that the high grounds are almost granitic in material and the low lands are basaltic calls for a differentiation mechanism which can only be in nonequilibrium. That nonequilibrium condition must be driven by the first heat wave. The second heat wave is the one of which Levin spoke this morning and the final energy input is the one which concerns those who put their feet down on it—the little changing at the 200-m crater dimension that fills up the small craters and puts a uniform dust layer everywhere on the surface through which we have to study what is below.

HAPKE: I would like to say that I tentatively support the comments of the previous speaker as far as the possibility that the highlands are more granitic than the lowlands. But I would not be quite as emphatic as he is. I think there is still quite a bit of room for doubt about this. In particular, in my experiments I have been having a good deal of difficulty in making granite powders bombarded by protons take all the photometric properties of the moon.

KENKNIGHT: Just one thing needs to be added to clarify that point. The color differences which do occur on the moon are really small. They are smaller than the color differences which Hapke and Dollfus and I are measuring. The way to understand that is in terms of the things that Shoemaker talks about all the time—that this mixing phenomenon involving rather short-range events—meaning kilometer trajectories, and so on —produces a general mixing, but one which strongly favors the local material staying where it is.

SHOEMAKER: I think it might be pertinent to point out what some of the different colors are here in terms of, say, source, so to speak, and time sequence. It is extremely difficult to find any correlation in this. For example, the brown or reddish patch in the Jura Mountains is the rim unit of Sinus Iridum. And if this is the impact basin, which I personally believe it is, then this is the material ejected from this spot right here from some depth. This brown patch is not associated with Aristarchus. It is much older than Aristarchus in part and is a very complicated area. There are many things in here but there are some very old pre-mare material units in here of the same general age as the units back around in the Apennines Mountains. It probably is related in general to the kinds of material that form the rest of the circumferential mountain range. You can notice a distinct color difference in the throw-out rim unit around Plato. It is a little bit redder than the surrounding, as well.

There is a very strange and interesting area on the mare, which is loaded with little volcano-shaped features. These are little domes, and John

McCauley has found that about 75 per cent of them have summit craters. If there is a good volcanic field of the terrestrial type anywhere on the moon, this is it; it is like much of the rest of the mare material. Up here we get complete differences in two craters which are both formed in the shallow part of the mare rim. That is a relatively thin layer of mare rim right in here.

Timocharis, you know, is light in here, which means it is blue. Pytheas is dark, which means it is reddish. It is apparently excavating down into material which in other respects seems to be the same. I have a hard time seeing any sense in this pattern.

GREEN: The idea of a gradually subsiding tectonic basin that is gradually filled with lava flows is appealing because of ocean basins on the earth. I wonder if Eggleston's point of view, which he discussed in papers given some time ago, on the filling of Alphonsus with Imbrium ejecta, is still adhered to in view of the recent Ranger IX photos of Alphonsus and if it corresponds to Imbrium debris; I do not agree with this point of view.

Also, I wish to state my points of disagreement with John O'Keefe about the mare areas being mostly ash flows. There are many surface features of ash flows that do not correspond with the features I think we see on the mare areas: the pressure ridges, the little craterlets on pressure ridges, the destruction of crater walls facing the maria areas, and things of that type, to which ignimbrite surfaces simply do not correspond, in my opinion.

VOICE: There has been a lot of talk today about dark halo craters. I would just like to point out that there are several dark halo craters on this photograph and they show up light. They are actually blue, which contradicts what John O'Keefe said a few minutes ago: he felt that the red was indicative of young; these actually are very young craters.

There is another one next to Timocharis too. They have blue rims and are identical in almost every respect to those craters that we see in the floor of Alphonsus. They have very diffuse dark halos, darker than the mare, and they show up blue.

ÖPIK: I wish to make two points. First, the surface of the moon is covered with dust and the dust is migrating and subject to radiation damage, darkening, and it collects in the lower portions of the maria and the different portions of the maria divide and slide down the slopes. Then there will be a systematic difference in color and appearance, locally and generally.

Second, the material of the maria may be more solid ground because the difference between them is found from the level differences existing. Maps which Baldwin and others have indicate that the maria are at a lower level than the highlands by 2 and 3 km, and with isostatic considerations it would mean that there, at least, they are considerably denser. I would consider a model where the mare is overflowed by a layer of, say, 6 km of lava, which is the amount produced in melting by a projectile. But the lava is consolidated as compared with the rubble of the continent. It may be exactly the same material. It is twice as heavy, and therefore I suspect from isostatic considerations it will be lower by 2 or 3 km.

GOLD: On the isostatic question, if the mare basins are made by large impacts, as most of us believe, then I think we have to worry about this situation: if any density differentiation had taken place prior to the formation of the mare basins, then, of course, as the mare basin is excavated, it is quite deep at the instant of excavation and immediately after the pressure is released from the explosion, the floor must jump up because the moon cannot maintain such an enormous departure from equilibrium.

Now, that means that the floor will come up and bring closer to the surface material that initially was much lower down in the moon. Therefore, if any general stratification had existed in the moon, which would make the lower layers denser—and that, after all, is probable—then any excavation of a large basin would always mean that it would not jump back to level and the equilibrium would jump back to less than level.

I would imagine this is an adequate explanation of why the basin was left low and why, even afterwards, it could fill with a lighter material than the surroundings and yet still be near equilibrium with the level below that of the mountains.

KOPAL: I think the view that maria are generally lower than highlands should be taken with several grains of salt, as these facts are not borne out by the latest hypsometric data which has been processed but only partially published so far; they are partly from the Air Force and partly from the Army. They have recently been analyzed harmonically and the relationship between the shape

of hypsometric curves and the relations between albedo curves between the maria and the highlands are rather indifferent.

For instance, the last points of the maria lying near the western limb of the moon are quite high. I think the statement that the maria lie low and highlands are high is true to a very limited extent, if it is true at all.

Secondly, in regard to what Dr. Gold mentioned here, I should like to stress the point that I made this morning. The differences between levels on the moon amount to between 4 and 5 km, and how these differences are maintained is a marvel. Even if we made the moon solid throughout and assigned to the whole bulk of the moon the modulus of rigidity of the highest level of solid granite, we would have difficulty in maintaining these differences on a cosmic time scale.

SAGAN: On the question of the relations between darkness, age, and color, perhaps Dr. Hapke could help enlighten us. If we take some material which is pulverized and pretty bright—it has not been irradiated yet—and we irradiate it with solar wind for an equivalent length of a hundred thousand or a million years, then we have darkened it and, if I understand right, we have also reddened it. Is that right?

HAPKE: That is generally correct for silicates.

SAGAN: Do you reach a saturation value of the albedo and color index?

HAPKE: Yes.

SAGAN: Or if we were to continue for tens or hundreds of millions of years, would we get even darker and redder material?

HAPKE: No, I don't think so. You see what happens is that at the same time we are bombarding and darkening it by putting this layer on the material, we are also eroding it by sputtering. Wehner's work would indicate that the time it takes to sputter away one particle of the material is of the same order of magnitude. So we would reach some sort of an equilibrium between the darkening and the sputtering rates.

SAGAN: So by the time the particle is as dark as the moon, it is not there.

HAPKE: No, this is just an equilibrium point I am talking about.

KENKNIGHT: May I clarify that point, please. The particles in our data are not appreciably sputtered by the time they are far darker than the moon. They have not been sputtered away even

.01 part by the time they are already quite darker than the moon, whatever their makeup.

SAGAN: Well, this clarification has not helped me understand it very well. Let me ask a related question. Is the difference in albedo and redness between, let us say, a bright ray material and some dark mare material within the range of your asymptotic redness and darkness, or if we were to irradiate something with the albedo of ray material, would we start approaching the darkness of mare material?

HAPKE: If we irradiated something which was a ray material, it would go toward the albedo of the maria. If it is a silicate, it would get redder in general. Now, just the opposite is observed on the moon, apparently.

SAGAN: Right. We did see red rays.

HAPKE: That is the rays are redder and the darker materials are bluer.

SAGAN: Right. So isn't there a contradiction if it turns out that the bright material is also redder yet irradiation darkens and reddens?

HAPKE: Yes, but there is a way out of this. There is one general class of materials which after irradiation gets bluer instead of redder, and that is materials which are rich in ferric oxide. Now, it is my understanding that, on the earth, lavas and igneous rocks which are rich in ferric oxide and which are reddish have gotten this way because they were oxidized by contact with the atmosphere. Can somebody enlighten me on this? Maybe this isn't the case. If we can figure out a way to make moon material which is rich in ferric oxide, like some of the volcanic ash from the Hawaiian volcanos, then they would get bluer instead of redder.

SHOEMAKER: The oxidation can also take place from the contained molecules in the lava.

HAPKE: So you don't need an oxygen atmosphere to do this?

SHOEMAKER: It is not clear that this would be necessary.

SAGAN: Well, I just wanted to call attention to this paradox between progressive reddening and progressive darkening.

HAPKE: If the lunar material were richer in water vapor, for instance, or water material, would this do it? Would this be sufficient to make it enriched?

SHOEMAKER: That is all one can say. It would be enlightening.

HAPKE: Would it be consistent with the lower density?

WHIPPLE: I think that the conclusion, from Dr. Kopal's and others' comments on the gravity, is that we very much need the circumlunar satellite to determine the gravitational field. On the earth this has proved highly successful, as we now have, from observing satellites orbiting about the earth, the tesseral harmonics well determined at least to the fourth order, and the zonal harmonics in the latitude to about the fourteenth harmonic. On the moon it will be extremely valuable to see what these heights mean in terms of underlying gravitational attraction.

KOPAL: I should very much like to echo your words, because it seems to me that one of the most important tools which the powers of NASA could possibly lend toward advancing the lunar status in this field would be to provide a selenodetic satellite to precede the photographic orbital mission. If one of these could be spared for purely tracking purposes at the beginning of the mission, it might very greatly increase the probable lifetime and chances of success for the subsequent ones.

If we have to launch the first photographic satellite first, we cannot guarantee how long it will remain in orbit or whether or not it will crash. One single selenodetic satellite launched in advance of the main program could answer a great many questions.

WHIPPLE: In this connection, I have heard it said that in the Ranger firings, the moon seemed to be about a kilometer or a mile—I have forgotten which—too small. The landing was about a second late. Does anybody have an explanation of this?

SHOEMAKER: What it means is that the center of mass of the moon, which is what we are working against in reference to the tracking, is displaced, according to the data that is available now from Ranger VII and VIII.

We may have a determination from Ranger IX by now. But in any event, from those two points the moon's surface is low—it is low by about 2 km, relative to the average lunar radius obtained from the purely external geometry of the moon.

WHIPPLE: I think this proves Dr. Hopmann's and your point, about the need for a circumlunar satellite to determine this, the actual gravitation versus the geometry.

KOPAL: I should like to interject a question—

whether to include how to determine if the center of symmetry and center of gravity of the moon coincide, and within what limits.

O'Keefe: One of the things that is known about the difference between the center of the moon's surface and the center of mass is that the center of mass lies above the center of the visible surface. This supports the remark that Gold made, that the maria must, at the bottom, be composed of heavier material than the other because the north side of the moon is composed mostly of maria, the south side is mostly non-mare. Also from the Russian work we know that the far side of the moon has fewer maria than the near side and this again should have led us to anticipate that the center of mass of the moon will be closer to our side than to the far side, if Gold is right again.

Whipple: There is one piece of research on the moon that was mentioned this morning by Dr. Levin, and we have not heard a word about it except indirectly through him. Professor Eckert, would you be willing to give us a quick description of your studies about the distribution of mass or density in the moon?

Eckert: There are two recent developments that have sharpened the question that has been around for half a century. I refer to the indicated large moment of inertia of the moon as shown by its orbital motion. This is measured by the motion of the moon's perigee and node. The total motion of the moon's perigee and node are 14 and 7 million seconds per century. This is almost entirely contributed by the attraction of the sun. The planets— I will speak mostly of the node, because this is the most influential one in what I want to say—the planets give 140 seconds per century. The dynamical flattening of the earth gives about 650 and 600. There are 2 seconds from relativity. The observations in both cases, we believe, are easily accurate to a second of arc and it is difficult to quibble about this.

In here I should put the effect of the shape of the moon. And in each of these cases we have a mathematical theory. We also have some observed constants. In the first case the observed constants are the orbital elements of the moon. In this case it is the dynamical flattening of the earth; then in this case it is primarily the radial distribution of mass in the moon.

In 1908, Professor Brown said, "Well, the distribution of mass in the moon must be very similar to that in the earth. Therefore, I compute 14 seconds." When he did this, the numbers did not add up right. He said, "Well, we don't know the dynamical flattening of the earth very well, so I will choose a value corresponding to an oblateness of 294." At that time other people thought the oblateness was higher. And so that was the way he escaped.

In 1937, Spencer Jones rediscussed the observations and he said, "The precision of these is high. We are not at liberty any longer to say 294. We must say 297. And whatever is left must be attributed to the distribution of mass in the moon."

What was left was 25 seconds. Since then I think it has largely gone unnoticed, in the case of the Sputniks and the determination of this quantity. We now say that it has to be 298.25 and in so doing, we have added 5 more seconds per century to our problem. So that is the difficulty.

Now, what are the assumptions we can make about the distribution of mass in the moon? First, that it is like the earth, 14 seconds. Second, that it is homogeneous, which gives us about 17 seconds. And third, even for the worst case of a shell, we have about 29. And this is what the current state of the observations show if we put on 5 seconds more for this change due to the satellites.

Now, when we compute the theoretical part, some assumptions and some uncertainties come in. For instance, we assume a little bit about libration theory, and so forth, and when we allow for the fact that this difference is made up of several pieces, one doesn't insist on this discrepancy of 10 seconds. But I would say that one could easily admit a second or two with good probability or a few seconds with middle probability, and 10 seconds with very low probability.

So at the moment I think we must assume that the evidence from this particular line of reasoning points to excessive mass toward the surface. Of course one would next say, "Well, let's look into some of these other areas," and that is proposed, of course.

Urey: How much excess mass toward the surface?

Eckert: If you attribute it all to that you will get the absurd value of a shell.

Eckert: As I say, one would naturally assume that the bigger the correction the less probable it would be.

Kopal: What safe bottom would you assume?

ECKERT: We had to assume the dynamical flattening of the moon at, I think, .000629.

KOPAL: Why did you have to assume it?

ECKERT: Because the number that we get here for the motion of the node is a product of two factors. One is the principal moments that we get from an assumption of Cassini's law and the other is the radial distribution.

KOPAL: What you have referred to as the dynamical flattening singles out the contribution of second harmonic only.

ECKERT: Yes.

KOPAL: But this one ignores, by virtue of the Eulerian equations, analogous effects of any other harmonics.

ECKERT: Yes.

KOPAL: Now, we have some evidence on the surface that other harmonics besides the second are as large as or larger than the second. In particular, the fourth harmonics in the shape of the surface seem absolutely larger than the second. I would like to ask what effect would inclusion of higher harmonics have in your deductions?

ECKERT: You are speaking of the moon now?

KOPAL: Yes, of the moon.

ECKERT: I have not investigated this.

KOPAL: I fondly and strongly hope that, when you do that, the need for compressing too much mass in the outer part will disappear.

O'KEEFE: There is a classic investigation by Bessel of the effect of the pear-shaped harmonic of the earth on the motion of the moon which would give us an upper limit to what we would expect in this case—whether it is, in effect, a third harmonic which is the next most serious thing on the motion of the moon.

Now, any reasonable values of the third harmonic would be completely negligible in their effect on the motion of the moon. And I think it is logical, too, that the third harmonic in the figure of the moon would be negligible in its effect on the dynamical properties of the moon.

KOPAL: I cannot say anything about the third harmonic, but the fourth harmonic acts in much the same way as the second and if the second is a significant one, so should be the fourth. In the shape of the moon the fourth is larger and what puzzles me beyond expression is if we were to accept, for the sake of argument, the finding to which Professor Levin referred, and if we were to also accept that the bulk of the lunar mass is

molten, how could heavier solid matter be floating on molten and lighter matter? Could we ask Professor Levin for the explanation? Perhaps he sees it.

O'KEEFE: The second harmonic dies out as the inverse cube of the distance and the fourth harmonic dies out as the inverse fifth power.

KOPAL: It is the old question, which is alluded to repeatedly: How would you make solid silicates float on molten ones?

GOLD: We don't know that the inside of the moon is molten; that is an additional assumption.

O'KEEFE: Professor Eckert, can we ask this question? I think that Professor Levin found this difference was concerned with the difference between the solid and the liquid state. This would amount, I think, to about 2 per cent in density. Would this be of any help in solving this problem of yours?

GOLD: No.

ECKERT: If you want to take the numbers as they come and not change any of the theory, the correction is very big. It means it is a shell. We actually get 29 seconds.

WHIPPLE: Two per cent probably would not produce one second, would you say?

ECKERT: Well, from the earth to a homogeneous sphere is between about 2 and 3 seconds per century.

WHIPPLE: Thank you very much, indeed. That was very enlightening.

GOLD: I suppose if it is a hollow shell, then these dimples are easily explained.

WHIPPLE: You have read my mind—I thought it was time to discuss dimples relating to the moon. Does anyone wish to say anything on this subject?

HOPMANN: I would like to illustrate the necessity of making further visual observations, with large instruments (and possibly with positional micrometers), on the landscapes which were investigated by Rangers VII, VIII, and IX. The Ranger pictures are, to a certain extent, "snapshots" taken with evening illumination. They were photographed with about 20 degrees of sun height. Visual observations should be made at a sun height of about 5 degrees at waxing and waning phase in order to determine the details which could not be recognized on the Ranger pictures. The following examples show what even small instruments (6–8 inches) have already achieved.

The Ranger VII photograph 166 (Figure 19–3), which was taken from a height of 423 km, shows

TABLE 19–1: Telescope Height versus Heights from Ranger Photographs

No.	ξ	η	λ	β	x	y	h(H)	h(R)	Remarks
1	−0.247	−0.157	−14.5°	− 9.0°	—	—	0.24	—	Outside the photo
2	− 346	− 230	−20.9	−13.3	115	58	0.50	0.47	Darney δ
3	− 347	− 162	−20.6	− 9.3	135	160	0.34	0.41	Bonpland γ
4	− 379	− 169	−22.6	− 9.7	75	167	0.46	0.98	Bonpland E[a]
5	− 409	− 163	−24.5	− 9.5	17	178	0.80	—	Langer Höhenzug[b]

NOTES: The columns give the orthographic co-ordinates, the longitude and latitude, the position on Ranger VII photo number 166 in millimeters, and the relative height according to Hopmann's (H) earlier observations and the Ranger (R) observations.

[a] From (H) external shadows of crater walls; from (R) shadows in crater centers.

[b] See text.

for example some mountain ranges which are also partly delineated in the *IAU-Moon Atlas* or *Catalog* (by A. Muller, Vienna, and M. Blagg, England), and indeed, even on the map by Mädler (1835). In the text accompanying this Atlas there is, on page 310, a detailed description of a bright plateau in the south (compare Ranger VII, frame 88). On September 23, 1964, I was able to measure the length of its shadow, using the 8-inch instrument of the Vienna University Observatory. This shadow was also seen on the Ranger photos.

Table 19–1 gives the position of the object in the Ranger pictures. This is counted in millimeters from the left bottom corner if North is up. On the Ranger picture the shadows are shorter and less sharp than in the visual observations. The heights which were identified agree very well. In number 5, a long mountain range should be pointed out which hardly appears on the Ranger picture but is clear, for instance, on Plate E-5a of Kuiper's *Orthographic Atlas*, and in the rectified Atlas Plate 18b. According to the "peak light-edge procedure"[1] an absolute height of terrain of about 2.5 km above the mean radius of the moon (which height should also be valid for Mare Cognitum) was determined for the region about Guerike.

Ranger VIII picture EP 32226/30 (Figure 19–4) was taken from a height of 243 km. The most prominent objects are the craters Ritter and Sabine, and especially the two parallel broad furrows (lower right, if North is above) and a very narrow rille (about 9 cm to the right and 17 cm up from the lower left corner). As early as 1817 Gruithuizen, using a 2-foot long telescope, saw these furrows and, since that time they have been

[1] J. Hopmann, *Peak Light-Edge Procedure* [*Spitze Lichtengrenze—Verfahren*], Vienna: Middle University Observatory (1964), p. 67.

shown on all moon maps (Lohrmann, Mädler, Schmidt, etc.). From 1890 to 1900 the Austrian amateur astronomer Krieger, in Dalmatia, using a 6-inch telescope, discovered many regions of the moon, each on about 20 favorable evenings, in such a way that he was able to insert the finest details into a suitable copy of the *Paris Moon Atlas*. His work was completed by König as a great three-volume opus with almost 100 tables. The text for the region "Ritter and Sabine" contains 11 large pages of excellent detail. One of these has the description of the two furrows, and four pages are devoted to the 39 rilles.

In the region of the Ranger pictures, about 30 small craters and 5 rilles, in all, are shown, of which, however, only the most prominent, and these only in part, are recognizable. The others are not recognizable because they appear only at morning light and partly because, in the Ranger snapshots, the sun was already too high and there were no longer any shadows in the rilles. A more detailed comparison of the Ranger pictures with Krieger-König's atlas will be made elsewhere.

We can make comparison of the new photographs in Figures 19–5 and 19–6 (Ranger IX shots 539 and 573, taken from heights of 420 and 225 km respectively) with older representations. All of them contain the mountain range which stretches through the middle of Alphonsus. Mädler and Schmidt (1835) assigned the same height to the central mountain as that which is yielded by the shadows on the Ranger pictures; they observed with a 4-inch telescope. The great map of Schmidt (1870) yields, in some ways, more details than sheet D5a of the *Orthographic Lunar Atlas*. Schmidt shows, on the inside of Alphonsus, 25 craters and 3 rilles. All this is surpassed by the description in the work of Krieger-König, in which there are

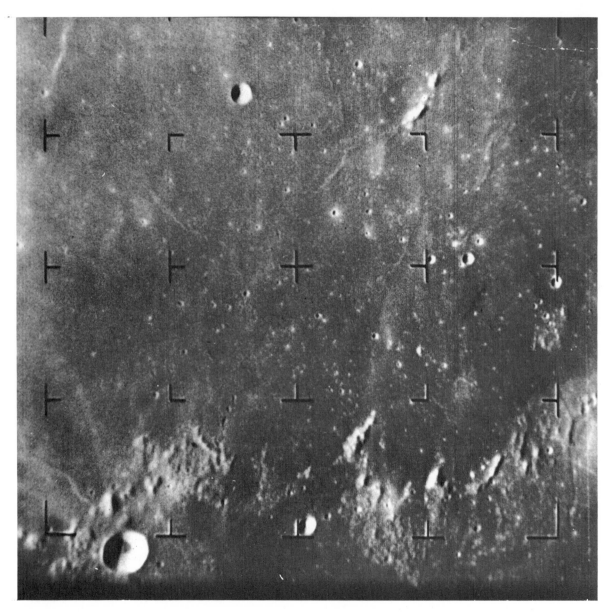

FIGURE 19–3: Ranger VII photograph 166, taken from a height of 423 km.

FIGURE 19–4: Ranger VIII photograph EP 32226/30, taken from a height of 243 km.

three text pages for Alphonsus, and his map shows 100 smaller craters and about 20 rilles. Only the strongest or most prominent of these details are recognizable on the Ranger photos and a more detailed comparison should be undertaken in this respect. The important dark places, already known to Mädler (page 304), which are important for the interpretation of the physics of the moon's surface, are discernible only with difficulty on the Ranger pictures.

If one were able to recognize so much through visual observation on 4-, 6-, and 8-inch instruments, it appears that, for a correct cartography of the impact regions of the three Rangers, it would be absolutely necessary to investigate them visually, possibly micrometrically, with much larger instruments and at the maximum number of different phases.

WHIPPLE: Thank you, Professor Hopmann. It shows that even the Ranger doesn't disclose everything. Now, Mr. Nicks was going to report on plans for a circumlunar orbit.

NICKS: I thought it might be relevant to mention that the first lunar orbiter will contain also a Doppler tracking equipment to provide the very kind of information that was suggested. It will probably be put into a large orbit initially and tracked long enough to get this kind of information before a closer approach. In addition to the track-

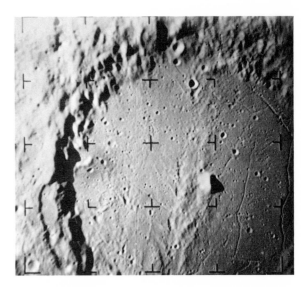

FIGURE 19–6: Ranger IX photograph 573, taken from a height of 225 km.

ing information, the stereo capabilities of the cameras and their coverage should provide additional information.

KOPAL: If we want to get as much information as we can on the gravitational field, we must chance at least one satellite to make it around as close as possible, because the effects of high harmonics grow as inverse powers of the distance from the center of the moon.

NICKS: It will change orbits from high orbits down to low orbits.

KOPAL: That will be good.

WHIPPLE: There was a question before the house as to whether Mare Imbrium debris was thrown across Alphonsus. Does anyone wish to respond to that question? It may be relevant to the dimples.

VOICE: I don't think it is a slight bit relevant to the dimples, but Jack Green insists that we believed the Imbrium debris was there, and many of Eugene Shoemaker's colleagues have always felt there was none. I think the Ranger photographs show that there was none.

KENKNIGHT: We have so many comments these days about Ranger IX having been really the first to see that these highlands are so very, very unusual in lacking these certain size craters. Was this true?

WHIPPLE: As I remember from the pictures, Ranger VIII showed some indication, but not as good a closeup near Tranquillitatis.

KENKNIGHT: So that you can say this is not peculiar to that area?

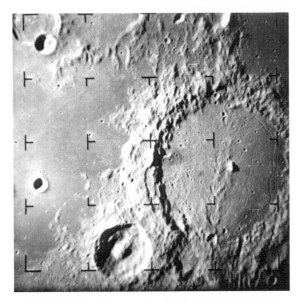

FIGURE 19–5: Ranger IX photograph 539, taken from a height of 420 km.

WHIPPLE: I don't think that can be answered. Can anyone try?

KENKNIGHT: Well, no, it is not peculiar to a very small area, because we saw quite a lot of it. But was there really any data that anybody could mention that indicated that the earlier shots saw that these highlands were peculiar?

O'KEEFE: The black ridge on Ranger VII seemed to some people—I think Ewen Whitaker mentioned this—to be somewhat crater-free.

GOLD: I wish to raise another point on the strings of craters which I thought should have been brought up. On the Ranger pictures and, of course, as known long before, there are six strings of craters that are obviously highly correlated with each other in more or less a straight line pattern. Many of those are connected with a rille and seem to imply some internal actions: some crevice along which something has happened, like sinkage or explosion or something of internal origin. There is another class of lines of craters that look completely like externally impressed explosion craters and have no reference in their line to any other of the linear structures of the moon. These go over hill and dale, irrespective of highland and low land and nevertheless trace out more or less a straight line. One such is seen in Ranger IX (Figure 19–7) a very prominent one that goes on the low land but then eventually climbs up on the mountains, and there are two or three big craters on the mountains.

I would like to hear the Chairman's comment on this. I believe that it is perfectly explicable to have primary craters that come in a string in a linear arrangement of that kind provided that meteoric material that comes in is, in the first place, made up of, say, rocks with ices that gradually evaporate in space and therefore leave a bundle of rocks, lightly glued together in the course of time, floating around in space. So long as there is no force to pull them apart, very small forces will suffice to keep such a bundle going along together. Then when it finally has the misfortune to approach the earth-moon system it will be exposed to the gravitational force of the earth or the moon prior to falling into the moon, and it is reasonable for it to become strung out in a line, that is, the line defined by the gradient of the gravitational field and the plane of the orbit.

Consequently I would believe it perfectly possible to have these as primary craters; therefore it is not

FIGURE 19–7: Ranger IX photograph, showing chain of small craters (*top*).

safe to conclude merely from a linear pattern that it is a feature of inner origin.

WHIPPLE: I would agree that that is possible of comets. I think perhaps Professor Öpik is the astronomer who pointed out that comets may be multiple. We do know that splitting is a very frequent phenomenon. In fact, a student of mine is now studying it and it looks as though a great many new comets split.

In terms of asteroidal debris, meteorites, Dr. Öpik says they are the same, but I believe that in the case where the material must be removed by shock from a solid body, it is very unlikely that they come in pairs. This has been argued, for example, with regard to the Canyon Diablo, the great crater. There were some other pieces with an accompanying piece. Supposedly two were coming together, but I have always considered this to be unlikely. Now the radiometric or the isotopic data have confirmed that they were all in one piece. It seems to me, however, that with comet nuclei it would be possible.

GOLD: But comets would often hit the moon.

WHIPPLE: I would think so. My own thinking on it is that if we would admit some heat in the moon and subsequent cooling, following Dr. Levin's general idea that the moon perhaps is not quite as hot now as it was, then the cracks which would occur in the maria and in the highlands provide a means whereby the volcanic action from below can reach the surface, so that if a crack is formed by the shrinkage, an opportunity is provided for a vent to occur and for a form of volcanic activity to take place. I rather prefer that idea to the multiple collision theory, although I certainly won't say the other was impossible.

GOLD: I think you have misunderstood me. I know there are some features of internal origin; I have no doubt about that.

WHIPPLE: You are saying without a rille?

GOLD: There are others without a rille which have also been interpreted as necessarily being of internal origin. I would deny that and say those can be of external origins. In some cases they appear to be.

WHIPPLE: I would agree that they could be, yes.

O'KEEFE: I think the question is easily answered in practice, because at least sometimes these craters line up with the tectonic grid of which we have heard a good deal of discussion. In other words, there are systematic crack systems that go all over the moon. If a line of craters like this parallels the general system of cracks in the moon, then it is probably of internal origin.

I would also like to say that it is very difficult to arrange to have the forces of gravity large enough to tear it apart before it hits. Could some of Shoemaker's people say anything about these particular chains?

SHOEMAKER: I want to make a comment that, although we haven't yet performed any good statistical studies on it and it just seems to crop up visually, apparently there are a good many more pairs of craters, where two craters occur close together, than we have any right to expect just from random distribution. These include all sizes of craters from the big ones that are easily seen in the telescope on down to many of those seen only in the Ranger pictures.

GOLD: That is right.

SHOEMAKER: This certainly leads me to believe that there are pairs of objects going around together in space.

GOLD: Yes, I agree with that very much, and those pairs of craters are characteristically of a similar age and appearance.

WHIPPLE: I would like to comment about their occurring close together and in a line. The major effect as they would come in would be for the differential forces to pull them out in the orbit plane along the direction of the orbit. Of course the orbit is one thing and their coming in is another. Consequently it would be quite possible for them to be separated, especially if they collided at small angles, by an appreciable distance, that is, a number of miles but not hundreds of miles.

LEVIN: Perhaps some of the strings of craters are similar in origin to bright rays. Perhaps it is a secondary ejection but from pieces of greater strength, so that they form not rays but a string of craters.

Commenting on the appearance of double craters, I would like to say that one Soviet commentator has published several papers on interesting correlations between craters that form strings which are not linear but have very complicated forms. There is a regularity in the change of dimensions of these craters from one part of the string to the other. He had found several examples of such regular patterns.

WHIPPLE: Thank you.

SHOEMAKER: I want to say that the pairs of craters of which I was speaking are the kinds we call primary, that are primary in shape, to distinguish them from the well-defined chains or strings of secondaries.

VOICE: If I may go back to the dimple chain craters, if there is such a thing as a dimple crater, why don't we see, in the chain craters which are extensions of rilles that are suggested as collapsed features, more of these having a dimple form?

WHIPPLE: I thought some of them did.

VOICE: Very few, I think, in comparison to the total number that you see. I would expect that more of them would have a dimple form.

SHOEMAKER: Some of them on Ranger IX do. There are some chains of little dimple craters. But do you mean why aren't the bigger ones that way? A dimple crater is a shape below a certain size range. That is an empirical answer and it doesn't say why it occurs that way. We just don't see dimple craters much bigger than about 300 m across. There are well-defined chains of these things in Darney, as you can see in Ranger VII pictures.

VOICE: In the case of Dr. Gold's pairs of craters, we have the Clearwater Lakes in Quebec, which are two large lakes, 14 and 20 miles across, both with many impact features to support them.

GOLD: And the Carolina basins.

VOICE: The Clearwater Lakes themselves were investigated by Tanner a couple of years ago for the orbital stability of the meteorites that may have formed them. He concluded that they would be stable under planetary perturbation of a sufficient size to change this and that the increase in the separation of the two objects as they came into the earth's atmosphere would be perhaps on the order of 30 per cent. These two circles are essentially touching each other with diameters on the order of 14 to 20 miles.

SHOEMAKER: Where is Tanner's result published?

VOICE: In the *Journal of the Royal Astronomical Society of Canada* about two years ago.

WHIPPLE: I am really amazed that very many of these could occur in meteorites. If they were made by comets, I won't be bothered. But you don't know in this case whether it was meteoritic or cometary in origin, do you?

VOICE: I don't know.

WHIPPLE: Well, you might if you picked up any meteorites around. Have you?

VOICE: No. As in the case of all these, these two have been well drilled. There are lots of cores available. I don't know whether there is any immediate evidence available from them; there might be in time.

GOLD: But could one not have comet material? Are you opposed to having comet material partly ice and partly rock?

WHIPPLE: If I am going to make these out of solid meteoritic material.

GOLD: Then I agree, but material that is partly ice, partly rock?

WHIPPLE: No, I have no absolute objection to that, but I expect a mixture from comets.

GOLD: If that doesn't exist, then of course there is always the chance of it evaporating the ice and leaving the rocks that are going together in the same orbit but not attached.

WHIPPLE: Quite possible, yes, but I expect finite velocities of separation.

VOICE: A dimple crater is a new word in the vocabulary of the subject of the moon. Can we get a definition now as to what the panel means by a "dimple" crater?

WHIPPLE: A pimple can have a depression at the top, whereas a dimple is all depression. In Oceanus Procellarum there is an area with a great many of these, and there are also some near Copernicus, which look like secondaries.

SAGAN: It seems that there are two independent hypotheses, each of which explains a sequence of craters. Is it possible to illuminate one of them? The tectonic hypothesis is that somehow there are vents that form the craters in sequence and the rilles as well. Is the opposite also possible? Is it possible to form the rille by the impact of this linear sequence of craters that Gold spoke of, or is that out of the question?

GOLD: I think it is out of the question because in many cases the rilles have splits in them which are V-shaped or Y-shaped, and I really cannot believe there are objects floating around in space making a pattern in the shape of a letter Y or a letter V when they hit. I would emphasize that those two kinds of linear arrangements of craters look quite different, and that if one believed a linear arrangement is a possibility coming from space, one would easily classify them into those two quite distinct kinds. Figure 19–7 shows that string of craters which climbs up on the mountain and down a row of ridges. I believe that to be of external origin while all the many features that we saw in the crater, not very well in this picture but on bigger ones, are believed to be of internal origin. But I think there is no indication of any other kind of a tectonic feature connected with it. It is just a series of holes, a very dense one.

LEVIN: Have you evidence that they are of primary origin and not secondary?

GOLD: It is an awful lot of material to be of secondary origin and be ejected as one chunk. I don't think we have any other cases where we have patterns of this nature and of such large size as secondaries strung out like that.

SHOEMAKER: There are secondaries of this size and number, but where is the primary that is associated with it?

GOLD: Close to it. A big primary one could account for it. But far from it, I don't think that one is.

LEVIN: But can we find something similar to the origin of rays? Perhaps the rays are formed by the trajectories of low inclinations and such strings of craters by the trajectories of the high inclinations.

WHIPPLE: The only trouble is that they don't know what the origin is. They could originate from the Moscow Sea on the other side, perhaps.

UREY: There is another one right here. One wonders what the probability is of having two events like that so close together.

WHIPPLE: Some of you may have special subjects you want to introduce.

DUBIN: I would like to ask a question of everybody. Several years ago Fritz Zwicky indicated to me that an impact on the moon of a few kilograms would be visible by telescope; this was in regard to whether a meteorite could be observed on the moon by telescope. The impact of Ranger was around three kilometers a second, and had a mass of a couple of hundred kilograms near the terminator. I think attempts were made to observe it. Based on impact theory and on the lunar conditions (that mass, near the terminator), should the secondary ejecta have been observed, and was it observed?

SHOEMAKER: Attempts were made to observe all three shots, VII, VIII, and IX. The seeing for Ranger VII across the country was generally too poor to arrive at a definitive conclusion. On Ranger VIII there was excellent seeing, both in Arizona and in California, and some experienced observers were on telescopes there, Herring at the 84-inch telescope at Kitt Peak and Don Wilhelms on a 36-inch refractor at Lick. They had superior seeing conditions, they knew the exact time of impacts, the information was coming over the radio, they had the exact impact point, and nothing was observed. I think that we can conclusively say now that it is most unlikely, at least for an impact in mare material, that this can be observed. Ranger IX, however, was in the daylight so we were not able to get a good check.

WHIPPLE: Thank you. I would like to ask a question of Dr. Levin because there were reports that Lunik II, which landed on this side was observed. Is it your opinion that people did see a dustcloud in conjunction with that?

LEVIN: It was a report from agrarian observers— amateurs looking with small telescopes. So I am not very convinced.

WHIPPLE: Thank you. That has been a question we were all interested in.

SAGAN: There is one topic I would like to raise, and that is the general question of the albedos of objects in this corner of the solar system. Hapke's results have suggested that the reason the moon has such a low visual albedo of about .07 is because it is exposed to the progressive ravages of the solar wind.

Now, the planet Mercury, for example, has a slightly less visual albedo. If the lower albedo of Mercury is attributed to the same cause, this immediately tells us some interesting things about Mercury which we don't know about on any other ground.

For example, it says something about the magnetic field strength near the surface of Mercury and puts it at less than a few hundredths of a gauss. Some people have suggested that Mercury was a bare planetary core with a very strong magnetic field. If this argument is right, it is an argument for a very low magnetic field strength. It also says something about the atmospheric pressure at the surface of Mercury because even with 10 KeV protons, if there is as much atmospheric pressure as, let's say, 10 millibars, these protons will not reach the surface, so it sets a limit on the possible atmospheric pressure on the surface of Mercury.

I think one can also ask about the low albedo of Mars, whose atmospheric pressure and magnetic field strength we have some idea of. There the albedo is not nearly as low, and while this is not necessary to explain a visual albedo of the order of 15 percent, particularly in the case of Mercury, these experiments with lunar intent seem to have Mercurian content.

WHIPPLE: Any other comments.

GOLD: Yes, naturally we have thought about the case of Mercury in this connection. Mercury not only has an albedo the same as the moon, but it has a very similar phase function.

VOICE: And polarization?

GOLD: And polarization; and one is inclined to wonder whether the surface structure is somewhat similar. Of course, all this is enormously at variance with the radio-thermal data showing an absence of a phase around Mercury and consequently this is such a great puzzle that, at the moment, I don't wish to say that we can infer very much about the surface of Mercury. The back seems to be warm and, if there were a lunar-type surface and no atmosphere, the back would be cold.

SAGAN: But what do you think about the intermediary case, where we have some millibars of pressure?

GOLD: If we put enough atmosphere there to take enough heat from the front to the back, then we have no reason to expect this kind of a surface to continue to be maintained either from bombardment or from the structural properties of the powder.

Consequently, I am very puzzled about how I can reconcile the temperature situation with a moonlike surface. On the other hand, there are various satellites of the major planets that have all the properties, as near as one can tell, of the moon, and I am inclined to think that they have just the same kind of material on them.

SAGAN: I am a little puzzled by the remark about Mercury. Are you saying that if there is enough atmosphere to carry the heat around to the dark side, there is too much to allow the protons to get to the surface?

GOLD: Too much, yes.

SAGAN: That implies a calculation of the advective heat transport from the bright to the dark side.

GOLD: Yes.

SAGAN: My understanding was that this was a generally unsolved problem. I am delighted to hear it has been solved.

GOLD: It has been solved to the nearest two orders of magnitude but this is enough to make this point.

SHOEMAKER: I would like to make some additions to the discussion about the data, that I don't think have been clearly brought out, and I would like to second Dr. Gold's proposition that it is not just the albedo, but also the polarization and the photometric functions that constitute the powerful arguments. Now two things are involved here, not just the solar wind or irradiation, but also the small particle bombardment. These two things have to be considered together, and we should not match the albedo of powder formed by just the simple process of irradiating the powder once on the surface. This is not what we will see on the moon's surface. It is a case of constantly mixing this irradiated material with fresh material which is being thrown in and having it reach a steady state after a while. That steady state will have a higher albedo than the actual albedo that we would get at a steady state albedo from proton bombardment alone. Now these two things are going on together; therefore we should be very, very careful when we match the albedo, let us say, or any of these other properties with something that is done in the laboratory in a much more simple situation. We must consider what the true steady state is on the surface.

VOICE: I understand there was a discrepancy between the experiments. Hapke said that he was approaching lunar albedo and Wehner says that if he continues proton irradiation he gets albedos lower than those on the moon.

SHOEMAKER: Hapke gets albedos on some materials that are lower too. What you should be matching is something that will in fact be darker than the moon if you don't mix it in with fresh material all the time.

GOLD: The fact is that for a great many rocks, results are obtained which give darker final values than the moon. In particular, for example, chondrites go very black indeed. It is conceivable that the mixture would then return to the actual albedo of the moon.

The reason Hapke was hesitating about this is that we have not contrived to make the entire photometric law, polarization, color, and albedo correct for any such mixture, but that may be merely because we have not yet hit on the right mixtures. We have made it exactly right, however, for primary materials that are fully saturation dark and that do not happen to darken too much. For those, we have arrived at a complete fit, as you can see.

O'KEEFE: Let me press this point again, If I may. During your paper you spoke of erosion at the rate of 1 μ per year, which would mean in 10 years you would lose 10 μ. Now, Hapke talks about taking 100,000 years to darken a grain occupying about 10 μ. How is it possible to keep the moon surface darkened at all if you remove 10 μ per year? Let us not detour around the question by saying that there is mixing, because clearly it makes no difference whether you darken 1 cm in 100 million years or whether you darken 10 μ in 100,000 years. It is still difficult to remove more than 1 cm per 100 million years if you darken at the rate that your experiments indicate.

GOLD: I darken the grain in a period, which I don't think we can say very precisely, but which is perhaps about 10,000 years.

WHIPPLE: I don't think that we can answer the question properly without a laboratory, or a good theory, and I haven't seen any evidence for either on this one point yet.

DUBIN: I think there is a potential neglect though, and that it should be included here as a point of record. Assuming that the cosmic dust in space is also subject to proton bombardment and accordingly could be darkened by a similar process, although not exactly in the same way, the incoming dust that went to the moon and Mercury might be dark already and cover the surface, giving a low albedo.

GOLD: Yes.

VOICE: If I understand your mechanism correctly, it is sputtering and one would presume that the incoming dust would not in fact have been darkened.

WHIPPLE: The proton bombardment is darkened.

VOICE: That is right. It's sputtering. It darkens the underside of it.

O'KEEFE: And it doesn't darken the particle it hits. It darkens the ones near it, and you are right.

UREY: May I make a remark? It seems to me that the particles coming in from space would come at such a high velocity that they would be completely destroyed in the process and raised to a very very high temperature and nothing in the way of darkening would persist beyond the process.

WHIPPLE: I agree.

GOLD: That is correct, but the added point to be made is that the condensation of the material which is so evaporated by the incoming dust itself will have an effect on the coloration which we do not at the moment understand, but it will be a large effect.

KENKNIGHT: One difficulty that arises in the laboratory achievement of our goal is that indeed the bombardment produces some effects in a powder that won't be present on the moon, where the disturbance of the particles is going on at all times and maintaining the underdense structure.

We find, indeed, that after we have bombarded the surface for a while it becomes quite stable, although not really solidified, whereas in the early part of the experiment we must be extremely careful not to jostle the sample or the photometric function will be changed. Later on we can even drop the whole thing and it won't fly apart.

WHIPPLE: Can you answer the question whether the darkening comes directly from the proton bombardment or by the indirect effects of sputtering?

KENKNIGHT: I can say that it is essential that there is a sputtering back and forth, but indeed we don't see any evidence that the tops of our small particles are lacking in color. However, there you see it is very complicated, because you can transfer material up on the top too, and we don't see this.

GOLD: We have a great deal of experience with bombarding all manners of substances and it is clear that the dominant effect is that of a deposition by sputtering on the underside of particles, a far, far stronger effect. There are other darkening effects, too. One of these effects is the straightforward reduction due to hydrogen being put into the crystal lattice, that is, a straightforward reduction of metals in the material. That occurs in the solid sample just as much, of course, and is a much smaller effect.

WHIPPLE: I see. Thank you. Now, is there any remaining subject that somebody in the audience wants to bring up?

VOICE: I would like to direct a question to Professor Levin, if I might. He gave an early thermal history of the moon, similar to the earth, that is, melting by radioactivity, differentiation, and then formation of a crust. On the earth, of course, as we go into the interior we get an increase in density and yet I understand from his presentation this morning that he expects that as we go into the interior of the moon we would get a decrease in density from this same thermal cycle. I wonder if he would explain that.

WHIPPLE: May I see if I understand it?

LEVIN: Yes.

WHIPPLE: The point is that because of the small gravity there is very little compression, so the major effect is that of temperature being highest at the center and, therefore, the density being lower at the center. Is that correct?

LEVIN: Yes, that is correct.

WHIPPLE: Dr. Green?

GREEN: This question is to Dr. Kopal. What causes the luminescence in enstatite achondrites; is it the iron-free enstatite, or in the case reported, the oldhamite impurities that were present in the Bustee and the Bishopville and perhaps the Khor Temiki?

KOPAL: I believe the question could be more appropriately answered by Professor Urey; he knows more about it.

UREY: I don't know the answer.

VOICE: Using ultraviolet radiation, not higher energy radiation, we have shown that it is not the enstatite or the oldhamite but the mineral forsterite, which occurs in about 5 per cent abundances in Norton County. You can actually irradiate and pick out these grains and do whatever you will with them.

WHIPPLE: Can you give us the composition, roughly, of forsterite?

VOICE: Forsterite is just Mg_2SiO_4.

WHIPPLE: Thank you.

SAGAN: Regarding the albedo of material in

space, I thought we might have a number for the albedo on the Mars cloud?

WHIPPLE: The only evidence is that it is low. The fairly high polarization indicates that the albedo is quite low and it might be lower than the moon. I quote only from Southworth, who has given it some consideration.

UREY: May I make a remark? In regard to this fluorescence of the enstatite achondrites on the moon, we got a number of samples at La Jolla from the Engels and passed them over to Steele at General Atomics, so that he could test them to see whether other things would fluoresce. We found one that would fluoresce beautifully in the red. It is dolomite, $MgCa(CO_3)_2$.

WHIPPLE: Very interesting.

UREY: I don't know what particular thing fluoresced.

WHIPPLE: Is there anyone on the panel who has a last comment to make?

KOPAL: May I make one remark about Professor Levin being with us? He doesn't come very often to these parts. I should just like to voice a little caution not to take liberties with internal temperatures of the moon.

WHIPPLE: Do you wish to respond?

LEVIN: I have heard enough.

WHIPPLE: Professor Urey, you began this meeting. Would you like to conclude it with some pertinent comment?

UREY: I would like to remark about one thing that nobody has asked about, and that is the very different presentation by Dr. Levin and myself in regard to the possible origin of the moon.

I said it might have escaped from the earth. I did not favor this, but many people think so. In the second place, it might have been captured by the earth.

Levin, on the other hand, accumulated the moon in the neighborhood of the earth. The two were completely mutually exclusive. Now, I would like to state why this occurs and maybe Levin will have some correction to make.

I assume that the abundance of iron in the sun is what the astronomers say it is. I also assume that the abundance of iron in the earth is what the geophysicists say it is. The two disagree very markedly. For that reason I have to find some way

to account for the different chemical composition of the moon and the earth. If I understand Levin correctly, he believes that the geophysicists are wrong about the composition of the earth and the core does not consist of metal.

LEVIN: Yes.

UREY: So that accounts for the difference. I may say there is another group of people in the field who believe that astronomers are wrong, and that the earth and the sun have the same composition and the moon is off.

You can take your choice about it. I am inclined to take the opinion of the experts in various sciences in which I am not competent, and to work on that. I might say that Professor Francis Birch has published a paper recently in which he has estimated that the per cent of iron, perhaps plus nickel, in the earth is 38 per cent, which is higher than that of any of the meteorites that I know of, except the metal ones. And he bases his estimate, that the core of the earth consists of metal, on the tests that have been made with atomic bombs on the elements in this neighborhood. He maintains that it is not possible to get the properties of the core of the earth from elements in the neighborhood of aluminum, magnesium, and material like that. This is the reason for my view.

WHIPPLE: Now, Dr. Levin, as our distinguished visitor, you may have the last word.

LEVIN: The American scientist Libby has published papers in which he has explained why the experiment with shock compression does not give a phase transition into the metallic phase, but under the static pressure, the silicates can transform into metallic states. The process on the earth can be similar to that in diamonds. For them the required static pressure is about 20,000 atmospheres. However, under shock it can be that several million atmospheres are needed to transform them into diamonds.

WHIPPLE: Thank you very much. That, I think, is a good note to conclude on—diamonds. I would like to turn the microphone back to Dr. Hess. Do you have a final word?

HESS: I would like to thank all of the participants in the meeting for NASA. Thank you very much. I enjoyed it.

Subject Index

Acetylene: from interior of moon, 290. *See also* Alphonsus

Age of lunar formations, 290

Albedo: difference and thermal anomalies, 282; differentiation between highlands and lowlands, 117; normal, 141, 142; normal and particle size, 149; of moon, 310; of material in space, 312, 313; on slopes, 73; optical, 107; and proton irradiation, 90; variations in, 57

Alignments, 85. *See also* Grid

Alpetragius, 256

Alphonsus: central peak, 119, 256; central peak, a recent formation, 260; central peak, relation to Mare Imbrium, 259; crater, 18, 119; craters on floor, 289; crater walls, 265, 266; dark craters in, 293; dark patches in, 118; dark spots, 290; density of craters in, 92; filling with Imbrium ejecta, 82; floor of, 289; floor, cracks in, 255; and Imbrium collision, 16, 17; maria-like crater floor, 290; mountain range in, 303; outgassing, 139; peak of, 288; recent volcanism in, 261; relation to ring complex, 266; surface able to form craters, 259; typical volcano, 255; walls of, smooth appearance, 93, 94

Aristarchus: active in historic times, 297; anomalous reddening, 178; appears blue, 80; brown patch in, 298; concentration of suspected luminophors, 182; cooling curve, 205; glowing spots, 178; luminescence, 176; reddish-orange spots, 177, 178; spots near, 180; transient luminous phenomena in, 179

Aristoteles, 24

Ashen light: used to study depolarization, 157, 159, 160

Ash: volcanic, 97; will weld, 294

Ash flows: differential compaction, 261; dilute, as thin clouds, 266; dilute phase, 261; discussion of, 244; in western United States, 252, 254; how level, 257; mounding, 257; ponding at foot of hills, 266; softening by, 262, 263; temperature range, 257. *See also* Bandelier ash flow

Aso caldera: 245; compared to Albategnius, 255

Bandelier ash flow, 245, 247

Bearing strength: 23, 79, 105, 293, 294; effect of grain shape, 293; from angles of repose, 294, 295; from observed slopes, 295, 296; mare ground unstable, 118; one to five kg per cm², 103; toughness, 107

Blackening: depth of penetration of, 265; of highland areas, 265

Blocks: absence of, 71

Bonpland-H: thermal enhancement in, 216

Brightness: from microdensitometer, 74; of lunar eclipses, 174; of moon, fluctuations, 174; and phase angle, 141; of zodiacal cloud, correlated with planetary magnetic number, 182

Brightness temperature: and two-layer models, 207

Bullialdus: 26, 36; of asteroidal origin, 102; secondaries from, 8; secondary swarm, 68; swarm associated with, 29, 36

Caldera group: in Africa, 252

Calderas: and ash flows, 264; caused by eruption of ash and pumice, 241; in western United States, 252, 254; Kilauean type, 243, 247, 250, 252, 257; Krakatoan type, 244, 245, 247; Long Valley, 247; on faults, 255; other resurgent, 247; size of, 245; terrestrial, 241; volumes of material from, 254

Carbonaceous chondrites: from the moon, 20, 21; and origin of life, 21

Cauldrons: resurgent, 250, 256, 257

Cavities: and lunar craters, 92

Center of mass: 300; closer to our side, 301; lies above center of visible surface, 301

Central dome, 256

Central Mountain, 245, 250

Chondrites: darkening by proton bombardment, 311; do not match photometric properties of moon, 154; high-iron group, 4; low-iron group, 4

Circumlunar satellite, 300

Clearwater Lakes, 308, 309

Coherence, 23

Collapse features, 9, 117, 120

Color: 141; brown associated with high silica, 297; effect of irradiation, 153, 298, 299; and flow phenomena, 290; and time sequence, 298

Color boundaries: 69, 99; and ash flows, 81; by filter photography, 79; connection with ridge, 80, 81; differences, 79, 80, 298; and dust, 85; sharpness, 99

Color index: saturation value, 299

Comets: material of, 309; splitting of, 307

Composition, chemical: 3, 4, 21; differences, 297; effect of, on photometric characteristics, 153, 154; effect of, on irradiated igneous rock, 149; of highlands, 154; of maria, 154

Concentric fractures, 255, 257

Condensation of material evaporated by dust, 312

Conductivity of moon's surface, 231

Contamination: of earth from moon, 9; of moon from earth, 21

Cooling curves: 218, 219; computed, 202; Copernicus, 205; effect of changing parameters, 204; effect of changing thermal diffusivity in lower layer, 204, 205; effect of changing thermal diffusivity in upper layer, 204; explanations for, 207; for crater Tycho, calculation of, 202; kink in, 113

Copernican system, 24

Copernicus: 24, 57, 107, 136; cooling curves of, 199, 200, 205; high radar reflectivity, 289, 290; luminescence, 176; rays, 100; secondaries of, 68, 137

Craters: black halo, 9; and blocky ejecta, 134; buried, 76; composite, 70; dark-halo, 93, 102; deficiency of small, 259, 260, 288; dependence on material, 132, 133, 134; distribution, 53, 54; effect of repeated formation, size, 66, 67; frequency below 1-km limit, 103; frequency distribution, 288; formed by missile impact, 132; geometry of, 125; halo, 17; impact origin, 288; irregular features of, 51; linear arrangements, 307, 309; mostly of impact origin, 102; modification of shape by erosion, 66; older, 112, 115; on summits, 102; pairs, 308; pairs and comets, 309; pattern of secondary swarm, 70; percentage which show shadow, 259; pre-mare, 102; protuberances in, 51, 53; shallow, 56, 100; shapes of, 23, 24; small, 36, 66, 67, 68; smallest ray, 58; soft-edged, 79; softened appearance, 259; steady-state distribution of, 289; streaks in the walls, 58; strings of, and

Name Index

The Nature of the Lunar Surface

edited by

Wilmot N. Hess, Donald H. Menzel, John A. O'Keefe

designer : Edward King
typesetter : Baltimore Type and Composition Corporation
typefaces : Baskerville (text), Craw Clarendon (display)
printer : Universal Lithographers, Inc.
paper : Warren's Silkote Opaque
binder : Moore and Company, Inc.
cover material : Columbia Riverside Vellum